Introduction to

Intelligent High-Speed Railways

Introduction to **Intelligent High-Speed Railways**

Tongjun Wang
China State Railway Group Co., Ltd., China

高等教育出版社
HIGHER EDUCATION PRESS

World Scientific

Published by

Higher Education Press Limited Company
4 Dewai Dajie, Xicheng District
Beijing 100120, P. R. China

and

World Scientific Publishing Co. Pte. Ltd.

5 Toh Tuck Link, Singapore 596224

USA office: 27 Warren Street, Suite 401-402, Hackensack, NJ 07601

UK office: 57 Shelton Street, Covent Garden, London WC2H 9HE

Library of Congress Cataloging-in-Publication Data
Names: Wang, Tongjun, 1964– editor.
Title: Introduction to intelligent high-speed railways / Tongjun Wang.
Description: Hackensack, NJ : World scientific, [2025] | Includes index.
Identifiers: LCCN 2024009331 | ISBN 9789811287442 (hardcover) |
ISBN 9789811287459 (ebook) | ISBN 9789811287466 (ebook other)
Subjects: LCSH: High speed trains--Technological innovations--China.
Classification: LCC TF1460 .I58 2025 | DDC 625.20951--dc23/eng/20240610
LC record available at https://lccn.loc.gov/2024009331

British Library Cataloguing-in-Publication Data
A catalogue record for this book is available from the British Library.

For any available supplementary material, please visit
https://www.worldscientific.com/worldscibooks/10.1142/13706#t=suppl

Desk Editor: Sanjay Varadharajan/Amanda Yun

Typeset by Stallion Press
Email: enquiries@stallionpress.com

Preface

High-speed railways (HSRs) are an important sign of transportation modernization and a concentrated reflection of a country's level of industrialization. China has built the world's largest and most modern HSR network. By means of introduction, digestion, absorption, re-innovation, and independent innovation, China has established a complete HSR technology system covering the three fields of engineering construction, equipment manufacturing, and operation management. The HSR technology in China has generally stridden into the top rank in the world and has reached the world's leading level in some fields.

Since the 21st century, new-generation information technologies such as building information modeling (BIM), cloud computing, big data, the Internet of Things (IoT), 5G, and artificial intelligence (AI) have witnessed their accelerated development, promoting the digital transformation of traditional industries, such as manufacturing, energy, transportation, and finance. Railways in major developed countries and regions in the world are actively using new information technologies to make strategic deployments for railway digitalization and intelligentization. In order to realize the continuous leadership of China's HSR technology, promote the coordination of the whole life cycle and the whole industrial chain, and comprehensively improve the service quality, efficiency, benefits, and safety level of China's HSR, the development of intelligent HSR has become an inevitable course.

This book systematically clarifies the scientific connotation, management methods, and system architecture of intelligent HSR. It summarizes key technologies such as intelligent construction, intelligent equipment, and intelligent operation and their innovative applications based on the intelligent engineering practice used by China's HSR. The work comprehensively reflects the frontier dynamics, top-level design, and innovations in intelligent HSR achieved in China and fills the gap in teaching material for colleges and universities teaching related courses in this field. This book has important theoretical and practical guiding significance. Typical features and innovations featured in this book are as follows:

1. The definition and scientific connotation of intelligent HSR are revealed in depth.

   Intelligent HSR is a new-generation intellectualized HSR system in which new technologies such as cloud computing, big data, IoT, mobile interconnection, AI, Beidou navigation, and BIM are widely used, and resources are efficiently utilized to achieve the comprehensive perception of information among mobile HSR equipment, fixed infrastructure, and internal and external environments. Used in conjunction with ubiquitous interconnection, fusion processing, active learning, and scientific decision-making, they enable the realization of an integrated lifecycle management system.

   The typical characteristics of an intelligent HSR mainly include three aspects:

   i. Driving business decision-making based on the integration of models and data. That is, supporting the scientific decision-making of HSR construction (manufacturing), operation and maintenance, service, and safety through the integrated application of the characterization model, mechanism model, and massive dynamic and static data.
   ii. Realizing the lifecycle management of infrastructure and equipment. That is, realizing the unified management of the whole life cycle from design to construction (manufacturing) and operation.
   iii. Using intelligent technology to reduce costs, raise efficiency, improve services, and ensure safety. Intelligent HSRs are a means of transport aimed at making traveling more convenient, faster, safer, more reliable, more comfortable, more energy-efficient, more environmentally friendly, more economical, and more efficient by means of the deep integration of intelligent technology with the HSR business, instead of through the simple accumulation and application of intelligent technology.

2. The management method of "model–data drive and axis–plane coordination" for an intelligent HSR is put forward for the first time.

   Based on the integration of models and data, the overall efficiency of the system is optimized to realize the lossless transmission of forward information and reverse iterative optimization in the HSR's whole life cycle, as well as the overall collaboration of business elements at different time sections in the whole life cycle. This includes coordination of geology, subgrade, bridge, tunnel, track, communication, signaling, power and electrification, station building, and engineering economy in the design stage; coordination of quality, progress, safety, investment, environmental protection, materials, and the external environment in the infrastructure construction stage; coordination of planning, materials, production capacity, quality, resources, and environment at the equipment manufacturing stage; coordination of track maintenance, communication and signaling, the vehicles, passenger flow, transportation, and the external environment in the operation stage. The "model–data drive and axis–plane coordination" management method guides the intelligent practice of the Beijing–Zhangjiakou High-Speed Railway, and may be applied to other similar complex giant systems.

3. The intelligent HSR system's architecture, which integrates technology, data, and standards, has been built in an all-round way.

   The technical system clarifies the system composition of an intelligent HSR and the interrelationship between the components, mainly including three sections (intelligent construction, intelligent equipment, and intelligent operation), 10 fields, 18 directions, and N innovations. The standard system clarifies the overall composition and classification of intelligent HSR standards, as well as the boundaries and relationships among various standards at all levels, mainly including general basic and management standards, intelligent HSR application standards, and platform and supporting technology standards. The intelligent HSR data system specifies where data come from, where they are stored, and how to use them, establishes a life-cycle data model and storage structure, and plans such three levels as data collection, storage and processing, and analysis and application, according to the architecture mode of platform + application. The intelligent HSR system architecture has been released as a technical standard by China State Railway Group Co., Ltd. (China Railway), providing guidance and basis for the intelligent construction of new HSRs and the intelligent improvement of operating HSRs.

4. The latest achievements of intelligent HSR, under the guidance of the management method and system architecture, are systematically demonstrated.
   i. In the field of intelligent construction, an intelligent space–air–ground integrated survey and BIM-based collaborative design have been realized; digital factories such as an intelligent beam yard, intelligent track slab yard, intelligent double-block sleeper yard, and intelligent machine-made sand processing yard have been built; complete intelligent construction technologies for bridges, tunnels, subgrade, tracks, passenger stations, and communication, signaling, power, and electrification have been established; and a BIM+GIS-based engineering management platform has been developed and applied.
   ii. In the field of intelligent equipment, intelligent EMUs with functions such as intelligent driving, intelligent safety, and intelligent service have been developed. The automatic driving of 350 km/h high-speed trains based on CTCS3+ATO has been realized for the first time. Restrictions for new train operation control technology with functions such as moving blocks and station-area integrated remote control have been shattered. An intelligent traction power supply system with holographic perception, multi-dimensional fusion, reconstruction and self-recovery, and intelligent operation and maintenance has been built. An intelligent communication system composed of 5G-R, bearer networks, and multimedia dispatching communication has been proposed, and an intelligent inspection and monitoring system covering infrastructure, mobile devices, perimeter intrusion, the external environment, and natural disasters has been established.
   iii. In the field of intelligent operation, intelligent passenger stations with functions such as an automated information service and intelligent management control have been achieved. An intelligent ticketing system with the characteristics of perceptual automation, service networking, and autonomous execution has also been developed. Key technical problems for the intelligent preparation of train diagrams, intelligent centralized traffic control, and intelligent

comprehensive dispatching have been solved, and the intelligent infrastructure operation and maintenance system, prognostics and health management (PHM), and intelligent electric multiple unit (EMU) operation and maintenance system have been built.

iv. In terms of a basic platform, the first main data center in the railway industry based on cloud computing has been built; the railway data service platform has been developed and deployed; the data lake of "temperature–granularity–sensitivity" integration and the data analysis algorithm of "general–private" combination have been innovated; and the railway BIM service platform has been established.

This book comprises six chapters covering the definition and architecture for intelligent HSRs, management methods for the complex giant systems of intelligent HSRs, the intelligent construction of HSRs, the intelligent equipment of HSRs, the intelligent operation of HSRs, and a demonstration of the innovative application of these advances in the Beijing–Zhangjiakou High-Speed Railway.

We would like to thank Academician Lu Chunfang and Academician He Huawu of the Chinese Academy of Engineering; Ye Yangsheng, chief engineer of China Railway; Luo Qingzhong, deputy general manager, and Li Ping, chief researcher, of the China Academy of Railway Sciences Corporation Limited; and Zhang Xingchen, director of the Teaching Steering Committee for Transportation Majors of Institutions of Higher Education of the Ministry of Education, for reviewing this book during their busy schedules. We would like to thank Higher Education Press for its careful planning and full support for the publication of this book!

We would like to thank China Railway Engineering Corporation, China Railway Construction Corporation Limited, CRRC Corporation Limited, China Railway Signal & Communication Corporation Limited, and China State Construction Engineering Corporation Limited, as well as relevant universities and scientific research institutions, for their contributions to the intelligent development of railways in China!

This book has been funded by scientific research projects of the Ministry of Science and Technology of the People's Republic of China, the Chinese Academy of Engineering, and China Railway, and extensive domestic and foreign literature has been consulted during the preparation process. We would like to express our gratitude here!

Due to limited time and cognitive limitations, there may be omissions and deviations in this book. Any comments and suggestions for improvement for this book will be greatly appreciated. In the future, with the development of new technologies and the continuous iteration of applications, the contents of this book will be revised and improved!

About the Author

Wang Tongjun, born in 1964, graduated from Beijing Jiaotong University, China, with a major in railway transportation. He is currently the vice-general manager of China State Railway Group Co., Ltd., and engages in the research of railway engineering construction and informationization and intelligent railway innovation. He has published 41 papers, authored 4 books, and holds 11 patents. He has received the Construction Award, Handover Award, and Special Contribution Award from buildingSmart International (bSI), as well as the Digital Award from the Union of International Railways.

Professor Wang currently holds several positions, including the director of the China Association of Railway Engineering Construction, the deputy director of the China Civil Engineering Society and Railway Expert Advisory Committee, and the director of the Railway BIM Alliance.

Contents

Chapter 1

Definition and Architecture of Intelligent High-Speed Railways

1.1 Intelligent and Digital Railway Development around the World

In recent years, railways all over the world have been facing challenges such as the rapid growth of passenger and freight transport demands, the need for customized transport services, and ubiquitous transport safety concerns. As the supply capacity of railway transport infrastructure remains stable, to significantly improve the efficiency and cost-effectiveness of railway transport organizations, the service quality of passenger and freight transport must be optimized, and the safety of railway transport must be improved by adopting the new generation of information technologies, which constitutes the only way of railway development in all countries. Keeping up with the global sci-tech development driven by a new round of sci-tech revolution and industrial reform, major railway countries all over the world established their strategies and ideas on intelligent technology development suitable for themselves in order to meet the new requirements of economic and social development and transport market competition for railways and to further improve the efficiency and cost-effectiveness of railway transport operations. Guided by new strategies and ideas and based on advanced technologies, such as cloud computing, the Internet of Things (IoT), big data, artificial intelligence (AI), mobile internet, satellite navigation and positioning, and building information modeling (BIM), many countries have issued plans for intelligent development in light of their national conditions and the development status and needs of enterprises. These plans strive to build intelligent railway equipment capable of automatic inspection, diagnosis, and decision-making, to construct intelligent railways oriented toward digital management throughout the whole life cycle, and to provide personalized transport services, real-time and predictable safe operation and maintenance, and more targeted operation management, thus improving the intelligent level of railway operation management, passenger and freight services, engineering construction, rolling stock maintenance, and infrastructure maintenance in all aspects.

Broadly, the railway intelligence process is a process where cutting-edge digital and information technologies are applied in the railway industry to improve the efficiency and performance of railways so as to reduce costs and improve capacity, provide reliable connections to build a safe and efficient railway network, offer customers better value-added service to improve customer experience, and make full use of data to strengthen railway competitiveness. It is also called the railway digitalization process. In the following section, railway digitalization development strategies and application cases are examined on a country/region-by-country/region basis.

1.1.1 *Europe union*

1.1.1.1 *Development strategy*

Since the 1980s, the capacity of some important corridors of the European railway network has become saturated, and the emergence of transport modes such as low-cost airlines and shared transport has negatively affected the railway transport market to a certain extent. In such a context, the European Union (EU) is ambitious to strengthen the competitiveness of railways and increase the market share of railway transport, regarding digital technologies as one of the important means to overcome difficulties in railway development. Under the guidance of major strategies, such as "Europe 2020," the EU has issued a series of railway sci-tech development strategies and plans in recent years, including the 2011 White Paper on Transport, "Roadmap to a Single European Transport Area — Towards a competitive and resource efficient transport system," Rail Route 2050, and Shift2Rail.

(I) **Roadmap to a single European transport area — Towards a competitive and resource efficient transport system**

In order to achieve the goal of "building a more competitive and resource-efficient European transport system to solve major social problems such as rising transport demand, traffic congestion, energy supply security and climate change," set out under the EU 2020 strategy on March 28, 2011, the EU issued the 2011 White Paper on Transport "Roadmap to a Single European Transport Area — Towards a competitive and resource efficient transport system." The white paper proposes a forward-looking transport policy for the purpose of making the current European transport system competitive and resource efficient. To fulfill this vision, the EU plans to achieve the following objectives in the next 40 years: transforming the current oil-dependent transport pattern and cutting emissions by 60% (compared with 1990) without affecting transport efficiency and mobility; building an efficient backbone network for multimodal transport; cutting CO_2 emissions by 40% by 2050 (compared with 2005); establishing a clean urban transport system; and achieving 10 goals for a competitive and resource-efficient transport system to reach a 60% GHG emission reduction target.

In view of the inherent advantages of railway transport in environmental protection, land use, energy consumption, and safety, the white paper emphasizes the importance of railway transport in Europe and sets out six development goals directly related to HSR transport:

(1) By 2050, complete a European HSR network. Triple the length of the existing HSR network by 2030, and maintain a dense railway network in all member states. By 2050, the majority of medium-distance passenger transport should be by rail.
(2) Complete a fully functional and EU-wide multimodal TEN-T core network by 2030, and complete a high-quality, high-capacity network and a corresponding set of information services by 2050.
(3) By 2050, connect all core network airports to the rail network, preferably high-speed;
(4) By 2020, establish the framework for a European multimodal transport information, management, and payment system.

(II) **Rail route 2050**

To support the 2011 White Paper on Transport, the European Rail Research Advisory Council (ERRAC) has released the "Rail Route 2050" strategic plan, proposing a development route toward a competitive, resource-efficient, and intelligent rail transport system for 2050 in seven aspects: intelligent mobility, energy and environment, security, certification, competitiveness and enabling technologies, strategy and economics, and infrastructure.

To achieve this vision, ERRAC has prioritized the development of seven fields: intelligent mobility 2050, energy and environment 2050, security 2050, safety and certification 2050, competitiveness and enabling technologies 2050, strategy and economics 2050, and infrastructure 2050.

(III) **Shift2Rail**

The EU attaches great importance to the development of scientific research and innovation. The Framework Programmes for Research and Technological Development (also called Framework Programmes, or FPs) are an important part of the EU's sci-tech and innovation policy. The first program was launched in 1984. Up to now, there have been eight framework programs which have set out to achieve the EU's scientific and technological development objectives, priority research areas, and main activity guidelines for a certain period. The FP8 — "Horizon 2020," which was launched in January 2014, is the largest R&D and innovation program ever and a funding tool for the EU to implement innovation policy.

Relying on "Horizon 2020," on June 16, 2014, the Council of the European Union issued Council Regulation No. 642 Shift2Rail (S2R) with a total budget of EUR 920 million. The S2R Joint Undertaking was formally established from July 7 to 9, 2014, and the candidate for interim executive director was nominated. S2R is the first European rail initiative to seek focused research and innovation (R&I) and market-driven solutions by accelerating the integration of new and advanced technologies into innovative rail product solutions and developing necessary technologies to build the Single European Railway Area (SERA).

S2R promotes collaboration in the railway sector and cooperation and knowledge transfer in industries, academia, and research institutions, focusing on reducing life cycle costs, enhancing railway network capacity, improving service reliability, and punctuality, to achieve the objectives of integrating European railways, strengthen the attractiveness and competitiveness of European railways, and consolidate the leading position of European railways in the global market (as shown in Figure 1.1).

The S2R hopes to achieve the following key objectives by 2030: a 50% increase in the reliability and punctuality of rail services, a 100%

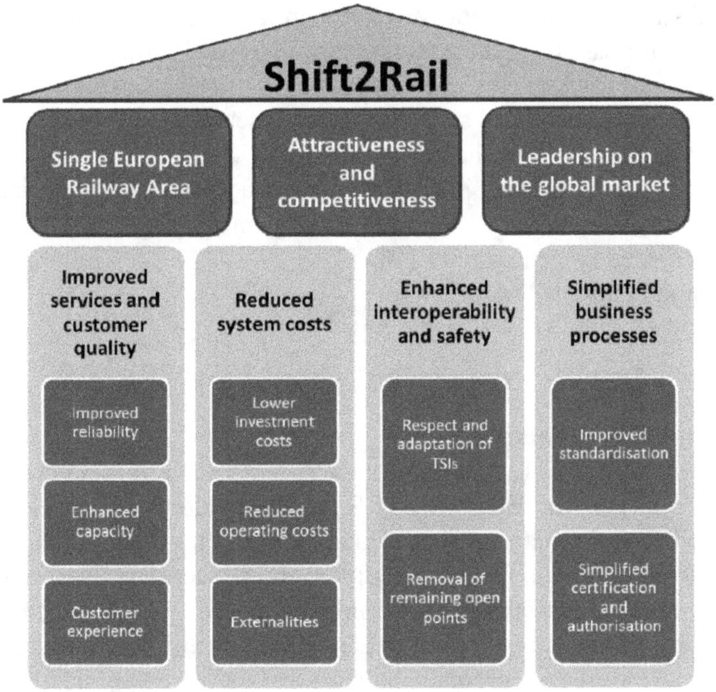

Figure 1.1. Strategic objectives of Shift2Rail.

increase in the capacity of the railway transport system, and a 50% reduction in the life cycle cost of railway transport (developing, building, operating, maintaining, renewing, and dismantling infrastructure and rolling stock). In order to fulfill the overall objectives above, S2R organizes its work, aiming for improved service quality, reduced system costs, enhanced interoperability and safety, and simplified business procedures.

Every S2R innovation program (IP) or cross-cutting activity (CCA) is further divided into various numbers of technology demonstrators (TDs) as the research directions. For implementation, each IP usually has one lighthouse project and several special research projects implemented over the years:

IP1: cost-efficient and reliable trains, including high-capacity and high-speed trains;
IP2: advanced transport management and control system;
IP3: low-cost, sustainable, and reliable high-performance infrastructure;
IP4: attractive IT solutions for railway services.

The CCAs focus on their innovative research on various IPs and methodologies for assessing the achievement of the S2R objectives. It has six work areas (WAs):

WA1: long-term needs and socioeconomic research;
WA2: key performance indicators (KPIs);
WA3: safety, standardization, intelligent maintenance, and intelligent materials;
WA4: system integration, safety, and interoperability;
WA5: energy and sustainability (WA5);
WA6: human capital (WA6).

1.1.1.2 *Application cases*

(I) **Intelligent equipment**

(1) *Hybrid ERTMS/ETCS-3 system*

The European railway ERTMS/ETCS-3 system adopts the moving block system, which eliminates ground signals and track circuits or axle counters for checking the cleared/occupied track sections. It can greatly shorten the time interval between trains, significantly improve the carrying capacity of lines, and further enhance the flexibility and reliability of the system. It aims to overcome challenges faced in the application of the ETCS-3 system by using the existing equipment (such as a limited number of trackside positioning equipment). It can be upgraded on ETCS-2 lines without installing expensive auxiliary technical equipment outside the system. This not only avoids complicated new operation regulations but also significantly improves the carrying capacity of lines.

The main characteristics of the hybrid ERTMS3 system (Figure 1.2) are as follows: it adopts a virtual fixed block system; it comprehensively

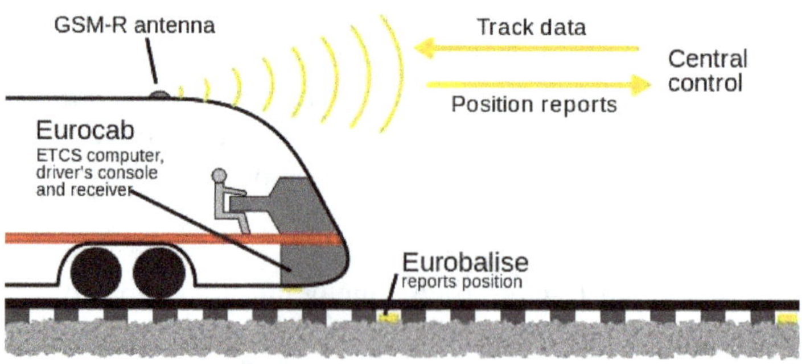

Figure 1.2. Hybrid ERTMS/ETCS-3 system.

uses information about onboard train positioning equipment and trackside positioning equipment; it defines many virtual subsections through trackside positioning equipment; it supports mixed operation of trains with and without a train integrity device monitoring system running on the same line and with safe and reliable operation; it has similar carrying capacity compared with the lines with a complete ERTMS/ETCS-3 system; and the more trains equipped with the train integrity monitoring system, the higher the carrying capacity of the lines.

(2) *Real-time remote diagnostic monitoring system*

The real-time remote diagnostic monitoring system (R2M) combines software with onboard equipment so that a failure alarm can be generated during effective maintenance. The R2M can establish real-time communication with the maintenance center to enable the maintenance team to make effective failure diagnoses and arrange related work. In this system, the NCU obtains the status data of vehicles without an onboard train management system by establishing connections with the onboard train signal and system. Before the vehicle arrives at the maintenance center, R2M can analyze the real-time status information of rolling stock and component systems based on these data, so that after the vehicle arrives at the maintenance center, targeted inspection and failure maintenance can be carried out efficiently and quickly.

(3) *ETCS-ATO onboard system*

The ETCS system involves a large number of technical specifications for interoperability. Therefore, the ERTMS Users Group, Unisig, and the EU Agency for Railways (ERA) jointly developed an interoperable ATO architecture based on ETCS. The proposed architecture was initially envisaged as two-level automation. Although ETCS and ATO have different functions, they share the same set of onboard equipment and the same train–ground communication system. The locomotive radio communication is connected to the ATO, ETCS radio block center, and interlocking system through an independent channel and IP network. In this architecture, the interlocking system, ETCS radio block center, trackside ATO, and other ground equipment are connected with the traffic management system of the control center. Onboard systems (including onboard ETCS and onboard ATO) implement driving operations and, at the same time, evaluate factors such as energy consumption, operation time, interval distance, and comfort level of passengers. ATO and ETCS share a completely standardized interface, and this is the key to realizing interoperability.

Mexico City–Toluca Railway, with a length of 57.7 km, a maximum design speed of 160 km/h, and a travel time of 39 min, will be open to traffic in 2018. For the project, ETCS-2 Baseline 2.3.0d and a semi-automatic train operation system to the GoA-2 grade are used. Drivers will be required to close the doors and operate the train during periods of disruption. Trains will operate at a 2.5 min headway at a maximum speed of 160 km/h.

1.1.2 *France*

1.1.2.1 *Development strategy*

In the future, the French National Railway Company (SNCF) will focus on the development of transportation capacity, competitiveness, passenger transport, and public funds. It plans to make technological breakthroughs in digitization, the IoT, AI, network security, energy, and other aspects, realize new business models such as low-cost, worldwide door-to-door service and shared cars, and explore the application of new technologies such as AI, energy storage, 3D printing, robotics, automatic driving, and new materials in specific railway scenarios. In 2015, the SNCF unveiled the "DIGITALSNCF" strategy. Relying on the Industrial Internet, using equipment and tools such as target-specific communication networks, data clouds, and various sensors and supplemented by manual transaction processing, the strategy plans to connect the three major sectors of trains, railway networks and station buildings through networks, thus making the equipment, trains, and station buildings all connected. On the one hand, the strategy pursues safer transportation, higher production efficiency, a better energy economy, and better work quality. On the other hand, it is dedicated to meeting the passengers' demands for punctuality and comfort. It is expected that in 2031–2040, a competitive, convenient, and sustainable railway system will be built for customers, which will be closely integrated with future transportation.

(1) **Connected equipment:** It is planned to use 3D printing, equipment sensors, energy sensors, robotics, digital testing, and other technologies to realize the digitalization of rolling stock, repairable parts, and information flow.

(2) **Connected trains:** It is planned to set data collecting sensors on trains in commercial operation. The collected train data will be intelligently analyzed by computers.

(3) **Connected railway network:** It is planned to improve the modernization of infrastructure overhaul and maintenance through real-time remote monitoring and digitization of the whole railway network.

(4) **Connected station buildings:** It is planned to install sensors in the station elevators, platforms, and other equipment and facilities so as to improve the utilization rate of equipment and save energy.

The specific roadmap for DIGITALSNCF is as follows:

(1) **Short term (until 2020):** Improve the existing railway system and introduce automatic driving on suburban railways. Promote 3D printing technology and reduce manufacturing time and cost of parts by 20%.

(2) **Medium term (2021–2030):** Build a disruptive and innovative railway system with functions including real-time train positioning, "door-to-door" transportation, intelligent control of passenger flow, self-adaptation of railway network capacity, etc.

(3) **Long term (2031–2040):** Provide customers with satisfactory, reliable, safe, and accessible services in real time; maximize the utilization rate of the railway network and inventory to reduce costs; simplify and standardize the systems and accelerate the application of new technologies; fulfill its mission in public service by optimizing resource utilization and limiting carbon emissions; incorporate the railway system into the global "door-to-door" transportation services; and turn the station into one that integrates various services and various modes of transportation.

1.1.2.2 *Application cases*

(I) **Intelligent equipment**

(1) *Hesop system*

Alstom (France) and Hamburger Hochbahn AG (Germany) jointly launched the Hesop system (Figure 1.3), which can obtain the braking energy of trains and redirect it to other places for reuse.

Figure 1.3. Hesop system.

As an advanced system, the Hesop system can not only provide traction voltage to the network but also recover braking energy from vehicles, which helps control the temperature in the tunnel. Furthermore, the system can be integrated into the existing system through transformation or built into the new lines in their construction stage. At present, there are about 128 Hesop systems operating in the metro networks of Milan, London, Dubai, Panama, and Riyadh and on trams in Sydney and Milan.

At Rauhes Haus station on Metro line U2 in Hamburg, the system can recover more than 99% of the traction energy generated during braking, thus reducing carbon dioxide emissions by reducing energy consumption.

(II) **Intelligent operation**

(1) *Optimet real-time train occupancy system and Optimet Orban map*

In recent years, Alstom has developed a high-tech predictive maintenance tool called HealthHub, which can automatically monitor the technical status of rolling stock, infrastructure, communication & signaling, and other equipment in real time and accurately identify which equipment needs to be replaced or repaired and the date of replacement.

(2) *HealthHub predictive maintenance tool*

SNCF recently signed an agreement with IBM (United States) using IBM's Watson IoT platform, big data cloud computing technology, and special industrial sensors developed by SNCF itself to remotely monitor trains and railway infrastructure and create conditions for predictive maintenance.

1.1.3 *Germany*

1.1.3.1 *Development strategy*

In 2016, DB signed a cooperation agreement on "Digital Railway Strategy" (Railway 4.0) with the German Federal Ministry of Transport and Digital Infrastructure and the German Railway Industry Association. Aiming at improving passengers' satisfaction, the "Digital Railway Strategy" involves deep technical changes in all aspects of the railway system, such as production, operation, maintenance, and customer interaction, thus fully supporting Germany's "Transport 4.0" strategy. The "Railway 4.0" strategy includes transportation 4.0, logistics 4.0, and infrastructure 4.0, supported by production 4.0, working environment 4.0, and information technology 4.0. To realize the "Railway 4.0" strategy, DB has initiated a series of projects and put a lot of effort into the digital development of railway transportation and operation. It mainly includes: establishing a multi-mode ride-hailing platform, developing mobile phone apps for logistics business, providing intelligent travel and train route planning services for passengers; introducing online diagnostic tools for locomotives, virtual planning, and control systems for infrastructure construction projects; and using digital equipment for maintenance work, thus gradually forming an intelligent operation system; promoting the development of railway research and innovation by establishing relevant laboratories and organizing "entrepreneurship summer camp."

With different goals being set for the short, medium, and long terms, "Railway 4.0" aims to realize deep technological changes in all aspects of the railway system, such as production, operation, maintenance, and customer interaction.

(1) **Short term (until 2025):** Realize the semi-automatic train wireless network distribution, provide the next-generation electronic travel service, and connect the passenger's mobile equipment directly with the base station signal through a special train design. Meanwhile, by the end of 2020, realize the planning strategy of applying BIM to all railway construction projects.

(2) **Medium term (2025–2035):** Realize automatic operation of trains, provide more flexible and personalized modes of transportation, and successfully develop robot cabs and put them into use.

(3) **Long term (2035–2045):** Establish new digital workshops; realize e-commerce, 3D printing maintenance, and full-automatic operation; use intelligent equipment as a regular tool for equipment maintenance.

1.1.3.2 *Application cases*

(I) **Intelligent manufacturing**

(1) *3D printing*

In the railway sector, 3D printing technology is mainly applied in the manufacturing of rolling stock parts. DB has worked together with Siemens and other companies to realize the successful application of this technology.

In mid-2015, DB began to implement the R&D of 3D printing technology and manufactured more than 200 components of 50 different types. At the same time, DB also decided to make full use of the experiences gained by aerospace enterprises in the field of 3D printing and jointly carry out R&D and promotion of 3D printing technology with Materialise, EOS, Laser Zentrum Nord, Knorr-Bremse, Siemens, SLM Solution, Leupold Legal, and other companies. When using 3D printing technology to produce spare parts, DB adopted additional technologies, such as selective laser sintering and selective laser melting, which connected the tiny parts on the photosensitive plate to form 3D pictures with existing shapes. Among its partners, Materialise creates 3D printing projects, EOS provides equipment, Laser Zentrum Nord provides printing services, Knorr-Bremse and Siemens provide the construction of rolling stock and ground facilities, SLM Solution develops and promotes the selective laser melting technology, and Leupold Legal takes care of the legal issues, including forming intellectual property rights for 3D printing technology.

Although 3D printing technology has been applied in the manufacturing of railway parts, due to the printing material limits, printing technology limits, cost limits, etc., 3D printing technology can only be applied to some small-scale manufacturing and high-end customized products, and there are considerable restrictions on its wide promotion and use.

(II) **Intelligent equipment**

(1) *Digital interlocking system*

The digital interlocking system enables the command of interlocking equipment to be transmitted to outdoor signal devices, such as switches and signals, through standardized interfaces in a digital manner for the first time, and the outdoor signal devices are powered by locally distributed power terminals; the IP transmission protocol can be used to transmit information through a data network. The control of outdoor signal equipment is implemented by an object controller (electronic programmable controller), and the control distance is far larger than that of electronic interlocking, which can greatly reduce the use of cables. DB Netz plans to realize the DSTW/Neupro architecture in three steps: demonstrative project tests, early-stage batch pilot projects, and batch construction projects.

For this purpose, eight kinds of interfaces were tested in several digital-interlocking demonstrative pilot projects, which are used to connect ETCS section control centers, adjacent interlocking equipment, crossing signal equipment, interlocking equipment monitoring consoles, general digital input and output signals, color light signals, track circuits, axle counter equipment, turnout switch machines, etc.

The early-stage batch pilot projects for the digital interlocking system are distributed in five stations or sections in DB's railway system, such as the Annaberg-Buchholz South Station and Trier-Koblenz sections. The interlocking equipment all adopts a standardized DSTW/Neupro interface. The goal of this stage is to make each manufacturer's products pass the examination and certification and provide a design basis for the subsequent batch production of system equipment.

DB is now promoting the construction and deployment of digital interlocking system and ETCS system simultaneously. The combination of ETCS system and digital interlocking technology can further improve the railway automation, achieving a real technological advance in railway operation and maximizing the benefits.

(III) **Intelligent operation**

(1) *EMU predictive maintenance*

DB and Siemens initiated a pilot project of predictive maintenance in the second half of 2016: the predictive maintenance and overhaul of Velaro D (ICE3) high-speed EMUs. Siemens will become the first railway equipment manufacturer in the railway industry to own and operate a data service center (mobile data service center). The center, located in Munich, will receive data from trains, analyze them in the central diagnostic system, and predict failures through calculation and processing. These prediction results will be used by experts as the basis for verifying the proposed measures and suggestions and will be directly notified to the technicians in DB's maintenance workshop so that these technicians can quickly monitor the current state of rolling stock and take appropriate measures.

1.1.4 *Switzerland*

1.1.4.1 *Development strategy*

In order to improve the operational efficiency of HSR through digitalization and the application of new technologies, SBB proposed the SmartRail 4.0 strategy in 2017, which puts forward the strategic objectives for enterprises and customers from five aspects: cost, capacity, availability, safety, and service. In terms of cost, SBB plans to reduce the annual operating cost by one-third to provide customers with better prices and better service quality. In terms of capacity, SBB plans to increase the transportation capacity by 15–30% and improve the flexibility of operation. In terms of availability, SBB plans to improve the signal system performance by 50%, reduce travel disturbances, and improve punctuality. In terms of safety, SBB plans to reduce 90% of railway operation failures and provide customers with a safer environment. In terms of services, SBB plans to realize internal interoperability and provide customers with a better travel experience.

At the same time, SBB divides the implementation of the SmartRail 4.0 strategy into three stages:

Stage 1 (until 2020): Simplification of the centralized interlocking equipment, standardization of key interfaces, realizing of European Train Control System Stage II, and so on;

Stage 2 (2020–2030): Modularization of ATO trains, improving the technical content of vehicle design;

Stage 3 (2030–2040): Seamless connection of other transportation modes with railways, 50% lower cost of the railway system, and so on.

1.1.4.2 *Application cases*

(I) **Intelligent operation**

(1) *Iltis system*

Siemens (Switzerland) and GGB (Switzerland) have jointly built the Iltis System, the world's first traffic control system based on "cloud computing." Specifically, GGB only needs to obtain permission to use all the functions of the Iltis traffic control system online by purchasing the license, instead of purchasing all kinds of hardware equipment and software systems previously needed for train traffic control. The biggest advantage of this system is that GGB no longer has to invest heavily in hardware and software related to traffic control, and the maintenance work is also done directly by Siemens without the need for technicians to visit the site.

(2) *ADL system*
The adaptive train control system (ADL system) is an independent part of the rail control system (RCS), which has two main functions. The first is to identify and optimize the conflicts in train operation. By providing slow-down suggestions to train drivers in advance, the system can prevent the deep deceleration of trains en route due to ground limit signals or stop signals so as to avoid braking and speeding up again and, at the same time, achieve the purpose of protecting the railway line and saving energy. The second is the EcoDrive function, which is used to reduce and avoid trains en route arriving at their destination more than one minute earlier.

1.1.5 *United Kingdom*

1.1.5.1 *Development strategy*

The UK formulated the Digital Railway Strategy in 2018 to integrate a series of key works and technologies for the greater benefits for passengers and freight customers, as shown in Figure 1.4.

In Digital Railway Strategy, there is a three-stage roadmap, including short-, medium-, and long-term goals as follows:

(1) **Short-term goals (until 2019):** Based on the current railway development achievements, update and expand the existing railway lines. At the same time, learn from the advanced technological achievements in similar fields at home and abroad and carry out the application of innovative technologies.

(2) **Medium-term goals (2019–2027):** Focus on the five major railway lines of Anglia, London North East, South East, Wessex, and Western, and deploy their development plans. Greatly improve the limited transport capacity through digitalization, which will bring wider benefits to railway users and the British economy with lower operating costs.

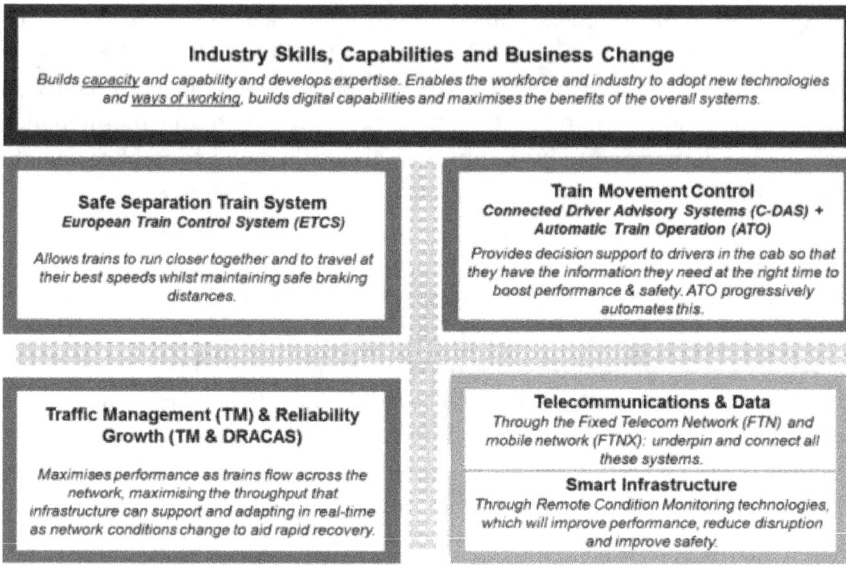

Figure 1.4. Key components of digital railway strategy.

(3) **Long-term goals (after 2027):** Further integrate the digital technologies with railway application scenarios. It is expected to reduce costs in digital signal, intelligent infrastructure, and train control. In the future, digital technologies will be widely applied to the daily operation of railways.

1.1.5.2 *Application cases*

(I) **Intelligent equipment**

(1) *3D scanner*

In Scotland, Network Rail used a 3D scanner for the first time to survey the track structure so as to realize accurate modeling of tunnel infrastructure.

Since April, the track geometry (ATG) engineering team of Network Rail has been using 3D scanning technology to collect detailed information about tracks and surrounding facilities, such as bridges, fences, and tunnels. The application of 3D scanning technology improved the safety of railway engineers, reduced the time spent by workers on tracks, and improved the detail and accuracy of data.

(2) *VR virtual modeling*
Network Rail has adopted VR technology in station reconstruction, which can save time and cost and improve the safety level of stations. With the assistance of VR technology, Network Rail can easily achieve 4D virtual modeling, thus making construction preparations in advance and reducing the disturbance to train services. At the same time, VR technology enables engineers to avoid duplication of work, identify potential risks and hazards caused by station reconstruction, and improve the safety level of stations.

(3) *BVR UAV*
In Britain, a survey was conducted on the first section of the HS2 line using SenSat UAV and cloud technology. Using a small UAV for railway line survey can not only reduce the cost and relieve the heavy financial burden of the HS2 line but also demonstrate the UAV's BVR capability, which lays a good foundation for its wide application in the future.

(II) **Intelligent operation**
(1) *Intelligent lighting equipment*
Intelligent lighting equipment refers to a modular interactive lighting post, which is a combination of LED energy-saving lighting technology, cameras, speakers, internet, and a series of charging applications.

Embedded CCTV cameras are installed in the intelligent lighting post, which can monitor, record, and feedback the real-time video to the accident control room. The security staff in the accident control room can make quick and accurate decisions according to the real-time situation. An intelligent lighting post has a built-in broadcasting system, which can be used by station managers to send out information about train delays or cancelations or to communicate directly with passengers in a certain area of the station. In case of an emergency, the station staff can use the lighting post to issue important safety announcements and remind passengers of emergency evacuation procedures. Integrated with the broadcasting system and LED lighting, the lighting post can guide passengers to the safe area via an effective route.

(2) *Tracksure project*
A British railway organization initiated a project known as Tracksure (Figure 1.5), aimed at developing a cheap and easy-to-fit sensor that can

Figure 1.5. A train equipped with Tracksure.

turn any rolling stock into a track monitor. The sensors are embedded in Siemens' GSM-R locomotive radio systems and collect data about the condition of rails and rail beds by picking up track vibrations. Variations in these data reveal the location of under-track voids and gaps between the sleeper and ballast that lead to an increased risk of rail breakage and generally result in reduced vehicle performance.

1.1.6 *China*

1.1.6.1 *Development strategy*

China has been implementing its IHSR development strategy in three stages:

In stage 1 (2018–2020), China has constructed IHSR demonstration projects: Beijing–Zhangjiakou and Beijing–Xiongan IHSR Demonstration Projects.

In stage 2 (2021–2025), China will strive to make great breakthroughs in IHSR theory and technology and fully master the technology of the whole industrial chain of IHSR from design and construction to operation. This mainly includes the establishment of BIM intelligent building standard systems, self-learning self-adaptive genealogically intelligent EMUs, and comprehensive-perception driverless train operation (DTO).

In stage 3 (2026–2035), China will upgrade the IHSR application from assistance to automatic control so as to realize full automatic control of railway operation. This mainly includes the wide application of intelligent building technology, self-repairing intelligent EMUs, and exploration of unattended train operation (UTO).

1.1.6.2 *Application cases*

(I) **Intelligent construction**

(1) *BIM*

Trial applications of BIM technology in China's railway construction have been gradually carried out, and progress has been made in compiling relevant standards, which has significantly improved China's high-speed rail construction management level.

In terms of strengthening process management, China follows the principle of putting testing first and carrying out process simulation for difficult and important projects. In terms of optimizing the implementation scheme, BIM modeling and VR simulation technologies have been successfully applied to optimize the design scheme of large railway passenger stations. In terms of optimizing construction organization, the 3D model combined with a collaborative management platform is used to examine the construction organization design of bridges crossing large rivers and long tunnels under complex conditions and optimize technical schemes, engineering measures, and resource allocation; at the same time, in terms of schedule control, QR code technology is employed to compare the actual schedule with the plan in real time so as to realize the dynamic management and timely warning of construction schedule.

(II) **Intelligent equipment**

(1) *Unmanned aerial vehicle (UAV)*

UAVs are playing an important role in surveying railway lines, checking for changes in the surrounding environment, and monitoring disasters along the lines and have therefore been recognized by some railway units in China.

During the construction of Harbin International Container Center, UAV was applied to solve the contradiction of a large construction area against the limited vision of monitoring personnel. China Railway Shanghai Group Co., Ltd. regards UAVs as an important means to investigate the potential safety hazards of flood control points. In combination with other measures, such as flood control channels, it has established a 3D flood control and emergency rescue system. China Railway Xi'an

Group Co., Ltd. carries out mapping and investigation of topography and geological conditions along the railway line by UAV, satellite survey, and other technical means, and through integrating meteorological, hydrological, rainfall, foreign body monitoring, and alarm information, the company has built a flood control management information system platform to continuously improve the level of risk prevention and control.

China Railway Guangzhou Group Co., Ltd. uses UAVs to carry out safety monitoring of HSR power supply and has introduced the 6C system (HSR power supply safety detection and monitoring system) in the Guangzhou power supply section, where UAVs and unmanned substations are important components. Applying the remote wireless monitoring and transmission systems, including UAV detection, the relevant staff can directly understand the train status of the online running EMU of the company, which realizes all-around inspection to ensure that railway staff can discover and eliminate potential safety hazards of equipment in time, thus ensuring the safety of train operation.

(2) *Automatic train operation*

The automatic train operation (ATO) system boasts five functions: automatic train departure, automatic sectional operation, automatic stop at the station, automatic door opening after stopping, and automatic joint control between the train door and platform safety door. Meanwhile, the ATO system can automatically adjust the maximum speed of the train according to the train running section. An unmanned cab is shown in Figure 1.6.

Figure 1.6. An unmanned cab.

(3) *Intelligent management system for railway train operation safety*
A railway-dedicated "Intelligent Management System for Railway Train Operation Safety" has been developed successfully, and the main elements of the system include Anquanbao, a safety watch, and an onboard controller. The main functions include fatigue degree monitoring, action and behavior recognition, user identification, abnormal state early warning, online real-time monitoring, and cloud intelligent management. It can not only effectively monitor the physiological indicators and action tracks of the wearer but also handle abnormal alarms and can be connected with the data center in real time to realize comprehensive data processing so that management personnel can control the working state of train drivers synchronously.

(III) **Intelligent operation**
(1) *E-payment*
With the help of smart terminals such as mobile phones, e-payment, including Alipay, WeChat payment, UnionPay, Apple Pay, QR code, and NFC, has become an integral part of our daily lives.

Guangzhou Metro applies the IC credit card, a kind of UnionPay card capable of making small-amount payments without a password, which is very convenient for the passengers to take the metro without the need to recharge in advance. Chengdu's entire metro network has officially launched the mobile electronic payment system, and passengers can use WeChat, Alipay, or other means to buy tickets or recharge "Tianfu Tong" cards via mobile phones to take Lines 1, 2, 3, and 4. Qingdao Metro Line 3 trialed "cloud ticket purchase" when it was put into service, using which passengers can purchase tickets online through the mobile APP and then scan the code at the station to fetch the tickets.

Shanghai has begun to try out the "online ticket service" in metro stations. Passengers can purchase tickets electronically through the "Alipay" or "Fufeitong" APP on their mobile phones.

1.2 Definition and Connotation of IHSR

1.2.1 *Background of IHSR*

Since the beginning of the new century, we have been witnessing remarkable breakthroughs in information and intelligence technologies. While a new round of technological revolution and industrial transformation is emerging, as well as the application of new technologies such as AI, big data, cloud computing, IoT, BIM, and satellite navigation flourishing, society rapidly ushers in the intelligent era and embarks on the process of upgrading the traditional industries with high technologies. As a result, intelligent transport is being developed in all aspects, and many innovations and applications are being made in the intelligent grid, smart city, smart agriculture, and smart medical care. The intelligent era also puts forward more stringent requirements for the innovation and development of railways: for instance, how to improve the efficiency of transport organization to solve the contradiction between transport demand and capacity supply; how to accurately understand the demands of clients to provide more targeted and high-quality customer service and to design passenger and freight products adapted to market demands; how to realize the automatic identification of potential risks to improve the ability to prevent accidents and faults and adapt to the high-speed and heavy-duty transport mode; as a heavy asset industry, how to master the deterioration mechanism of equipment to improve the delicacy management of equipment, accurately predict the equipment maintenance time and items, and greatly reduce the equipment maintenance cost. In this background, the concept of intelligent railways emerges, and research on intelligent railways has started worldwide in an all-round manner.

Recently, many countries, including Germany, France, the UK, the US, and Japan, have formulated their railway digitalization development strategies so as to optimize transport service quality, enhance transport safety and transport organization efficiency, reduce transport costs, and improve operation efficiency by promoting the efficient integration of new technologies and railway services. Undoubtedly, future competition in the HSR field worldwide will hinge on the progress of digitalization and intelligentization.

Developing IHSR is an inevitable choice in the general trend of developing information and intelligence technologies around the world. Today, a new round of technological revolution is upcoming, and technologies such as information, new materials, manufacturing, and new energy are boosting and widely applied to almost all fields, which has facilitated major technological changes characterized by environmental friendliness, intelligence, and ubiquity and helped social production and consumption shift from industrialization to automation and intelligentization. Consequently, new production modes, industrial patterns, business models, and economic growth areas continue to spring up.

The core of the new round of technological revolutions is the deep integration of information and intelligence technologies in the manufacturing industry. Now, intelligent manufacturing based on cyber-physical systems, such as intelligent equipment and intelligent factories, is leading the change of manufacturing mode; network crowdsourcing, co-design, mass customization, precise supply chain management, full life cycle management, and e-commerce are reshaping the industrial value chain system; and intelligent terminal products, such as wearable products and home appliances, are expanding new manufacturing fields. Meanwhile, information and intelligence technologies are greatly influencing the development of transport industry in various countries, presenting the tendency of automation, intelligence, and smartness. **From the perspective of the development of the intelligent railway technology around the world, the application and innovation of new technologies, such as IoT, big data, and AI, in key railway technical fields are being continuously promoted. Countries are speeding up the progress of transport informationization and intelligentization to provide passengers with convenient services.**

Developing IHSR is an effective way to enable railway transport to meet the requirements of economic and social development. As the artery of the national economy, the key infrastructure that ensures the people's well-being, and the backbone and major transport mode of the comprehensive transport system, the railway plays a crucial role in the economic and social development for various countries. In the new era, the railway industry should fulfill the mission of facilitating technological innovations according to the requirements of quality development, giving support to economic and social development, and implementing major national strategies in all countries to boost the construction of an innovative country and enable the people to feel more satisfied. To this end, the railway industry must vigorously push ahead with the innovation and application of information and intelligence technologies and actively promote the development of IHSR. Therefore, **developing IHSR is an important supporting measure to make China an innovative country. Developing IHSR is a significant means for the railway to better serve the people.**

Developing IHSR is vital to realizing the sustainable development of the railway itself. In recent years, with the continuous construction of large-scale railways, especially HSR, in various countries, safety and efficiency have become more and more important for the sustainable development of railways. On the one hand, due to the rapid extension and expansion of railway networks and the application of new technologies and equipment, the construction and operation of railways, especially HSR, are facing a more severe safety situation. On the other hand, due to the fiercer competition among various modes of transport, convenience, shorter time, and comfortable experience, which directly determine their market attraction and market shares, become the focus of competition. It is urgent to further improve the efficiency of transport organizations under the condition that HSR are connected and form a network, enhancing passenger travel efficiency, passenger travel experience, and the market competitiveness of railways. In addition, HSR has an obvious comparative advantage in energy conservation and environmental protection. However, as other transport modes pay more attention to energy conservation and environmental protection and continuously improve their technical equipment, HSR needs to consolidate its advantage by virtue of the latest technological innovations. Therefore, **developing IHSR can effectively improve the safety level, effectively enhance transport efficiency and benefits, and further reduce energy consumption.**

1.2.2 *Concept of IHSR*

In this background, IHSR is defined as follows based on the characteristics of intelligence and the internal and external requirements for IHSR:

IHSR is a new-generation HSR system to which new technologies such as cloud computing, big data, IoT, mobile internet, AI, satellite navigation, and BIM are widely applied so that resources can be comprehensively and efficiently used to realize the comprehensive perception, ubiquitous interconnection, fusion processing, active learning, and scientific decision-making of HSR mobile equipment, fixed infrastructure, and internal and external environment information and to implement integrated management of the full life cycle.

IHSR is a new-generation HSR system which deeply integrates information and intelligence technologies with all the HSR fields. Information and intelligence technologies endow IHSR with new functions and characteristics in addition to the traditional characteristics of HSR, so that IHSR can comprehensively perceive the railway system elements in real time, adapt to the environment, and actively learn to realize the deep fusion and wide sharing of all data, realize fault self-diagnosis of facilities and equipment, and provide scientific and reasonable support for

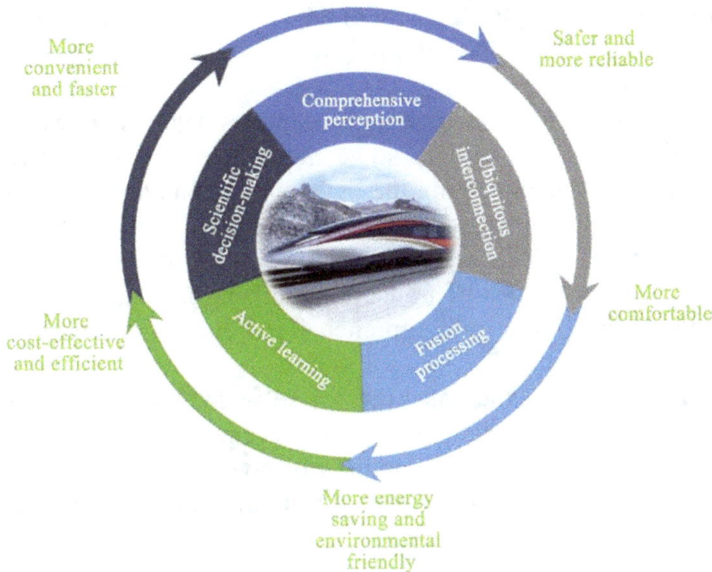

Figure 1.7. Characteristics and development goals of IHSR.

decision-making by mining, analyzing, and reasoning with mass data and information. To sum up, IHSR is characterized by comprehensive perception, ubiquitous interconnection, fusion processing, active learning, and scientific decision-making, as shown in Figure 1.7.

Comprehensive perception: By building a sensor network characterized by multi-dimensional perception and wide coverage, effective communication among fixed facilities, mobile equipment, and the natural environment of HSR is implemented to comprehensively and thoroughly perceive all railway elements, such as the railway network, train, climate, environment, and passengers, so as to provide information support for the operations management of HSR.

Ubiquitous interconnection: By relying on such technologies as IoT, wireless communication, sensing, and detection, wide, deep, and safe exchanges and sharing of various types of information of fixed facilities and mobile equipment are implemented. Further, intelligent dispatching and safety control of trains are carried out on the basis of full spatial-time and dynamic HSR information collection and integration so as to realize the coordinated management of the railway network and trains and improve the operation efficiency of HSR.

Fusion processing: Multi-source heterogeneous data resources in different time and space are used to solve the problem of inconsistent and incomplete data. Accompanying the development and application of new technologies, the new-generation IHSR system will realize deep integration across various fields, such as the sharing and fusion of information of meteorology, other modes of transport, tourism, and business, so as to provide a sufficient basis for comprehensive decision-making.

Active learning: With the capability of self-adaptation to the environment and learning, IHSR can continuously accumulate a large amount of data and information, select the optimal behavior strategy by perceiving the external environment, and constantly iterate, improve, optimize, and update the database of the system.

Scientific decision-making: Based on big data analysis and knowledge reasoning, decision-making information is optimized by mining, refining, computing, analyzing, reasoning, forecasting, and matching massive heterogeneous information in the whole life cycle so as to provide more efficient and swift support for auxiliary operations management and business decision-making by relevant organizations.

IHSR will achieve the development goals featuring greater safety and reliability, cost-effectiveness, efficiency, comfort, convenience, energy savings, and environmental friendliness.

Safer and more reliable: Through the perception of the status of fixed facilities, mobile equipment, transport processes, and the natural environments of HSR, the prediction and warning of equipment faults and traffic events are carried out for advanced prevention so as to improve the capacity of HSR to ensure safe operation.

More cost-effective and efficient: Through the intelligent optimization of HSR transport organizations, transport efficiency is improved. Through the management of the full life cycle of HSR equipment and facilities, the transformation from planned maintenance to condition-based maintenance is realized so as to reduce the maintenance cost. Through the lean operations management of HSR, operational benefits are enhanced.

More comfortable: A number of user-friendly designs are applied to EMUs and stations so as to provide passengers with all-round and whole-process services, meet their diverse and personalized service requirements, and facilitate their travel.

More convenient and faster: By innovating in modes of transport and rebuilding business processes, a one-stop transport service from door to door is provided for passengers so as to provide transparent travel information, diversified travel services, and a convenient service experience.

More energy savings and environment friendly: The power structure and the mode of train operation control of EMUs are optimized so as to realize on-line monitoring, intelligent analysis, and energy-saving control of power consumption in all links and thus reduce the energy consumption of HSR. The building structure and the equipment performance are optimized so as to minimize environmental and noise pollution and accelerate the green and sustainable development of HSR.

1.3 System Architecture of IHSR

The intelligent HSR system architecture specifies the composition of intelligent HSR and the interrelationships between the components in aspects such as technology, data, and standards, and it can provide important top-level design guidance for the design, construction, and operation of intelligent HSR. The intelligent HSR system architecture is mainly composed of a technical system framework, a data system framework, and a standard system framework, which effectively guides the intelligent innovation practice in the Beijing–Zhangjiakou HSR and the Beijing–Xiongan Intercity Railway. It fills the gap in the top-level design of HSR intelligence in the world.

1.3.1 *Dimensional design of IHSR system architecture*

Intelligent HSR is a complex cyber-physical system covering multiple disciplines, fields, and new technologies. Its construction is not accomplished overnight but needs to be promoted in stages and steps under the guidance of a unified system architecture. The intelligent HSR system architecture is a guiding framework for the construction and operation of

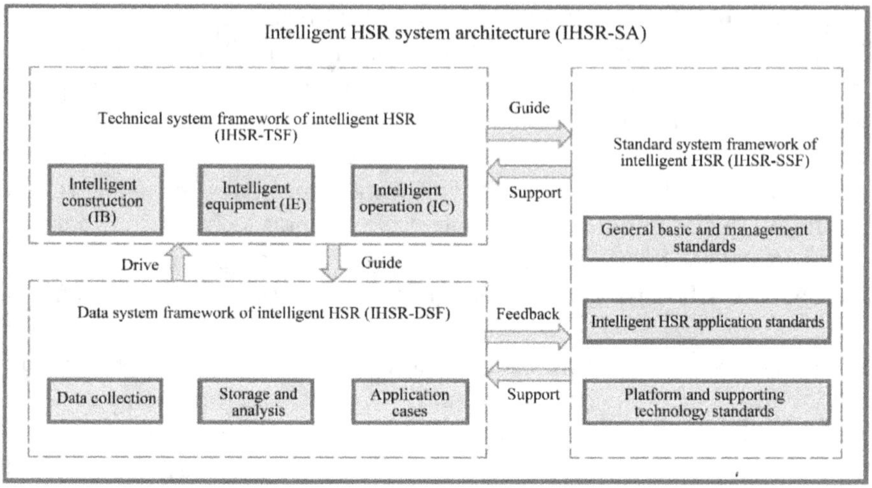

Figure 1.8. Intelligent HSR system architecture.

intelligent HSR based on the systematic summary of the intelligent innovation practices in the Beijing–Zhangjiakou HSR and the Beijing–Xiongan Intercity Railway, covering the technical system framework, standard system framework, and data system framework of intelligent HSR. With the continuous development of technologies and application needs related to intelligent HSR, its system architecture can be continuously optimized and expanded according to actual needs in the future. The specific content is shown in Figure 1.8.

The technical system framework of intelligent HSR is the core of the intelligent HSR system architecture. It defines the technical composition of three sections, namely intelligent construction, intelligent equipment, and intelligent operation, and provides guidance for the formulation of the data system and standard system frameworks.

The standard system framework of intelligent HSR provides standard support for the implementation and application of the technical system and data system frameworks in terms of data, technology, and management. It defines the general basic and management standards, technical standards of intelligent HSR, platform and supporting technology standards, etc.

The data system framework of intelligent HSR defines services such as data collection, storage and analysis, and application display for the data generated by the innovative applications of intelligent construction, intelligent equipment, and intelligent operation defined in the technical system framework and other related internal and external data, with the life cycle management of data as the focus, to provide data driving for the technical system framework and feedback for the standard system framework.

1.3.2 *Technology system of IHSR*

1.3.2.1 *Overall architecture*

The technical system framework of intelligent HSR is designed based on the principle of classification and hierarchical design, and it includes five levels from top to bottom, namely section, field, direction, innovation, and support platform, totally involving 3 sections, 10 fields, 18 directions, *N* innovations, and 1 platform. The detailed composition is shown in Table 1.1.

The 3 sections refer to intelligent construction, intelligent equipment, and intelligent operation.

The 10 fields refer to survey and design, engineering construction, construction management, mobile equipment, communication and signaling, traction power supply, inspection and monitoring, passenger service, transport organization, maintenance, and repair under the framework of the 3 sections.

The 18 directions refer to the space–air–ground integrated engineering survey, BIM-based engineering design, intelligent construction of bridge, tunnel, subgrade, and track works, intelligent construction of passenger station works, intelligent construction of communication, signaling, power and electrification works, BIM+GIS-based engineering construction management, intelligent EMU, intelligent comprehensive inspection train, signaling, communication, intelligent traction power supply, intelligent inspection and monitoring, intelligent passenger transport, intelligent ticketing, intelligent comprehensive dispatching, intelligent traffic control, integrated operation and maintenance PHM of track, communication, signaling, and power supply, and intelligent operation and maintenance of EMU under the framework of the 3 sections and 10 fields.

N innovations refer to intelligent innovations such as GIS-based intelligent exploration, BIM modeling, intelligent beam yard, intelligent construction of passenger stations, intelligent construction of electrification works, BIM-based virtual construction, intelligent monitoring of EMUs, intelligent inspection data analysis, intelligent traction

Table 1.1. Technical system framework of intelligent HSR.

Technical System Framework of Intelligent HSR			
3 Sections	**10 Fields**	**18 Directions**	**N Innovations**
Intelligent construction	Survey and design	Space-air-ground integrated engineering survey	GIS-based intelligent exploration; space-air-ground integrated intelligent survey and mapping; digital survey delivery
		BIM-based engineering design	BIM modeling; collaborative design; digital design delivery
	Engineering construction	Intelligent construction of bridge, tunnel, subgrade, and track works	Intelligent beam yard; intelligent shield tunneling; intelligent subgrade filling; intelligent slab yard
		Intelligent construction of passenger station works	Intelligent construction of passenger stations; coordination for construction of integrated transport systems
		Intelligent construction of communication, signaling, power and electrification works	Intelligent construction of electrification works; intelligent construction of communication works; intelligent construction of signaling works; intelligent construction of information works
	Construction management	BIM+GIS-based engineering construction management	BIM-based virtual construction; full-process digital management; digital completion delivery

Intelligent equipment	Mobile equipment	Intelligent EMUs	Intelligent monitoring of EMUs; intelligent diagnosis of EMUs; intelligent services of EMUs
		Intelligent comprehensive inspection trains	Intelligent inspection equipment; analysis of intelligent inspection data
	Communication & signaling	Signaling	Automatic control of station route; train operation control; automatic train operation (ATO)
		Communication	Intelligent communication bearer network; intelligent mobile communication; intelligent dispatching communication
	Traction power supply	Intelligent traction power supply	Intelligent traction substation; simple, unified and standardized OCS; intelligent power supply dispatching system
	Inspection and monitoring	Intelligent inspection and monitoring	Inspection and monitoring of infrastructure; monitoring and warning of natural disaster; intelligent monitoring of perimeter intrusion; intelligent monitoring of environment
Intelligent operation	Passenger service	Intelligent passenger transport	Integrated production command and management of passenger transport; intelligent management and monitoring of passenger transport equipment; key station and train service; passenger safety of stations and trains; intelligent station and train service

(*Continued*)

Table 1.1. (*Continued*)

Technical System Framework of Intelligent HSR			
3 Sections	**10 Fields**	**18 Directions**	**N Innovations**
		Intelligent ticketing	Electronic ticketing; intelligent product design and ticketing organization; travel planning; targeted marketing; integrated traffic information sharing
	Transport organization	Intelligent comprehensive dispatching;	Transportation situation awareness; planning integration and coordination; intelligent plan adjustment; intelligent emergency dispatching
		Intelligent traffic control;	**Automatic adjustment of train operation plans; route and command control; integrated simulation of train operation control**
	Maintenance and repair	Integrated operation and maintenance of track, communication, signaling, and power supply;	Digital log management; intelligent fault diagnosis; condition-based intelligent integrated warning; intelligent decision-making on operation and maintenance
		Intelligent operation and maintenance of EMU	Fault prediction; health assessment; operation and maintenance analysis; decision-making support
1 Platform	Basic platform of intelligent HSR	Master data service; Metadata service; Geographic information service; Big data analysis services; Artificial intelligence service.	
	Supporting technology	Network security assurance; Beidou satellite navigation; Internet of Things.	

substation, intelligent inspection and monitoring of infrastructure, intelligent station and train service, electronic ticketing, automatic adjustment of train operation plan, and intelligent decision-making of operation and maintenance under the framework of the 3 sections, 10 fields, and 18 directions.

One platform refers to the basic platform of intelligent HSR providing support and services for the technical innovation of intelligent HSR. The basic platform provides master data, metadata, geographic information, big data analysis, artificial intelligence, and other services for N innovations in 3 sections (namely intelligent construction, intelligent equipment, and intelligent operation), 10 fields, and 18 directions and adopts supporting technologies such as network security assurance, Beidou satellite navigation, and IoT.

1.3.2.2 *Framework of intelligent construction*

According to the construction process of infrastructure, the intelligent construction section is divided into three fields, namely, survey and design, engineering construction, and construction management.

The field of survey and design includes space–air–ground integrated engineering survey and BIM-based engineering design. The space–air–ground integrated engineering survey mainly includes GIS-based intelligent exploration, space–air–ground integrated intelligent survey and mapping, and digital survey delivery; BIM-based engineering design mainly includes BIM modeling, collaborative design, and digital design delivery.

The engineering construction field includes the intelligent construction of bridges, tunnels, subgrades, and tracks, the intelligent construction of passenger stations, and the intelligent construction of communication, signaling, power, and electrification projects. The intelligent construction of bridges, tunnels, subgrades, and tracks mainly includes intelligent beam yard, intelligent shield tunneling, intelligent subgrade filling, and intelligent slab yard; the intelligent construction of passenger stations mainly includes intelligent construction of passenger stations and coordination for the construction of integrated transport systems; the intelligent construction of communication, signaling, power, and electrification projects mainly includes intelligent construction of electrification projects, communication projects, signaling projects, and information projects.

The field of construction management includes BIM+GIS-based engineering construction management, mainly involving BIM-based virtual construction, full-process digital management, and digital completion delivery.

To sum up, the intelligent construction section can be summarized into 3 fields, 6 directions, and 19 innovations, and the innovation content can be expanded as needed in the future. The specific content is shown in Figure 1.9.

1.3.2.3 *Framework of intelligent equipment*

Intelligent equipment can be divided into mobile equipment, communication & signaling, traction power supply, and inspection & monitoring according to the classification of the main business objects. The field of mobile equipment includes intelligent EMUs and intelligent comprehensive inspection trains. The intelligent EMUs mainly include the intelligent monitoring of EMUs, the intelligent diagnosis of EMUs, and the intelligent services of EMUs; intelligent comprehensive inspection train mainly includes intelligent inspection equipment and intelligent inspection data analysis.

The communication & signaling field includes signaling and communication. The signaling field mainly includes automatic control of station routes, train operation control, and ATO; the communication field mainly includes an intelligent communication bearer network, intelligent mobile communication, and intelligent dispatching communication.

The field of traction power supply includes intelligent traction power supply and others, mainly involving innovations such as intelligent traction substations, a simple, unified, and standardized OCS, and an intelligent power supply dispatching system.

The field of inspection and monitoring includes intelligent inspection and monitoring and others, mainly involving inspection and monitoring of infrastructure, monitoring and warning of natural disasters, intelligent monitoring of perimeter intrusion, and intelligent monitoring of the environment.

To sum up, the intelligent equipment section can be summarized into 4 fields, 6 directions, and 18 innovations, and the innovation content can be expanded as needed in the future. The specific content is shown in Figure 1.10.

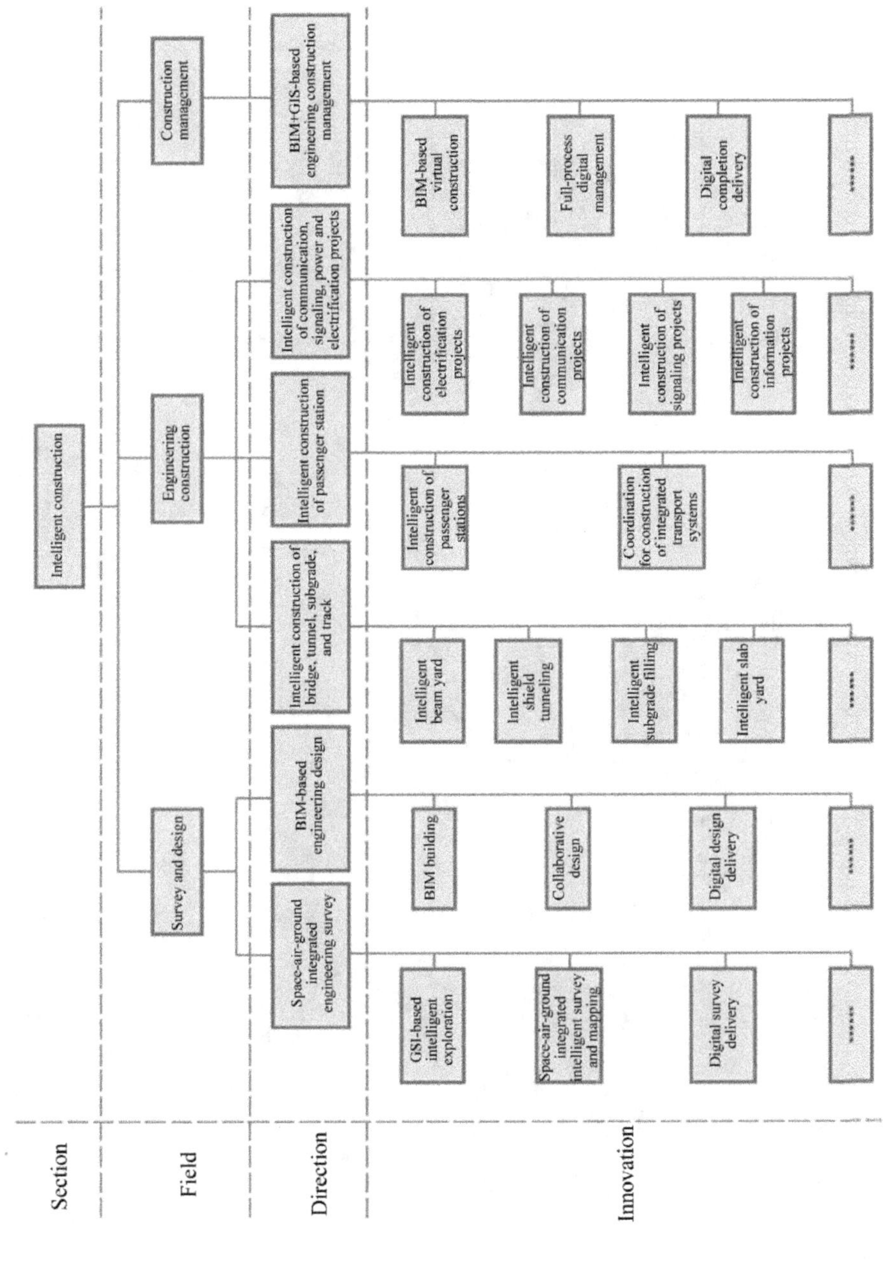

Figure 1.9. Framework of intelligent construction technology system.

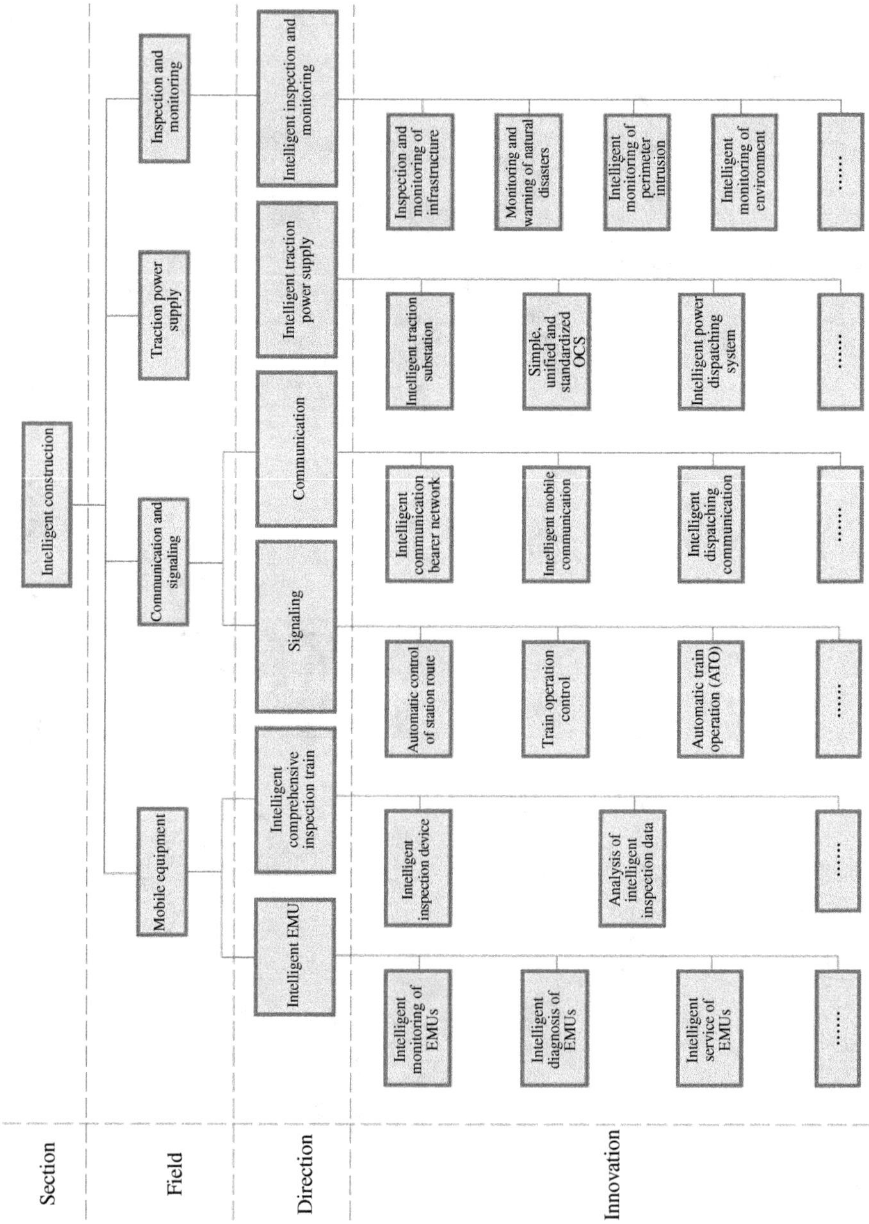

Figure 1.10. Framework of intelligent equipment technology system.

1.3.2.4 *Framework of intelligent operation*

Intelligent operation can be divided into passenger service, transport organization, maintenance, and repair according to the main business classification of operation.

Passenger service includes intelligent passenger transport and intelligent ticketing. The intelligent passenger transport mainly includes integrated production command and management of passenger transport, intelligent management and monitoring of passenger transport equipment, key station-and-train service, passenger safety of stations and trains, and intelligent station and train service; intelligent ticketing mainly includes electronic ticketing, intelligent product design and ticketing organization, travel planning, targeted marketing, and integrated traffic information sharing.

The field of transport organization includes intelligent comprehensive dispatching and intelligent traffic control. The intelligent comprehensive dispatching mainly includes transportation situation awareness, planning integration and coordination, intelligent plan adjustment, and intelligent emergency dispatching; intelligent traffic control mainly includes automatic adjustment of train operation plans, routes and command control, and integrated simulation of train operation control.

The maintenance and repair field includes the integrated operation and maintenance of track, communication, signaling, and power supply, and the intelligent operation and maintenance of EMUs. The integrated operation and maintenance of track, communication, signaling, and power supply mainly includes digital record management, intelligent fault diagnosis, integrated condition-based warning, and intelligent decision-making on operation and maintenance; the intelligent operation and maintenance of EMUs mainly includes fault prediction, health assessment, operation and maintenance analysis, and decision-making support.

To sum up, the intelligent operation section can be summarized into 3 fields, 6 directions, and 25 innovations, and the innovation content can be expanded as needed in the future. The specific content is shown in Figure 1.11.

1.3.2.5 *Framework of basic platform*

The basic platform of intelligent HSR is the data center, model center, and analysis center for the application systems of intelligent construction,

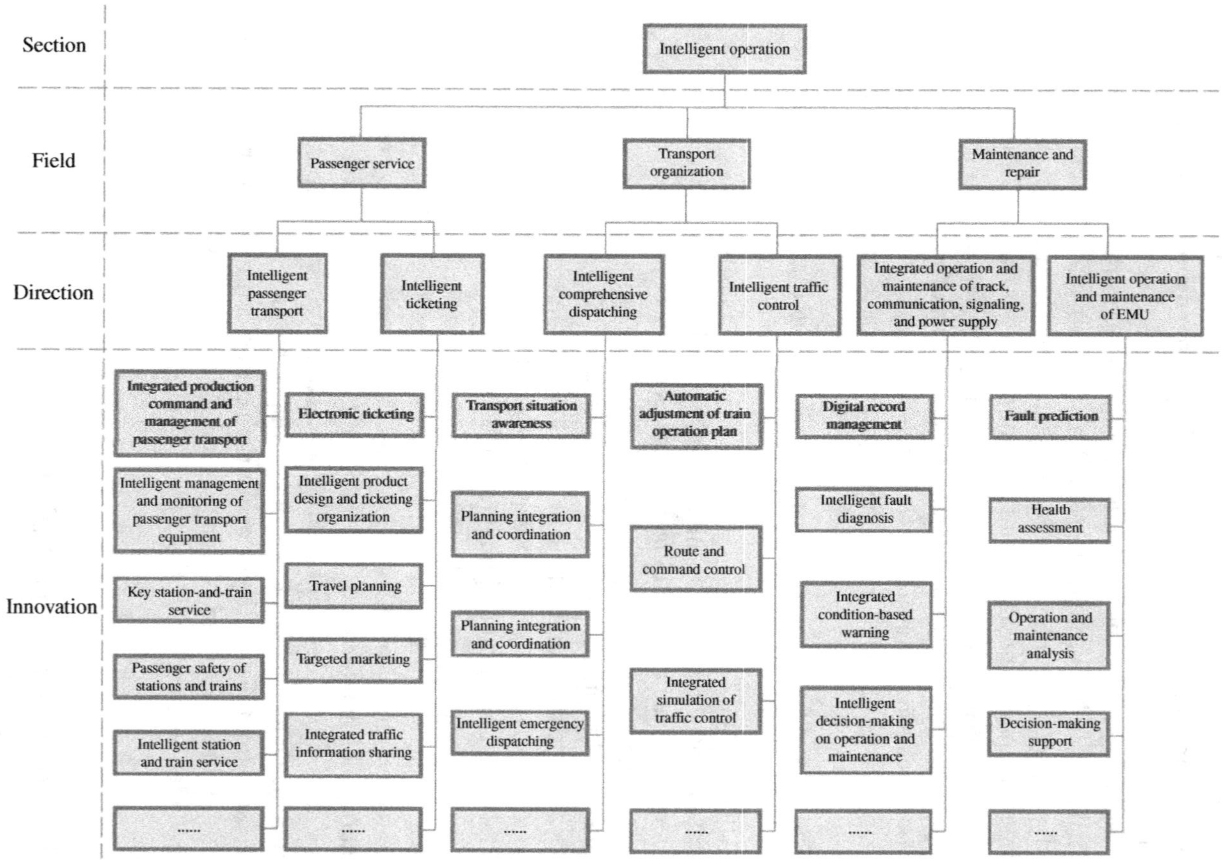

Figure 1.11. Framework of intelligent operation technology system.

intelligent equipment, and intelligent operation, and it uniformly provides services such as big data, model algorithms, AI, BIM+GIS, and Beidou information for all application systems:

(1) **Big data:** It provides services such as data access, storage, calculation, and sharing in terms of all data types, realizes centralized data management, offers cross-discipline data-sharing services to various systems, connects data links, and reduces the repeated construction of point-to-point data interfaces.

(2) **Model algorithms:** They are oriented toward the built-in mainstream statistical analysis and data mining model algorithms, which can be called by various application systems. For each application, it is only necessary to pay attention to specific application scenarios instead of the repeated development of general algorithms, which is conducive to reducing the development workload and shortening the development cycle.

(3) **AI:** For Computer Vision, speech recognition, unstructured text analysis, etc., it provides services such as data management, sample annotation, sample management, model training, and model deployment to support application scenarios such as personnel behavior analysis, equipment fault recognition, and semantic analysis for text from various disciplines.

(4) **BIM + GIS:** With the spatial information database for railways as the carrier and BIM technology and geographic information system (GIS) as the core, it provides BIM+GIS-based spatio-temporal information visualization service, element query service, network analysis, statistical analysis service, geometric service, 2D and 3D spatial analysis and visualization service for various business applications.

(5) **Beidou information:** With the Beidou satellite navigation technology as the core, it has the basic service capability of providing Beidou high-precision positioning, time service, and short-message communication functions to support the application of Beidou satellite navigation technology in railways nationwide. The main functions include online equipment management, high-precision position service, short-message communication service, time service, and operation management.

1.3.2.5.1 Relationship between basic platform and intelligent construction section

The basic platform collects information on raw materials, such as reinforcement and concrete, production data of the mixing plant, laboratory test results, surrounding rock measurements, settlement observations, and construction logs, from the intelligent construction section. At the same time, it provides modeling and data support for quality trend analysis of raw materials and semi-finished products, trend analysis of safety status in the whole construction process, project progress analysis, big data analysis for laboratories, big data analysis for settlement observation, etc.

1.3.2.5.2 Relationship between basic platform and intelligent equipment section

The basic platform collects data from the intelligent equipment section, such as disaster type involving alarm, kilometer post at monitoring point, alarm time, alarm level, speed limit section and speed limit value, onboard inspection and monitoring data, inspection and monitoring data for track inspection cars and manual inspection, basic account data of dynamic faults, static faults, and track maintenance equipment, EMU routing, maintenance, operation failure, EMU allocation, and operation overview to provide support for big data analysis for disaster monitoring, equipment safety warning analysis, etc.

1.3.2.5.3 Relationship between basic platform and intelligent operation section

The basic platform collects data from the intelligent operation section, such as ticket sales data, train punctuality and delay information, train stops information, passenger ticket stubs, internet order data, basic information of EMUs, train-set routing, actual routing, running kilometers, daily operation, and daily revenue to provide support for the benefit evaluation of passenger train operation, price monitoring of passenger transport market, failure rate analysis of EMU train operation equipment per million kilometers, etc.

1.3.3 *Standard system of IHSR*

The standard system framework for intelligent HSR consists of general basic and management standards, technical standards for intelligent HSR, and platform and supporting technology standards, mainly including 3 primary categories, 9 secondary categories, 26 tertiary categories, and several extensible categories, as shown in Table 1.2.

1.3.3.1 *General basic and management standards*

General basic and management standards include two secondary categories: general basic standards and management and service standards. The general basic standards consist of two tertiary categories: definitions and terms and general rules for preparation; the management and service standards consist of two tertiary categories: technology R&D and management standards and intelligence level evaluation standards.

1.3.3.2 *Technical standards*

The technical standards for intelligent HSR include three secondary categories: intelligent construction standards, intelligent equipment standards, and intelligent operation standards.

(1) **Intelligent construction standards consist of three tertiary categories:** survey and design standards, engineering construction standards, and construction management standards. The survey and design standards include extensible categories, such as standards for the implementation of railway engineering information modeling in the design stage. The engineering construction standards include extensible categories such as technical specification for intelligent construction of railway ballastless track slabs/sleepers, technical specification for intelligent filling of HSR subgrades, and technical specification for intelligent construction of railway shield tunnels. The construction management standards include extensible categories such as standards for BIM-based virtual construction and standards for full-process digital management.

(2) **Intelligent equipment standards consist of four tertiary categories:** mobile equipment standards, communication and signaling standards,

Table 1.2. Standard system framework of intelligent HSR.

Standard System Framework of Intelligent HSR			
3 Primary categories	**9 Secondary categories**	**26 Tertiary categories**	**Extensible categories**
General basic and management standards	General basic standards	Definitions and terms	
		General rules for preparation	
	Management and service standards	Technology R&D and management standards	
		Intelligence level evaluation standards	
Technical standards for intelligent HSR	Intelligent construction standards	Survey and design standards	Standards for implementation of railway engineering information modeling in design stage
		Engineering construction standards	Technical specification for intelligent construction of railway ballastless track slab/sleeper; technical specification for intelligent filling of high-speed railway subgrade; technical specification for intelligent construction of railway shield tunnel
		Construction management standards	Standards for BIM-based virtual construction; standards for full-process digital management

	Intelligent equipment standards	Mobile equipment standards	Intelligent EMU standards; standards for intelligent comprehensive inspection train
		Communication and signaling standards	Train operation control standards; ATO standards; standards for intelligent dispatching communication
		Traction power supply standards	Standards for intelligent traction substation; standards for simple, unified and standardized OCS
		Inspection and monitoring standards	Standards for inspection and monitoring of infrastructure; standards for monitoring and warning of natural disasters
	Intelligent operation standards	Passenger service standards	Intelligent passenger transport standards; intelligent ticketing standards
		Transport organization standards	Intelligent comprehensive dispatching standards; intelligent traffic control standards (CTC)
		Maintenance and repair standards	Standards for integrated operation and maintenance of track, communication, signaling, and power supply; standards for EMU intelligent operation and maintenance

(Continued)

Table 1.2. (*Continued*)

Standard System Framework of Intelligent HSR			
3 Primary categories	**9 Secondary categories**	**26 Tertiary categories**	**Extensible categories**
Platform and supporting technology standards	Basic platform standards	Data resource standards	Standards for classification and coding of railway engineering information modeling; data storage standards for railway engineering information modeling; standards for railway engineering information modeling expression
		Geographic information service standards	Technical specification for railway engineering 3D GIS data; geographic information delivery standards for railway engineering information modeling application
		Big data analysis standards	
		Artificial intelligence standards	
		AI platform standards	
	Network security standards	Evaluation standards for classified protection of cybersecurity	
		Standards for network security risk assessment	
		Network security management standards	
	Data center standards	Hardware resource standards	
		System software standards	
	Supporting technology standards	Standards for Beidou satellite navigation application	
		IoT standards	

traction power supply standards, and inspection and monitoring standards. The mobile equipment standards include extensible categories such as intelligent EMU standards and standards for intelligent comprehensive inspection trains. The communication and signaling standards include extensible categories such as train operation control standards, ATO standards, and standards for intelligent dispatching communication. The traction power supply standards include extensible categories such as standards for intelligent traction substations and standards for simple, unified, and standardized OCS. The inspection and monitoring standards include extensible categories such as standards for inspection and monitoring of infrastructure and standards for monitoring and warning of natural disasters.

(3) **The intelligent operation standards consist of three tertiary categories:** passenger service standards, transport organization standards, and maintenance and repair standards. The passenger service standards include extensible categories such as intelligent passenger transport standards and intelligent ticketing standards. The transport organization standards include extensible categories such as intelligent comprehensive dispatching standards and intelligent traffic control standards (CTC). The maintenance and repair standards include extensible categories such as standards for integrated operation and maintenance of track, communication, signaling, and power supply and standards for the intelligent operation and maintenance of EMUs.

1.3.3.3 *Platform and supporting technology standards*

Platform and supporting technology standards include four secondary categories: basic platform standards, network security standards, data center standards, and supporting technology standards.

Basic platform standards consist of five tertiary categories: data resource standards, geographic information service standards, big data analysis standards, AI standards, and AI platform standards. The data resource standards include extensible categories such as standards for classification and coding of railway engineering information modeling, data storage standards for railway engineering information modeling, and standards for railway engineering information modeling expression. The GIS standards include extensible categories such as technical specifications for railway engineering 3D GIS data and geographic information delivery standards for railway engineering information modeling application.

Network security standards consist of three tertiary categories: evaluation standards for classified protection of cybersecurity, standards for network security management, and standards for network security risk assessment.

The data center standards consist of two tertiary categories: hardware resource standards and system software standards.

The supporting technology standards consist of such tertiary categories as the standards for Beidou satellite navigation applications and IoT standards.

1.3.4 *Data system of IHSR*

1.3.4.1 *Overall architecture*

The architecture of the data system of an intelligent high-speed railway (IHSR) is divided into a data collection layer, a storage and analysis layer, and an application display layer from bottom to top. The data collection layer is used to collect data from the three major sectors of intelligent construction, intelligent equipment, and intelligent operation and the existing business information system, as well as other external related data, such as transportation mode, meteorology, and earthquake, so as to realize the collection of all-service and all-type IHSR data. Based on the basic platform of the IHSR, the storage and analysis layer standardizes and normalizes the data, establishes a full life cycle data organization and storage structure according to different subjects, such as construction, equipment, and operation, forms a set of standardized data resources that involve multidisciplinary integration and cross-business and cross-department sharing, and provides basic data management, data integration, data governance, and other services. The application display layer adopts the railway big data application mode of "platform + application" and carries out typical applications of big data around the fields of engineering construction, mobile equipment, infrastructure, transportation production, operation safety, passenger transport management, service, and big data sharing of the comprehensive transportation system. The details are shown in Table 1.3.

Table 1.3. Data system framework of intelligent HSR.

Big Data Application			
Data collection	Existing information system		
	Intelligent construction	Survey and design; Engineering construction; Construction management	Construction data
	Intelligent equipment	Mobile equipment; Communication & signaling; Traction power supply; Inspection and monitoring	Equipment data
	Intelligent operation	Passenger transport service; Transport organization; Maintenance and repair	Operation data
	External data	Aviation; Urban rail; Public transit; Meteorology; Internet; Public services	External data
Storage and analysis layer	External environment data		
	Basic ledger data of equipment and facilities		
	Intelligent HSR data lake		
	Intelligent construction data fields	Space-air-ground integrated engineering survey data; BIM-based engineering design data; intelligent construction of bridge, tunnel, subgrade, and track works; intelligent construction of passenger station works; intelligent construction of communication, signaling, power and electrification works; BIM+GIS-based engineering construction management data	

(Continued)

Table 1.3. (*Continued*)

Big Data Application			
	Intelligent equipment data fields	Intelligent EMUs data; intelligent comprehensive inspection train data; signal data; communication data; intelligent traction power supply data; inspection and monitoring data of infrastructure; monitoring and warning data of natural disasters; intelligent monitoring of perimeter intrusion; intelligent monitoring of environment	
	Intelligent operation data fields	Intelligent passenger transport data; intelligent ticketing data; intelligent comprehensive dispatching data; intelligent traffic dispatching data; integrated operation and maintenance of track, communication, signaling, and power supply; intelligent operation and maintenance of EMU	
	Basic platform of intelligent HSR		
	Basic data management	Master data; Geographic information; Metadata	
	Data integration	Data acquisition; Data cleaning; Data models	
	Data governance	Data standard; Data quality; Data security	
	Data sharing	Data assets; Data authorization; Data service	
	Big data and AI analysis	Data analysis; AI analysis	
Application display layer	Big data application for engineering construction Big data application for mobile equipment Big data application for infrastructure Big data application for transportation production Big data application for operation safety Big data application for passenger transport management, and service Big data sharing of the comprehensive transportation system		

1.3.4.2 *Data classification and organization*

Under the guidance of the IHSR data system framework, in order to further improve the overall data management level and data service capability of the IHSR, it is necessary to formulate the IHSR data subject classification and plan a unified and standardized basic data management method. Data classification is to distinguish and classify data according to certain principles and methods according to the attributes or characteristics of the data content and establish a certain classification system and arrangement order for data management and use. The purpose of data classification is to scientifically organize data and effectively discover and utilize data. Commonly used data classification methods include classification by business, data type, data generation frequency, processing in real time, and importance.

1.3.4.2.1 Classification by business

According to the concept, connotation, and internal and external demands of IHSR, from the perspective of the business division of IHSR, 13 data subject categories covering statistical analysis, planning and plans, construction management, marketing, customers, transport organization, passenger transportation, transport capacity resources, safety monitoring, finance, human resources, materials, and enterprise functions are proposed, as shown in Table 1.3.

1. **Data subject of statistical analysis:** The data subject of statistical analysis includes passenger transport statistics, operation statistics, financial statistics, decision support, and other data models, covering the statistical data models generated by the daily operation of IHSR, as well as the comprehensive analysis data models developed by summarizing and integrating operation data, as shown in Table 1.4.

2. **Data subject of planning and plans:** The data subjects of planning and plans include planning information, plan information, plan management, and other data models, covering the long-term planning information of intelligent HSR, as well as various plan information formulated according to the long-term planning and the management, implementation, and assessment of the plan, as shown in Table 1.5.

Table 1.4. Data models under the subject of statistical analysis.

Conceptual data model	Type	Description of data model
Passenger transport statistics	Analytical data	Statistics of passenger traffic volume, turnover, income, density, etc.
Transport capacity resources statistics	Analytical data	Statistics and efficiency of EMU operation and maintenance, etc.
Material equipment statistics	Analytical data	Statistics of quantity, budget, and application of materials and equipment
Environmental statistics	Analytical data	Environment-related statistics
Human resources statistics	Analytical data	Statistics related to the employees, types of work, and salary payment
Financial statistics	Analytical data	Statistics of financial-related statements
Energy consumption statistics	Analytical data	Statistical reports on energy consumption quantity, energy consumption plan, completion amount, etc.
Periodic operation statistics	Analytical data	Comprehensive statistical reports related to operation and production
Management and control support	Analytical data	Real-time analysis to support daily business operation control
Decision-making support	Analytical data	Comprehensive analysis to support intelligent decision-making

3. **Data subject of construction management:** The data subject of construction management includes data models such as construction projects and construction management, covering the businesses managed in the form of projects in HSR construction projects and other business fields. The project data include relevant data from the whole project life cycle and can be used for project analysis and decision-making, operation, and service after the completion of the project, as shown in Table 1.6.

4. **Data subject of marketing:** The data subject of marketing includes data models such as transportation products, pricing, and sales management, covering the management of sales processes, a timely grasp of market dynamics, passenger transport management level, service quality, and other information, as shown in Table 1.7.

Table 1.5. Data models under the subject of planning and plans.

Conceptual data model	Type	Description of data model
Planning information	Master data	Long-term macro planning information of IHSR, providing a reference basis for specific plans
Plan information	Master data	The specific information of each plan of IHSR, such as work breakdown and organization structure of the project
Planning management	Transaction data	It includes specific implementation information of the planning, such as the implementation of the plan/stage summary/coordination of various disciplines
Plan management	Transaction data	It includes specific implementation information of the plan, such as project screening/pre-feasibility study results/feasibility study and other information
Plan implementation	Transaction data	It includes the specific implementation information of each plan, such as the implementation progress and financial status of each plan
Plan assessment	Analytical data	Assessment information for the plan

5. **Data subject of customers:** The data subject of customers includes passenger, customer service, and other data models, covering various information about various customers, establishing detailed customer views, and providing data for formulating corresponding products and services, as shown in Table 1.8.

6. **Data subject of transport organization:** The data subject of transport organization includes dispatching plans, dispatching instructions, train operation, and other data models, covering the daily transportation management of various departments of railway transportation production. By formulating various plans and orders in transportation dispatching, it provides information services for other departments of HSR and guides the specific operations in each station and depot, as shown in Table 1.9.

Table 1.6. Data models under the subject of construction management.

Conceptual data model	Type	Description of data model
Construction project	Master data	Basic information about the construction project, such as project name and responsible department
Construction management	Transaction data	Specific management information during the construction project, such as work breakdown and project implementation stage of the construction project
Construction design	Transaction data	Specific design information of the construction project, such as construction drawings and design schemes
Construction implementation	Transaction data	Specific implementation information during the construction project, such as construction-related personnel and construction materials
Archiving of construction project	Transaction data	Archiving information of construction project
Review of construction project	Analytical data	Information on assessment and review for construction projects

Table 1.7. Data models under the subject of marketing.

Conceptual data model	Type	Description of data model
Price standard	Master data	Specific price information of products related to passenger transport and other value-added services
Market analysis	Analytical data	Analysis of recent trends in the passenger transport market
Transport products	Master data	Basic information on passenger transport products, such as passenger train timetable, seat configuration, and other information
Channel	Master data	Information on tickets and other sales and promotion channels
Pricing	Transaction data	Specific ticket price information
Sales management	Analytical data	Management information for the whole sales process of various transportation products

Table 1.8. Data models under the subject of customers.

Conceptual data model	Type	Description of data model
Passenger	Master data	Basic information of passengers
Customer loyalty	Transaction data	Management information on customer loyalty, such as customer points and frequent passenger information
Credit	Master data	Credit management information of customers' previous transactions
Customer service	Transaction data	Relevant information on customer service, such as customer consultation information and customer return visit information
Orders	Transaction data	Order information of customer transaction
Customer reference information	Analytical data	Information on the recommended products and services to customers based on their past transactions and other data

Table 1.9. Data models under the subject of transport organization.

Conceptual data model	Type	Description of data model
Dispatching plan	Transaction data	Various plans and actual information in daily transportation dispatching
Transport capacity plan	Transaction data	Transport dispatching plan further prepared based on information, such as transport capacity resources allocation and train operation arrangement
Dispatching order	Transaction data	Relevant data of specific dispatching orders, such as order number, order receiver, and the contents of order
Location information	Master data	Specific location and geographical data of equipment, such as specific data information of stations, depots, and lines
Driver and conductor management	Transaction data	Relevant management data of crew and driver
Centralized traffic control	Analytical data	Centralized traffic control data, such as CTC information and route instructions
Train operation	Transaction data	A series of data generated during train operation

7. **Data subject of passenger transport:** The data subject of passenger transport includes two data models: passenger station management and in-transit service, as shown in Table 1.10.

8. **Data subject of transport capacity resources:** The data subject of transport capacity resources includes data models such as basic information of resources, resource application, and resource detection, covering the whole life cycle data of transport capacity resources. Through the whole life cycle management of the use, management, and repair of transport capacity resources, fine management and control of assets are realized, as shown in Table 1.11.

Table 1.10. Data models under the subject of passenger transport.

Conceptual data model	Type	Description of data model
Passenger station management	Transaction data	Specific operation management data of passenger station
In-transit service	Transaction data	Relevant data of passengers in transit

Table 1.11. Data models under the subject of transport capacity resources.

Conceptual data model	Type	Description of data model
Resource planning	Master data	Long-term planning data of resources, such as investment and income planning information
Basic information of the resources	Master data	Basic information of resources, such as model and number
Technical details of the resources	Master data	Technical history information of transport capacity resources
Resource allocation	Master data	Equipment allocation information, such as allocated time and allocated units
Data of resource capability	Master data	Transport capacity information of transport capacity resources
Resource utilization	Transaction data	Relevant data generated in the process of resource utilization, such as status change information
Resource status	Transaction data	Equipment status data, such as receiving, repair, failure, and scrapping
Resource testing	Transaction data	Test data of equipment
Resource maintenance	Transaction data	Maintenance data of equipment

Table 1.12. Data models under the subject of safety monitoring.

Conceptual data model	Type	Description of data model
Security inspection	Transaction data	Security detection data, such as surveillance video
Supervision and inspection	Transaction data	Supervision and inspection data, such as regular inspection and safety supervision
Accident handling	Transaction data	Accident handling information, such as train operation accident handling
Early warning	Analytical data	Early warning information analyzed based on collected data
Emergency responses	Analytical data	Data on the treatment of accidents or emergencies, such as rescue information

9. **Data subject of safety monitoring:** The data subject of safety monitoring includes data models such as security detection, accident handling, and emergency responses, covering the whole life cycle of data from security detection to security management, as shown in Table 1.12.

10. **Data subject of finance:** The data subject of finance includes financial subjects, funds, budget, and liquidation, covering financial accounting, transportation liquidation, etc. Financial data can comprehensively reflect the production and operation status of units at all levels and provide strong support for correct decision-making and effective supervision, as shown in Table 1.13.

11. **Data subject of human resources:** The data subject of human resources includes organizations, personnel, etc., covering employee recruitment, employee development, and related salary management information, providing basic data for cultivating and developing employee capabilities and employee performance appraisal, as shown in Table 1.14.

12. **Data subject of materials:** The data subject of materials includes material categories, basic materials information, etc., covering the data required for materials warehousing, use, recycling, and other businesses to optimize the inventory structure and maintain reasonable materials reserves, as shown in Table 1.15.

Table 1.13. Data models under the subject of finance.

Conceptual data model	Type	Description of data model
Accounts of finance	Master data	Financial account information, such as account number, nature, and field
Funds	Transaction data	Fund management information, such as fund allocation
Accounting statement	Analytical data	Data of various accounting statements
Investment	Transaction data	Investment-related data, such as investment projects and investment income
Income	Transaction data	Revenue-relevant data
Financial analysis report	Analytical data	Financial analysis data
Budget	Transaction data	Budget-related data, such as budget plan and budget execution information
Liquidation	Transaction data	Transportation liquidation information, such as inter-administration liquidation information
Cost	Transaction data	Cost management information, such as the classified cost of each cost object
Financial performance	Analytical data	Financial performance data

Table 1.14. Data models under the subject of human resources.

Conceptual data model	Type	Description of data model
Organization	Master data	Information about the structure of organization
Post	Master data	Job function data
Personnel	Master data	Basic information of personnel
Qualification	Master data	Data related to personnel qualification
Recruitment	Transaction data	Recruitment information of enterprises
Training	Transaction data	Training information of personnel and employees
Salary	Transaction data	Data related to personnel compensation
Personnel performance	Transaction data	Data related to personnel performance

13. **Data subject of enterprise functions:** The data subject of enterprise functions includes data models such as legal affairs management, land for HSR, scientific and technological management, and office, as shown in Table 1.16.

Table 1.15. Data models under the subject of materials.

Conceptual data model	Type	Description of data model
Category of materials	Master data	Category information of materials
Basic information of materials	Master data	Basic information of materials, such as material name
Supplier	Master data	Supplier information, such as name and number of supplier
Procurement process management	Transaction data	Data generated in the procurement process, such as recorded information and quotation
Circulation of materials	Transaction data	Data related to material circulation
Use of materials	Transaction data	Relevant data during the use of materials
Disposition of materials	Transaction data	Data of material disposition status
Summary of materials	Analytical data	Material-related summary data
Materials procurement plan	Analytical data	Data of procurement plan of materials

Table 1.16. Data models under the subject of enterprise functions.

Conceptual data model	Type	Description of data model
Legal affairs management	Analytical data	Management data related to legal affairs
Land for HSR	Analytical data	Relevant management data of land for HSR
Publicity	Analytical data	Relevant publicity data
Scientific and technological development	Analytical data	Data related to scientific and technological development
Office	Analytical data	Office data
Process system	Master data	Relevant data of process system
Audit	Analytical data	Audit-related data
Legal documents	Master data	Legal affairs-related data

1.3.4.2.2 Classification by data type

Classification by data type refers to the classification of data according to the type of data content and the data storage form. It is mainly divided into structured data, unstructured data, and semi-structured data by data type, as shown in Figure 1.12.

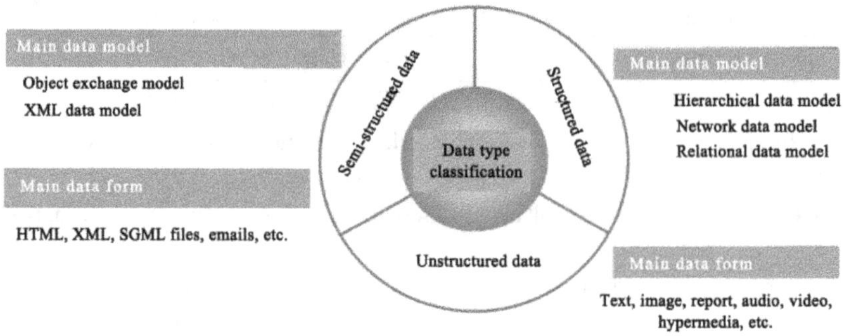

Figure 1.12. Classification by data type.

1. **Structured data:** Structured data refers to the data that are stored in a database, have a certain logical structure and physical structure, and can be logically expressed and realized using a two-dimensional table structure. Structured data have highly organized patterns, standard formats, and layouts and are stored, accessed, and processed in an organized and precise manner.

(a) *Data characteristics*: Data format and length specifications are strictly followed, usually with a predefined data model organizational structure, before data are generated. It is mainly stored in and managed with relational databases.

(b) *Main data model*: The logical data model of structured data is usually divided into the hierarchical data model, network data model, and relational data model. The hierarchical data model uses a tree (hierarchical) structure to represent entity types and relationships between entities; the network data model uses a directed graph to represent relationships between entities; the relational data model uses a two-dimensional logic table to represent relationships between entities.

2. **Unstructured data:** Unstructured data refer to data with irregular data structure and no predefined data model, and they usually cannot be represented by two-dimensional logical tables in a database. Machine log data, CCTV footage, X-ray images, and video are typical unstructured data.

(a) *Data characteristics*:

(i) The data have a variety of forms and are usually stored in different types of files rather than in a database.
(ii) The data have no data model, i.e., there is no predefined model or they are not organized in a predefined way. They have irregularities and fuzziness that are difficult to understand with traditional computer programs.

The main forms of unstructured data are text, image, report, audio, video, and hypermedia information in common formats.

3. **Semi-structured data:** Semi-structured data refer to data that may be irregular or incomplete and whose structure may change rapidly or unpredictably. Semi-structured data are specially structured data, which are quite different in structure and cannot be studied and applied in the current way of processing structured data. Therefore, semi-structured data are a type between fully structured data and completely unstructured data.

(a) *Data characteristics*

(i) They have a self-descriptive data structure, and the data structure and content are mixed together without obvious distinction.
(ii) The description of data structure is dynamic, and changes in data usually lead to changes in the structure pattern. It has a dynamic structure pattern.
(iii) The data structure is highly complex and difficult to integrate into the general description framework.
(iv) There is no order between elements, not all elements are necessarily present, and elements of the same name may take different forms.

(b) *Main data model*
The main data models of semi-structured data are the object exchange model and the XML data model.

(i) The object exchange model is a data model proposed for semi-structured data. It defines entities as objects, and each object is represented by a quaternion.
(ii) The XML data model mainly includes the Infoset model, XPath data model, DOM model, XQuery, and XPath data models. It is a variant of the object exchange model.

(c) *Main data forms*
Common date forms are HTML, XML, SGML files, e-mail, and other data with variable structure and content in file processing.

1.3.4.2.3 Classification by data generation frequency

Classification by data generation frequency refers to the classification of data according to the frequency of data generation (the expected amount of data generated per unit time or the frequency of arrival of a fixed amount of data).

1. **Demand-driven data:** Demand-driven data refer to the data that are generated irregularly, usually without a fixed generation rule and with a low generation frequency, such as petition information and emergency information.

(a) *Data characteristics*

(i) The generation of demand-driven data is uncertain and driven by demand. Emergencies and the impact of economic activities are demand-driven factors.
(ii) Generally, the frequency of data generation is low, and the production cycle can be calculated in days, weeks, or months. The data are discontinuous, and there are also obvious differences in similar data at different times, in terms of content, format, source, etc.

(b) *Data scope*
Demand-driven data mainly consists of social data, content media data, and other internet data. Data sources mainly include emergency data, website data, e-commerce, forums, application markets, news media, and other channels.

2. **Persistent data:** Persistent data refer to the data generated continuously according to their own rules per unit time. Persistent data are a data type with an intermediate production frequency between demand-driven data and real-time data.

(a) *Data characteristics*

(i) The data are generated regularly, usually at a fixed frequency for a long time, and a reasonable collection and analysis application scheme can be formulated according to the rule.

(ii) Due to the large amount of this type of data, the continuous accumulation of data generated, and their high-capacity characteristics, a distributed structure is required.
(iii) The frequency of data generation is higher than that of demand-driven data, and the production cycle can be calculated in minutes, hours, or days.
(iv) Persistent data are time-sensitive, but the timeliness requirement standard is not as high as that of real-time data.

(b) *Data scope*
Typical persistent data include statistical data, public data, and academic knowledge libraries.

3. **Real-time data:** Real-time data refer to the data generated, collected, and calculated in real time, and the intermediate data generated in the process of real-time calculation of data

(a) *Data characteristics*

(i) Real-time data are highly time-sensitive, and real-time data values become invalid over time. Outdated data can reflect the evolution of things and are usually stored as real-time historical data according to time and category.
(ii) Real-time data have a large data production volume per unit time and have the requirement of quick response.
(iii) Real-time data have the characteristics of real-time interaction. Data collection and analysis are all done simultaneously, and the data values are immediately reflected.
(iv) The frequency of data generation is high, and the common production cycle is calculated in seconds or minutes.

(b) *Data scope*
Typical real-time data include weather data, passenger transport transaction data, monitoring data of various disciplines, and equipment monitoring data.

4. **Time series data:** Time series data refer to the observed data series arranged in chronological order, that is, the data series recorded in chronological order for the same indicator.

(a) *Data characteristics*

(i) The generation frequency of time series data is the highest. New data are generated as things change, which can theoretically be calculated in the smallest time unit.
(ii) The data are strongly dependent on the time of generation (collection), and each piece of data is required to correspond to a unique time.
(iii) Data distribution is regular. Time series data are a series of observation values obtained in chronological order.
(iv) The amount of data is huge. Usually, the real-time monitoring system has tens of millions of monitoring points, and the data generated by the monitoring points are accumulated continuously every second.
(v) The data are highly comparable, and all data in the same data column are of the same caliber, including the number of periods and time points.
(vi) The data are highly predictable. Through the statistical characteristics and development rules of the time series within the data sample, a time series model can be built for out-of-sample prediction.

(b) *Data scope*
It consists of the data collected and generated by various types of real-time monitoring, inspection, and analysis equipment.

1.3.4.2.4 Classification by processing time

Classification by processing time refers to the classification of data according to the required time delay of data processing:

1. **Real-time processing of data:** The real-time data processing delay is required to be less than 200 ms. Streaming big data technology is mainly used to process data. At present, the widely used technologies include Tez, Spark Streaming, Storm, and Flink. These stream-processing frameworks load real-time data into memory one by one through stream processing for queries and other uses. Such a system can query the latest real-time data with efficient preset analysis and processing models and low data latency. However, due to the limited memory capacity, the system needs to discard the original historical data and cannot support *ad hoc* querying, analysis, and processing on a full big dataset.

2. **Batch processing of data:** The time delay for batch processing of data can be more than 200 ms. The batch-processing big data system needs to first aggregate the data into batches and then load them into the analytical data warehouse after pre-processing to achieve high-performance real-time queries. The batch-processing big data technology system includes parallel computing technologies, such as MapReduce and Spark. This kind of system can realize efficient *ad hoc* queries on complete large datasets, but it cannot query the latest real-time data and has problems such as high data latency.

1.3.4.2.5 Classification by importance

Data can be divided into high-sensitive data, sensitive data, internal data, and public data according to their importance:

1. **High-sensitive data:** The disclosure and destruction of high-level confidential information will have catastrophic consequences for the reputation, finances, or operations of the organization. It includes data such as detailed financial statements and core technologies.

2. **Sensitive data:** Confidential information about the organization, if improperly disclosed, may have a serious impact on the organization. It includes data such as sensitive customer data and internal product data.

3. **Internal data:** Content circulated within each functional department that does not involve confidential content and does not need to be disclosed. It includes data such as the telephone list, job information, etc.

4. **Public data:** Data that can be open to the public and society. It includes data such as marketing and publicity materials.

1.3.5 *Basic platform for IHSR with cloud–edge collaboration*

1.3.5.1 *Three-dimensional architecture*

The basic platform of intelligent HSR provides characterization models, mechanism models, dynamic and static data, and other model–data integration services for N innovations in 3 sections (intelligent construction, intelligent equipment, and intelligent operation), 10 fields, and 18 directions. The basic platform is deployed in the mode of cloud–edge collaboration for realizing the collaborative application of the cloud platform deployed in the main data center and the edge platforms deployed in stations, trains, and lines. The overall architecture of the basic platform of intelligent HSR integrates the physical, management, and collaborative dimensions, as shown in Figure 1.13.

1. **Physical dimension:** The cloud platform is deployed in the main data center of China Railway for realizing the functions of cloud data access, model management, model–data governance, model–data storage, model–data annotation, model training, model release, model deployment, resource management, cloud–edge collaborative management, cloud task orchestration and scheduling management, etc. The edge platforms are deployed in stations, trains, and beside tracks, mainly for realizing the functions of edge model management, data management, scheduling management, real-time analysis and application, edge–edge collaborative management, edge device management, transmission management, etc. Field devices mainly include sensor networks, IoT, and other kinds of sensor devices deployed in stations, trains, and beside tracks for acquiring data in real time and uploading them to edge and cloud platforms.

2. **Management dimension:** The basic platform will have a unified service framework based on model–data integration in the whole process of service development, service testing, and service deployment and operation to ensure the consistency of models and data in the whole life cycle. In the platform development stage, it is ensured that the basic coding, version management, and authority design of models and data are consistent, relevant, and dynamically related; in the platform test stage, it is ensured that the model–data integrated storage, model–data fabric, and model–data standards are consistent; in the platform deployment and operation stage, it is ensured that the model and data grow synchronously with the intelligent HSR system and are organically related.

3. **Collaborative dimension:** The cloud–edge collaboration mainly realizes the collaboration between the cloud platform and the edge platforms in terms of computing power, models, data, application, and security:

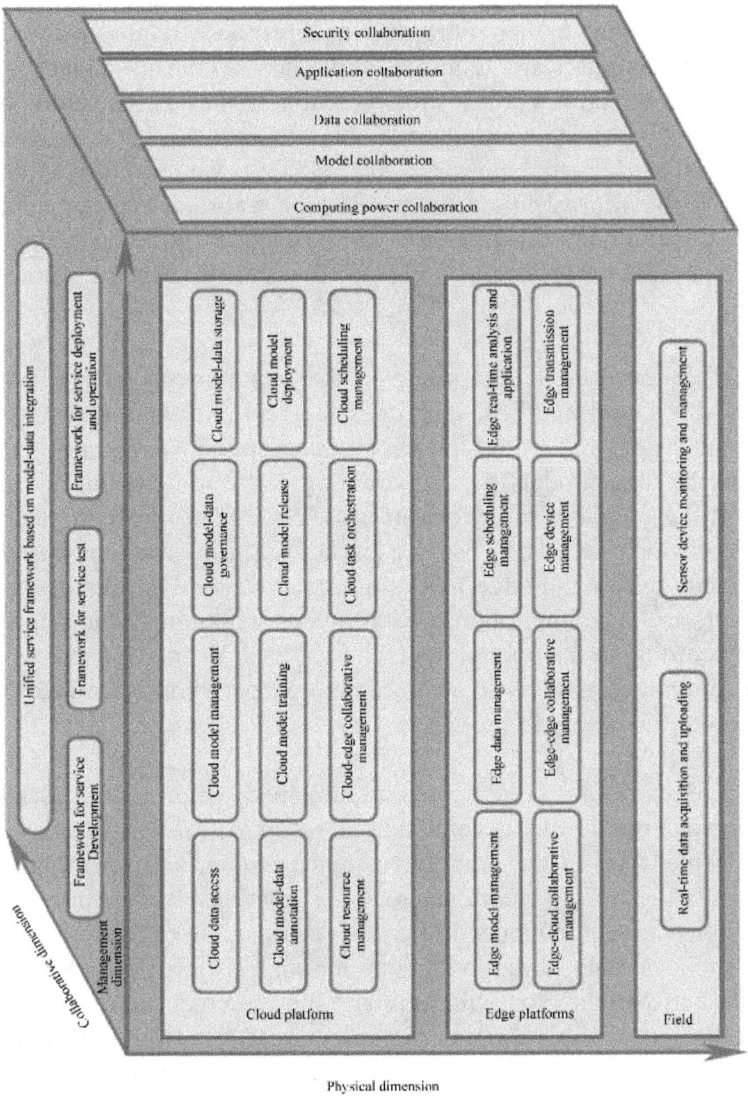

Figure 1.13. Overall architecture of basic platform of Intelligent HSR.

(1) *Computility collaboration*: Facing the computing requirements of a variety of complex task scenarios under the cloud–edge collaborative architecture, the cloud platform can allocate the overall computing power resources and provide task processing priority policies for different task scenarios according to their attributes and response requirements. The edge platform automatically identifies and processes the tasks that require high real-time detection but low model computing power. The cloud platform uses large-scale computing power to analyze tasks that require massive models or data to be processed and high accuracy. In addition, according to the priority policy, tasks with high real-time requirements are prioritized, while non-real-time tasks are processed in idle time, thus the overall computing power resources of the system can be optimally utilized.

(2) *Model collaboration*: According to the requirements of application scenarios, the cloud platform integrates the real-time data and massive historical data acquired by the field IoT and sensor networks, makes them stable and reliable models through training and optimization, and then distributes these models to edge platforms. Edge platforms complete the model deployment of edges, use models for real-time analysis based on the data acquired by the edge IoT, and upload the analysis results to the cloud platform. The cloud platform continuously optimizes and improves models according to the operation results fed back by models and distributes the optimized models to edge nodes to continuously improve the accuracy and timeliness of models.

(3) *Data collaboration*: Edge nodes can effectively acquire data related to production and monitoring generated by IoT terminals, as well as all kinds of data formed during the production, management, and operation processes, and then locally store the data in a limited space-time range. According to the requirements of the task scenario, the edge platform will first use the local data stored in its node for analysis. To achieve data collaboration between the cloud platform and the edge platforms, it is necessary to timely schedule the enormous amounts of data stored in the cloud platform for joint analysis and application when the accuracy is insufficient or more data are required to support the application of the algorithm.

(4) *Application collaboration*: The cloud platform is responsible for the integration and unified management of different model–data integrated

analysis models, provides deployment, invocation, and other interface services for edge platforms, supports the rapid deployment of simple and portable applications, such as container images, micro-services, and micro-applications on the edges, and provides a high-availability guarantee and live migration. In daily applications, the cloud platform can continuously enrich the sample library according to the data acquired by edges and iteratively train the machine learning algorithm and deep neural network models according to the error loss feedback of the models, thus improving the adaptability of each business scenario and the accuracy of intelligent analysis. Edge nodes have the conditions and operating environments for implementing network applications and can realize powerful intelligent analysis and fast output responses for various terminal applications. In case of downtime of the central server or interruption of network transmission, edge nodes can realize closed-loop management to ensure the normal operation of the edge business model and cooperate with the cloud in upgrading and maintaining the application system.

(5) *Security collaboration*: Security collaboration mainly includes collaboration in the areas of network, applications, data, and terminals. Network security collaboration refers to designing stable and secure transmission protocols for the data transmission of edge devices and edge gateways. In cloud resource management and data access services, not only data analysis, data storage, and other conventional functions should be provided but also intelligent sensing application functions, such as traffic cleaning and traffic analysis, shall be taken into account to deal with unknown traffic attacks.

Application security collaboration means the system needs to manage the users of both the cloud and edges in a unified way. Edges have the single sign-on (SSO) function, while the cloud stores the data generated by one user on multiple edges in a standardized manner and supports the asynchronous invocation function of edges.

Data security collaboration means that, on the one hand, edge nodes can intelligently select local caching strategy or remote transmission strategy to ensure data integrity according to network transmission conditions and data importance; and on the other hand, the cloud performs control over the edge data transmission through security policies and traffic cleaning and can timely identify and block the uploading of malicious traffic data from edges to ensure the security and reliability of the data collaboration process.

Terminal security collaboration entails strengthening the security of terminal devices through strategies such as terminal physical layer authentication, physical layer data encryption, terminal access identity authentication, and access permission management in order to control the risk factors of data breach or eavesdropping on the entire cloud platform caused by terminal intrusion or impersonation during the data interaction process of cloud–edge collaborative applications.

1.3.5.2 *Cloud–edge deployment architecture*

The deployment architecture of the basic platform of intelligent HSR is shown in Figure 1.14. The cloud platform, deployed in the main data center of China Railway or the cloud data center of each railway administration, is used to acquire all-discipline models and data from infrastructure, mobile equipment, and external environments throughout the whole life cycle, complete the analysis of compute-intensive and complex scenario requirements, and ensure the safe and efficient operation of HSR. Edge platforms mainly include station edge platforms, train edge platforms, and trackside edge platforms. The station edge platforms are used to monitor the status of station equipment and facilities, carry out integrated analysis of models and data, and guarantee that the stations are comfortable, energy-saving, and environment-friendly. The train edge platforms are used to acquire data from onboard sensors and study and judge the train operation safety status in real time to ensure train operation safety. The trackside edge platforms are mainly used to acquire the status information of infrastructure and mobile equipment and analyze the risks affecting operation safety based on machine vision data and structured data in real time to ensure operation safety.

The hardware configured for edge platforms mainly includes storage computing devices and data transmission devices. Storage computing devices mainly include edge AI computing nodes, edge gateways, edge controllers, and store-and-forward devices. The software configured for edge platforms mainly includes edge resource management, edge platform management, edge software service management, and the edge basic platform. Edge resource management mainly includes the local data resources transmitted from field devices and the models and

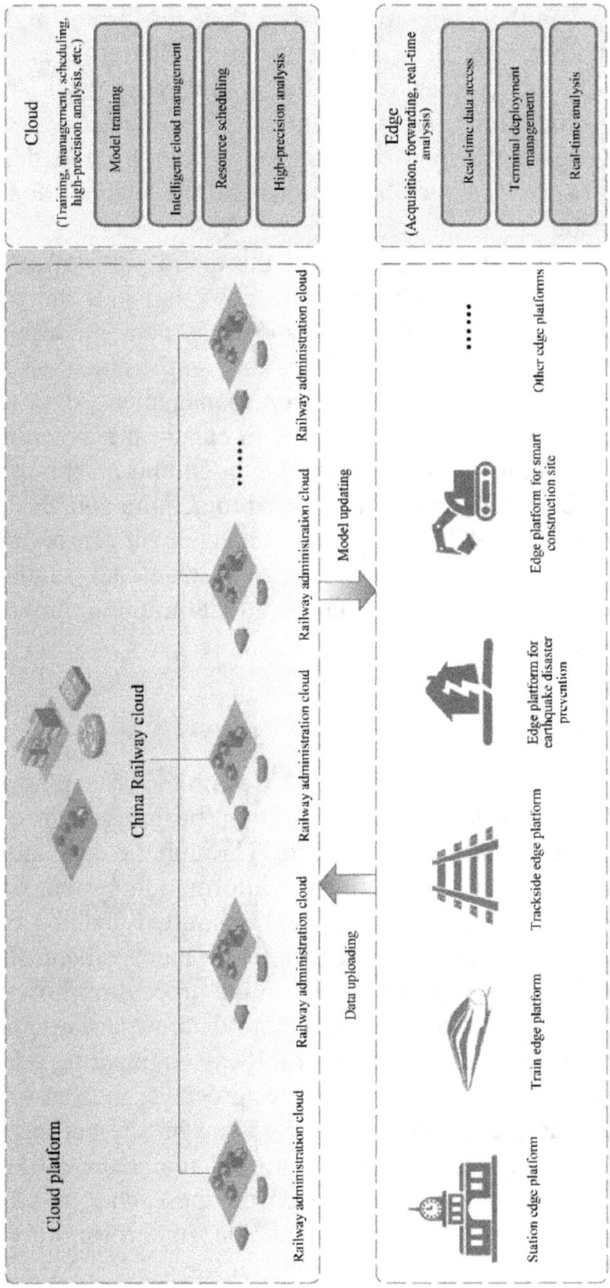

Figure 1.14. Deployment architecture of basic platform of intelligent HSR.

data resources distributed from the cloud. Meanwhile, edge platforms provide all edges with an operating system, a device driver, and a virtual mirror environment, and they provide the basic operating environment for all kinds of software services. Edge software services are mainly used for the scheduling management of edge devices and storage devices. The edge basic platform is responsible for the deployment management of specific edge applications and the analysis and processing of models and data.

The hardware configuration type of the cloud platform is basically the same as the functional attributes of edges, but it is more complex in terms of data scale and the corresponding data processing scale. Especially in terms of software configuration, on the one hand, the cloud platform realizes the tasks of application management, data resources, task scheduling, and model management to ensure the execution of the entire cloud–edge collaborative business model; and on the other hand, after the cloud basic platform is configured to acquire and store massive global production data, data storage and management can be realized on the cloud platform, and large-scale AI algorithm model training can be carried out, including data set annotation, model training, model release, and model deployment.

1.3.5.3 *Axis–plane coordination applications*

The axis-plane coordination application based on the basic platform is the integration of information systems in different business fields and stages throughout the life cycle of HSR projects. Through the data and data service provided by the basic platform, the information system data in the different stages of HSR projects can be transmitted and fed back, thus realizing information sharing and collaborative management of business applications within the whole life cycle. In the process of railway project construction, the aggregation of models and data, as an important information carrier, carries a large amount of railway engineering information, including the BIM model in the initial design stage, mechanism models and construction data continuously detailed and injected in the construction stage, and finally various operation and maintenance data in the operation and maintenance stage. Based on these data, a full railway life cycle model is formed as an important data support for railway operation and maintenance, as shown in Figure 1.15.

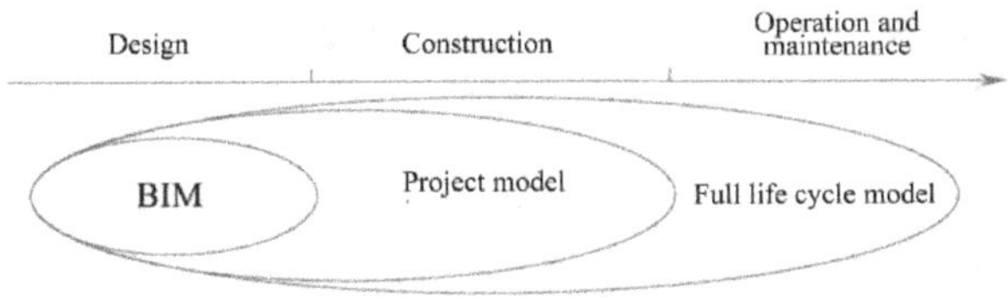

Figure 1.15. Relationship among full life cycle data models.

(1) **Internal collaboration at each stage**

(a) *Design stage*: The design stage involves the design of main elements, such as horizontal and longitudinal sections, geology, subgrades, bridges, tunnels, tracks, communication, signaling, power, electrification, station buildings, engineering economy, and other disciplines of a railway line. To realize design collaboration, BIM, GIS, and other relevant technologies need to be applied, the preliminary survey data, environmental information, BIM models, and other data provided by the basic platform need to be accessed, and the environmental and business data along the HSR line need to be aggregated, thus realizing the multi-angle, multiscale, multi-temporal, and multidisciplinary comprehensive collaborative design, visual management, and analysis of engineering construction data, with BIM models as the core. Through design resources management, process collaborative management, design data management, and design change management, process information can be recorded; models, data, and charts can be inquired and counted by all disciplines; and design standards are easier to implement, thus realizing project-based resource sharing, whole-process management of design documents, and collaboration among all participants, supporting multidisciplinary collaborative design and promoting collaboration among all discipline elements.

(b) *Construction stage*: The main business elements in the construction stage are quality, safety, progress, investment, materials, and environmental protection. By obtaining BIM models, mechanism models, design tables, and other relevant design data on the basic platform, efficient interaction among elements based on model information flow can be established. With BIM models as the core, by integrating information such as material mobilization, material processing, component prefabrication, factory production, and construction operations, the digital construction environment can be simulated from all aspects. Through unified coding, the linkage between real-time data and models can be realized to intuitively reflect the construction status. On this basis, the connection between processes and the interfaces among disciplines are controlled, and the model information flow runs through the whole construction process to realize the optimization and adjustment of the construction process and the collaboration of on-site elements and control elements of the HSR project.

(c) *Operation and maintenance stage*: The main business elements in the operation and maintenance stage are mobile equipment, track maintenance engineering, communication & signaling, traction power supply, transport organization, equipment operation and maintenance, security control, passenger service, external environment, etc. Based on the infrastructure design, construction, and completion acceptance information in the basic platform, the initial state of railway equipment and facilities during the operation period is managed to create a digital base map for operation and maintenance. In the operation and maintenance process, the above information is integrated with the dynamic and static inspection and monitoring information, maintenance and repair information, vehicle information, and passenger transport information so as to realize the unified management of all HSR elements, comprehensively analyze all kinds of information in an intelligent manner, put forward scientific suggestions for equipment maintenance and repair, transport organization, etc., and provide all-round information services for decision-making at all levels in the operation and maintenance process of HSR.

(2) **Data flow between stages**

Information flow management can ensure that the communication, discussion, and decision-making of HSR projects are carried out in a visualized manner at all stages, from project initialization and decision-making, survey, design, and construction to operation and maintenance, thus realizing the sharing and transmission of quality, security, environmental protection, progress, and investment information in the whole life cycle of a HSR project, supporting the knowledge exchange and collaboration among all participants, and effectively realizing dynamic control. The basic platform provides a unified platform for basic data (e.g., survey data, geological information, design information, construction process information, and operation and maintenance information), and the establishment of design delivery and completion delivery systems supports the data conversion and transmission from the design stage to the construction stage and then to the operation and maintenance stage. The basic platform and the design/completion delivery system jointly support the smooth transition and lossless transmission of data at each stage and further enhance the transmission and feedback optimization mechanisms of information flow, as shown in Figure 1.16.

(a) *From survey and design to construction*: Based on the data, models, and services of the basic platform, the functions of metadata management and the design delivery system are used to promote the flow of design drawings, design models, design tables, and other relevant information from the survey and design stage to the construction stage. This ensures the flow direction of information data and the integrity of the entire transmission process and also ensures the collaboration among management disciplines, thus optimizing the design, improving construction efficiency, and reducing rework. In light of the requirements of the integrated construction, operation, and maintenance application, the full element information model, documents, and description files are used as carriers, which contain information on the logical relationship between inter- and intra-disciplinary elements, as well as the iterative relationship of the result versions. Meanwhile, design results shall be stored on the basic platform for filing.

(b) *From design and construction to operation*: In the operation and maintenance stage, it is necessary to make full use of the survey, design, and construction deliverables to guide the allocation of funds, materials, manpower, and working hours. At the same time, it is required to perform real-time dynamic monitoring of infrastructures, analyze the infrastructure status, support the intelligent operation and maintenance of infrastructures, and keep infrastructure in good condition for use. The survey, design, and construction data (including design drawings, design models, design tables, construction models, construction process data, supervision data, supervision and inspection data, and acceptance data) are transmitted to the operation and maintenance stage through the completion delivery system to accumulate organizational process assets and provide basic data for operation and maintenance.

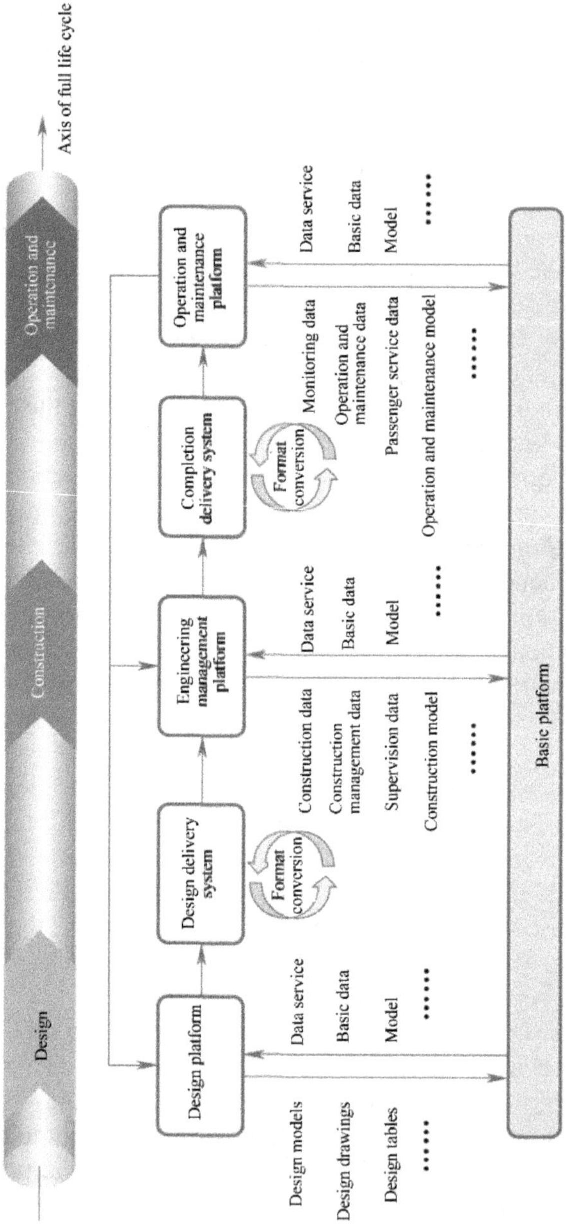

Figure 1.16. Collaborative application during the whole life cycle.

(c) *Optimize the survey and design according to construction feedback*: (i) Based on the actual site environment data, monitoring and measurement data, test data, and acceptance data during the construction process, models are created to analyze the internal rules and problems of the data, and through intelligent equipment and means, the information obtained is fed back to the design personnel. Through the dynamic design system, the design results can be adjusted and optimized. (ii) According to the monitoring and measurement data, comprehensive test analysis and forecast results, video monitoring, and images and photos in the construction process, the site and the design are actively compared to solve prominent problems, such as inconsistencies between construction and design, thus ensuring consistency between design and construction.

(d) *Optimize the survey and construction based on operation and maintenance feedback*: Through a unified data standard and basic platform, a bridge is built for data transmission throughout the whole life cycle so that operation and maintenance information can be correlated and compared with the design and construction data more smoothly, thus achieving the purposes of design optimization, process improvement, and experience summarization. (i) BIM model data are enriched and refined. In the operation and maintenance stage, the composition of model data is continuously perfected, thus providing a data basis for design optimization and improvement. (ii) The basic platform integrates the data of the whole life cycle of railway construction projects. The basic platform collects the entire life cycle data of construction projects across the whole national railway network, reflecting the internal relationship between the issues discovered during the operation period and the processes and methods used during the construction period to guide the selection of construction processes, methods, and materials for new projects.

Chapter 2

Management Methods for Complex and Giant Intelligent High-Speed Railway Systems

The intelligent high-speed railway (IHSR) is a new-generation high-speed railway (HSR) system that takes full life cycle management as its main focus and features comprehensive perception, ubiquitous interconnection, integrated processing, active learning, and scientific decision-making. It is a typical open complex giant system (OCGS) characterized by complexity, multiple levels, high risks, nonlinearity, and emergence. With a systematic analysis of management theories and methods related to the complex giant system (CGS) and on the basis of the concepts of qualitative–quantitative and model–data integration, this chapter innovatively puts forward an IHSR management method that is model–data driven and axis–plane coordinated.

2.1 Characteristics of a Complex and Giant ISHR System

A system is an organic combination of multiple elements that are interconnected and interacting. Qian Xuesen pointed out that systems can be classified into simple and giant systems according to the numbers and types of elements constituting a system as well as the complexity of their interrelations; additionally, systems can also be classified into simple and complex systems according to the degree of association between the subsystems within a system.

A simple system refers to a system that consists of a relatively small number of elements with relatively simple interrelations. For example, some non-living systems, such as measuring instruments, are small systems; a giant system refers to a system that consists of a vast number of elements, ranging from tens of thousands to trillions. If a system has many elements but only a few types and the interrelations between them are relatively simple, it is called a simple giant system, such as a laser system. If a system has many types of elements with a hierarchical structure and complex interrelations, it is called a complex giant system. The comparison between various systems is presented in Table 2.1.

A CGS is characterized by enormous scale and a wide variety of elements or subsystems, each with different natures and complex and variable interrelationships. It comprises multiple macro and micro levels, complex interconnections between different levels, and unclear action mechanisms. As a result, inferring its macroscopic behavior from microscopic descriptions through simple statistical synthesis methods is impossible. The human brain, body, organisms, ecology, geographical environment, and cosmic phenomena can all be considered CGSs. The social system is another type of CGS, with its constituent elements consisting of human beings. Due to the influence of human consciousness, the interrelations between the elements are not only complex but also highly uncertain.

The rise in system complexity has led to the emergence of theories related to complex systems, which encompass many subjects and involve

Table 2.1. Characteristics comparison between simple system, giant system, simple giant System, and complex giant system.

Description	Characteristics
Simple system	The number of elements or subsystems constituting the system is relatively small, and their interrelations are simple.
Giant system	The number of elements or subsystems constituting the system is very large (tens of thousands, tens of billions, and even trillions).
Simple giant system	There are not many types (several or dozens of types) of elements or subsystems constituting the system, and their interrelations are relatively simple.
Complex giant system	There are many types of elements or subsystems with a hierarchical structure constituting the system, and their interrelations are complex.

a variety of environments, coupled relations, and human factors. For instance, in the field of transport, the physical structure of transport systems is examined using network and system concepts in order to enhance transport efficiency and resilience. Similarly, in the field of energy, research is conducted on the network transmission of energy and the evolution of energy systems to optimize the relationship between energy supply and demand.

The IHSR represents the integration of technical, economic, management, social, and other elements and involves multi-aspect and multi-element coordinated control and management throughout the full life cycle of design, construction, and operation. All aspects, elements, and processes run interactively in a systematic manner, which results in the emergence of a giant complex system characterized by complexity, multiple levels, high risks, nonlinearity, and emergence:

(1) **Complexity:** The IHSR is a system that entails multiple elements, such as people, vehicles, railways, networks, and environments. It involves numerous areas, such as track maintenance, communication & signaling, traction power supply, mobile equipment, transport planning, dispatching and command, passenger transport service, and safety guarantee. The network structure and interrelations between elements are highly intricate, characterized by complex user needs, technology, R&D processes, testing and verification, as well as cross-department and cross-business coordination.

(2) **Multiple levels:** The construction and operation stages of IHSRs are characterized by lengthy time periods and a large number of subsystems, while the management hierarchy encompasses various levels, such as headquarters, railway administration, stations, depots, and workshops. The long work chain involves multiple aspects and participants, and high requirements are set for internal and external, multidisciplinary, multispecialty, and multilevel cooperation.

(3) **High risks:** The IHSR is characterized by a large construction scale, very complex construction scenarios, high operation speeds, and wide transport ranges. Therefore, it faces high technical and management risks. To ensure the safety of HSR construction and operation, requirements are put forward for reliability and safety throughout the entire life cycle.

(4) **Nonlinearity:** The IHSR system is a typical nonlinear system whose overall behavior or characteristics are not equivalent to the linear superposition of behaviors or characteristics of its internal elements or subsystems. For example, train operation control is a typical nonlinear complex system that has many interrelated elements, such as passenger flow, external environment, infrastructure operation and maintenance status, and vehicle operation status, which all exhibit complex nonlinear relationships and interactions.

(5) **Emergence:** After a certain period of interaction among the constituent subsystems of the IHSR system, several functions and characteristics have emerged in the whole system as a result of evolution, which do not exist in an individual subsystem.

2.2 Related Management Theory and Method for Complex and Giant Systems

The development of theories related to CGS has been synchronous with the advancement of complexity science. The evolution of complexity science can be broadly divided into three stages. The first stage emerged in the 1940s when Bertalanffy developed the general systems theory, which led to the emergence of the general systems theory, cybernetics, and artificial intelligence. The second stage began in the 1960s and continued through the 1970s, during which theories of dissipative structure, synergetics, hypercycle, catastrophism, chaos, fractals, and cellular automata emerged. The third stage, spanning the 1980s, witnessed the breaking down of previous discipline boundaries to facilitate comprehensive research. It was during this stage that the discipline that studies complex systems was named "complexity science." Qian Xuesen is recognized as the first person in China to recognize the significance of complexity science research. He put forward the OCGS theory and pioneered the study of complexity in China. The research on CGS centers primarily around three aspects: general properties and common behavior laws, system dynamics modeling and simulation, and the CGS being oriented toward specific applications. The current main theories and methods include traditional basic theories, such as general systems theory, cybernetics, and information theory, as well as emerging ones, such as data-driven and model-driven methods.

2.2.1 *Traditional basic theories: General systems theory, cybernetics, and information theory*

The general systems theory was proposed by the American scientist Ludwig von Bertalanffy in 1937. The core ideas of general systems theory mainly include the following:

(1) A system is an organic whole instead of a mechanical combination or a simple sum of various elements. The system governs and controls various elements through its overall action.
(2) The elements of a system do not exist in isolation. They are interrelated, constituting a whole. Elements determine the characteristics and functions of a system through interaction.
(3) Systems are constantly and dynamically developing and changing, and there are dynamic interactions and interrelations between internal elements, between elements and the whole system, and between the system and the environment. The emergence of general systems theory has completely changed people's way of thinking, gradually revealing the essential connections and internal laws of objective things.

Cybernetics, introduced in the 1940s by the American mathematician and electrical engineer Norbert Wiener, encompasses principles such as the control feedback principle and feedback method, possibility space and controllability principle, purposive behavior similarity principle and function simulation, input–output principle, and black box method. Cybernetics provides the tools for system-integrated adjustment and control of complex management objects, breaking down the boundaries between the natural science and social science and between engineering technology and biotechnology. It gained widespread use in management science after the 1960s. In his book, *Engineering Cybernetics*, published in 1954, Qian Xuesen presented the concept that the object of cybernetics is the system and was thus the first to systematically reveal the significance and application of cybernetics to scientific and technological domains, such as automation, aerospace, and electronic communication.

Information, proposed by the American scholar Shannon in the 1940s, serves as the foundation of cybernetics, enabling systematic management of the flows of people, logistics, finance, energy, and information in a system through the use of information technology. The main tenets of information theory include: (1) abstraction of purposive movements of systems into an information transformation process based on information; (2) synthesis of the information process of research systems, commencing directly from a holistic perspective based on interrelations and transformations; and (3) qualitative and quantitative analyses of the abstracted information. Information theory, modern networks, and computer technology provide important means for system integration control.

2.2.2 *Emerging theoretical methods: Data-driven and model-driven approaches*

Data-driven decision-making relies on heuristic rules to establish relationships between internal characteristics through the knowledge discovery of data. The discovered theorems or laws are then applied to business activities to achieve positive feedback on business effectiveness. Data-driven decision-making includes both data-driven thinking and a data-driven approach. Data-driven thinking enables systems to perform functions such as data-based forecasting, evaluation, dispatching, monitoring, diagnosis, decision-making, and optimization by using online and offline data from controlled systems. The data-driven approach mines information hidden in data based on the collected process data using various data processing and analysis methods, such as the multivariate statistical method, clustering analysis, spectrum analysis, and wavelet analysis, to provide guidance for operation and production. Figure 2.1 provides a visual representation of this process.

The model-driven approach (MDA) involves using models to understand, design, constrain, develop, execute, maintain, and modify systems. Model-based control aims to describe the physical operating characteristics of the controlled target using relatively accurate mathematical expressions and facilitates corresponding arithmetic control through a mathematical description of the controlled target. The model of the controlled target can be created using the parameter identification method or

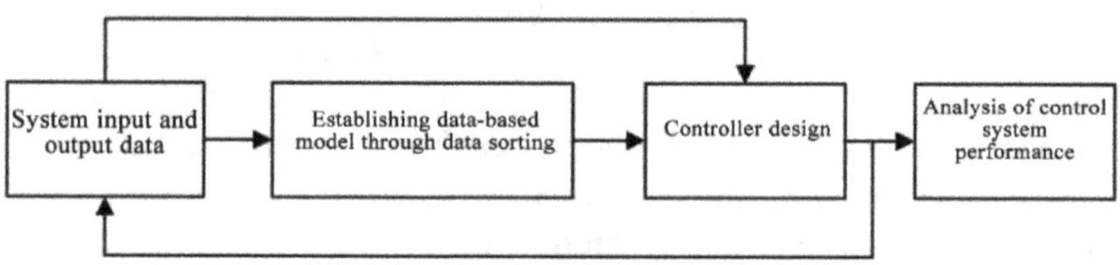

Figure 2.1. Data-driven approach.

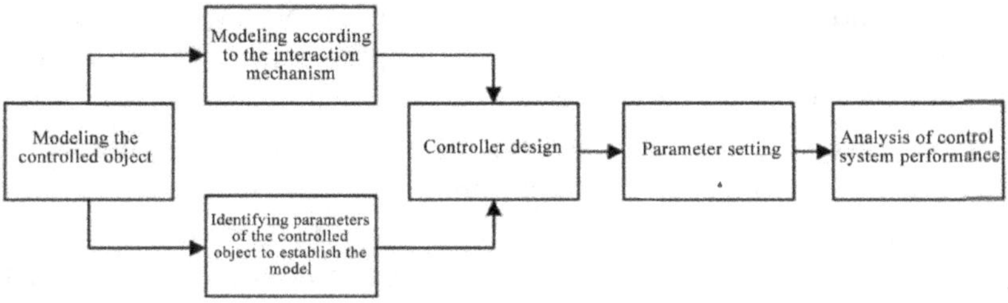

Figure 2.2. Model-driven approach.

be based on relevant physical parameters of the controlled target as well as its physical theorem. Figure 2.2 illustrates the flow chart of the model-driven research.

When applied to CGSs, traditional system theory, cybernetics, and information theory methods have significant limitations in terms of both theory and scopes of application. In the 1990s, Qian Xuesen pointed out that the comprehensive integration method combining qualitative and quantitative analyses is an effective way to address CGSs.

The IHSR system is an integration of complex engineering technology, fundamental theories, business knowledge, and coordinated decision-making. Therefore, it requires a comprehensive integration method that combines qualitative and quantitative analyses with model-driven and data-driven approaches. To be more specific, a multi-scenario business model is firstly created based on historical experience, data, and knowledge. This model serves to represent and identify the laws and mechanisms of HSR construction and operation. Next, a quantitative analysis is carried out on the model using actual data from each scenario. This allows for the establishment of a management method for the CGS of IHSR.

2.3 Model–Data Driven and Axis–Plane Coordination Management Theory (MDD&APC) for IHST

2.3.1 *System model architecture of model–data driven and axis–plane coordination method*

The IHSR is a CGS in which the entire life cycle is linked and all business elements are coordinated. Therefore, it is necessary to ensure data sharing and effective linkage at all stages, including design, construction, and operation. Additionally, comprehensive coordination of various business elements and the external environment involved in different stages of HSR is essential.

To achieve this goal, a construction and management method for IHSR GCS is proposed that combines qualitative and quantitative approaches, such as the general systems theory, data-driven method, and model-driven method. This method is a system model architecture that is "model–data driven and axis–plane coordinated" and aims to achieve optimal overall efficiency of the system. Refer to Figure 2.3 for a visual representation.

Model–data drive: "Model" refers to the representation or mechanism model of each business object of HSR; "data" refers to various dynamic and static data collected during the construction and operation of HSR. "Model–data drive" achieves the integration of model and data, including the representation model, mechanism model, and dynamic and static data. Centering on model and data integration, it facilitates forward lossless transmission and reverse iterative optimization throughout the full life cycle of IHSR, from design to construction and operation.

Axis–plane coordination: "Axis" refers to the main axis composed of the full life cycle of IHSR, covering design, construction, and operation; "plane" refers to the element plane composed of business elements and external environment involved in various time sections on the main axis. As shown in Figure 2.1, during the operation stage, business elements such as mobile equipment, communication & signaling, track maintenance, traction power supply, transport organization, equipment operation and maintenance, passenger service, safety assurance, and the external environment constitute the element plane. "Axis–plane coordination" refers to linkage throughout the full life cycle and coordination among all business elements of IHSR, aiming to achieve the optimal overall efficiency of IHSR. With the full life cycle as the main axis, it emphasizes the application of intelligent technology to all the assets, the entire industrial chain, and the full life cycle of IHSR. This is aimed at establishing a new mode for railway construction, operation, and management based on the full life cycle and promoting the seamless connection of tasks in different stages by integrating information systems in different business fields and different stages of the full life cycle of HSR. With all the elements of HSR as the plane, it emphasizes that on the static time sections of the full life cycle, support should be provided based on information flow for efficient interactions between all elements, such as different areas and external

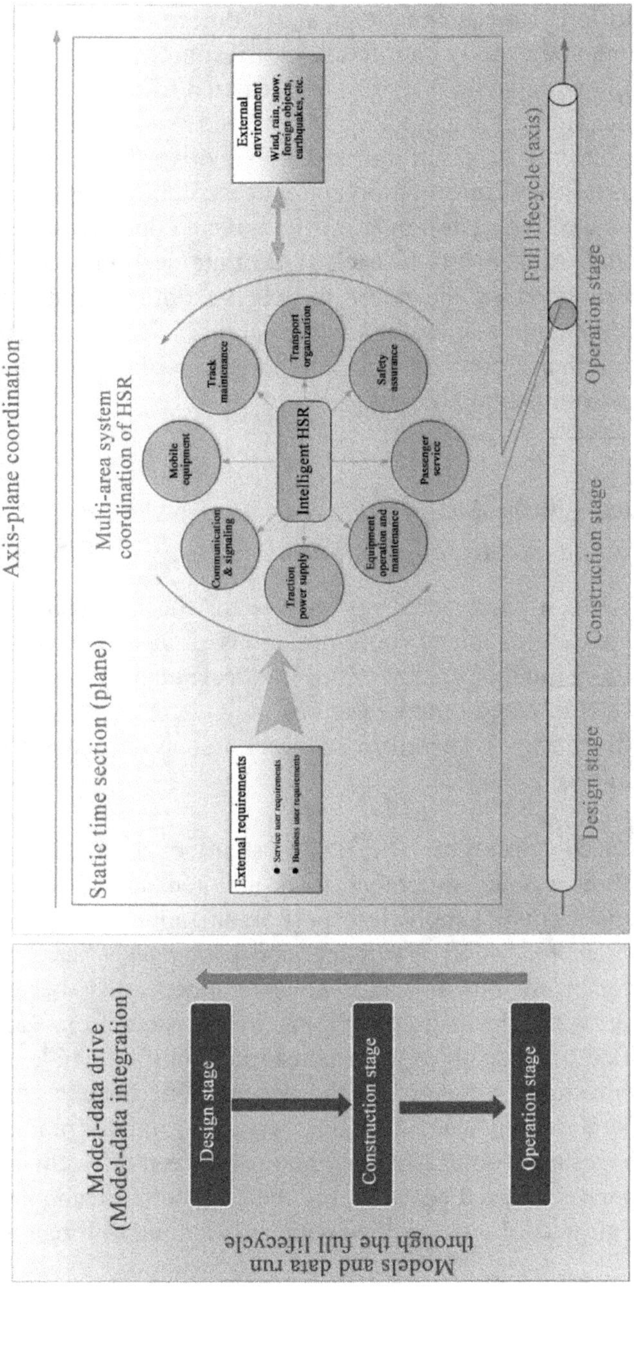

Figure 2.3. System model architecture of model–data driven and axis–plane coordination method.

environments, so as to achieve effective coordination between relevant elements of different areas and fields and promote the reasonable planning, allocation, and orderly coordination of resources.

To sum up, model–data integration is realized by the fusion of representation model, mechanism model, and data; the axis–plane coordination of IHSR is realized by means of the model–data drive. That is, the forward lossless transmission of information and reverse iterative optimization of the full life cycle covering design, construction, and operation are realized on the axis, and all elements in each static time section of the full life cycle are coordinated on the plane to achieve optimal comprehensive efficiency. This incorporates safety and reliability, cost-effectiveness and efficiency, warmth and comfort, convenience and rapidity, energy conservation, and environmental protection.

2.3.2 *Optimal system design featuring model–data driven and axis–plane coordination method*

"Model–data driven" is a necessary means for the management of the IHSR system, and "axis–plane coordination" is an ideal system operation mechanism. Based on the concept of full life cycle management, the IHSR system needs coordinated optimization of the "plane" and dynamic optimization of the "axis." Therefore, the optimal system design should include the following factors:

(1) **Multi-area coordination:** The HSR system covers multiple areas of expertise, such as mobile equipment, track maintenance, communication, signaling, traction power supply, transport organization, equipment operation and maintenance, safety assurance, and passenger service. All these areas are independent but unified. They are independent because their internal operation mechanisms are highly independent from each other. They are unified because all of them constitute a unified whole and work to achieve the overall optimal goal of the system. Different areas often need interactive operations, such as interface reservation, and this requires information exchanges and coordinated operations between different areas. The "model–data driven" based on BIM, big data, and other technologies can integrate models according to unified standards among different areas and

assist the efficient transmission and timely sharing of information among different areas through simulation and other functions.

(2) **Demand-focused:** First, it is necessary to meet the demands of service users, including passengers, other transport sectors, and partners. Passengers are the most important users of IHSRs. An IHSR must meet passengers' demands for intelligent travel service covering the whole process, from before traveling to after leaving the railway station. Other transport sectors, such as civil aviation, buses, and metro, work together with HSR to jointly provide travel services for passengers. Therefore, IHSRs must meet the demands for information sharing and the integrated development of joint services with other transport sectors. Partners include the hotel industry, tourism, the insurance industry, and other sectors. Intelligent HSRs must provide precise tailored information push services for passengers and share relevant information resources. Second, it is necessary to meet the demands of business users, including entities engaged in professional railway businesses and relevant professional users from engineering construction, equipment and facilities, transport organization, safety supervision, and other fields, to provide intelligent technical platform support for professional users.

(3) **Interactions with external environment:** HSR trains use new energy-saving materials, and the trains are designed to have a low-resistance streamlined head and smooth car body, which effectively reduces air resistance, thereby continuously reducing energy consumption during operation. Advanced braking systems and energy management tools are used to greatly reduce environmental pollution during operation, and automatic train operation technology is innovatively applied to comprehensively reduce the overall energy consumption of trains and realize benign interactions between the HSR system and the external environment.

(4) **Dynamic optimization throughout the full lifecycle:** The "model–data-driven" approach is used to unify data storage, semantic definition, and information transmission in all aspects throughout the full life cycle of HSR, covering construction and operation. Seamless connection is realized through lossless data transmission to effectively reduce management costs, greatly improve management efficiency, and promote the optimal overall efficiency of the full life cycle of IHSR.

2.3.3 *System operation mechanism of model–data driven and axis–plane coordination method*

The operation mechanism of "model–data drive and axis–plane coordination" for IHSRs takes the coordination throughout the full life cycle and the coordination of all elements in different time sections as the optimization goals, with the model–data integration as the driving element, as shown in Figure 2.4.

2.3.3.1 *Integrated model–data driven*

In the whole process of design, construction, and operation, sensor networks, the Internet of Things (IoT), new-generation communication technology, data processing techniques, and other means are integrated and applied to obtain the real-time status information of all equipment, facilities, personnel, and external environments of HSRs in a timely manner. Additionally, building information modeling (BIM) covering all stages of design, construction, and operation and the constitutive models of various infrastructures is utilized to realize the governance, storage, and analysis of model–data integration and provide a basis to ensure the reliability, availability, maintainability, and safety of infrastructures, thus promoting optimal overall efficiency throughout the full life cycle.

The forward lossless transmission and feedback iterative optimization mechanisms of model and data flows in the full lifecycle of design, construction, and operation are strengthened in a coordinated manner to realize the sharing and transmission of quality, safety, environmental protection, progress, investment, and other information driven by model–data integration in the full lifecycle of HSR infrastructure.

2.3.3.2 *Coordination throughout the full lifecycle*

(1) **Forward lossless transmission**

(a) *From design to construction*: The BIM technology platform is used to promote the lossless transmission of relevant models and data from design to construction. Focusing on the interface between the design results and the design and construction, relevant standards are established and

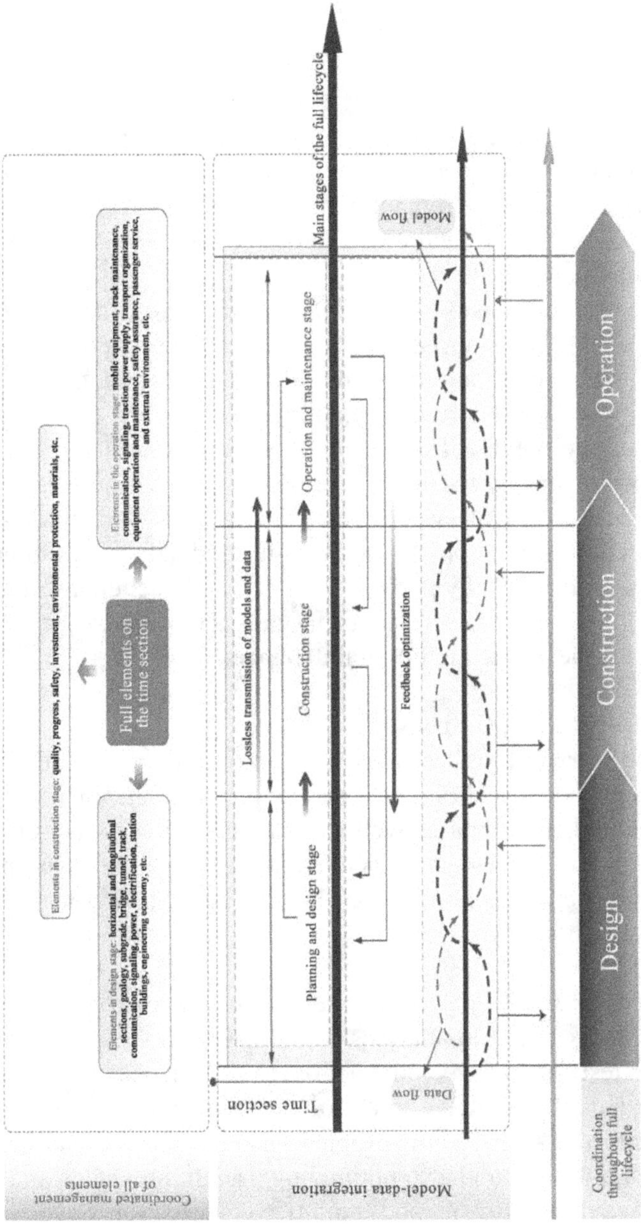

Figure 2.4. Operation mechanism of model–data driven and axis–plane coordination method.

improved to ensure the integrity of the model and data flow transmission in all areas.

(b) *From design and construction to operation*: In the operation stage, it is necessary to make full use of design and construction deliverables and information of each stage before operation, effectively allocate funds, materials, manpower, and operation time, strengthen infrastructure status analysis, carry out real-time dynamic monitoring, and keep the infrastructure in good condition to optimize operation standards, operation processes, and maintenance strategies.

The lossless transmission of models and data from design and construction to operation is promoted, and the construction management mode is comprehensively improved to build a progress control system for construction projects with project scheduling as the core, electronic engineering logs as the means of data collection, and construction organization and management as the main line. Additionally, there is a project quality and safety control system with laboratories and mixing plants under source control and construction quality and safety risk monitoring under process control to fully perceive, collect, and encode information in the construction process and transmit models and data to the operation stage to support the infrastructure maintenance.

(2) **Reverse iterative optimization**

On the basis of forward lossless transmission of data in the full life cycle of IHSR design, construction, and operation, with the increase in the amounts of models and data as well as changes in infrastructure service capacity, it is highly necessary to achieve the reverse iterative optimization of design by construction, design by operation, and construction by operation based on data analysis.

(a) *Feedback optimization of design by construction*: Based on the experience and problems found in analyses of construction sites, external environments, and test data during construction, timely feedback is given to the design stage. This is conducive to optimizing design deliverables such as construction drawings, preventing problems such as inconsistency between construction and design, and also optimizing design thresholds and parameters to improve design quality and minimize construction waste.

(b) *Feedback optimization of design by operation*: The BIM model in the design stage is an expression of the spatial relationship of engineering blueprints, while the BIM model in the operation stage carries complete data in the design and construction stages. The optimization of design can be promoted by comparing the data, such as mobile inspection, fixed monitoring, and manual inspection of the infrastructure in the operation stage, with the baseline data of the infrastructure in the design stage.

(c) *Feedback optimization of construction by operation*: The operation stage focuses on the functional and technical status of infrastructure, while the construction stage focuses more on construction methods and processes. The precise mapping relationship of infrastructure between the operation and construction stages should be established. A sound information transmission mechanism should be built through mapping conversion algorithms or tools, and data screening, conversion, and analysis should be carried out simultaneously with information transmission to solve problems, such as information loss and information translation, and to enable operation and maintenance information flow back to the construction stage more smoothly and thus improve the construction processes and methods and optimize the standards.

2.3.3.3 *Coordinated management of all elements*

(1) **Coordination of all elements in the design stage:** The BIM technology is used for multi-area coordinated design and scheme optimization. As a result, the design becomes more systematic, with higher quality and efficiency. HSR engineering design relies on the BIM-coordinated design platform, considers the coordination of multiple elements comprehensively, including horizontal and longitudinal sections, geology, subgrades, bridges, tunnels, tracks, communication, signaling, power, electrification, station buildings, and the engineering economy of railway lines, compares the feasibility study report, preliminary design, and construction drawing design, and considers in advance the reliability, availability, maintainability, and safety of the HSR after construction and operation to realize the dynamic optimization of the design scheme, minimize design changes during construction, and effectively control costs.

(2) **Coordination of all elements in the construction stage:** Focusing on personnel, machines, materials, methods, and environment for infrastructure construction, digitalization means are adopted to realize systematic coordination of elements, such as quality, progress, safety, investment, environmental protection, and materials, to

raise the construction management capacity and efficiency. Advanced perception technology is used to accurately identify high-risk factors in the whole construction process, dynamically correct the models and data information handed over at the design stage, and improve the overall economic benefits in the later stage of construction and operation; BIM technology is applied to carry out construction refinement based on visual 3D models, simulate and optimize the schemes of all areas, and thus effectively improve engineering construction efficiency and reduce costs associated with rework and changes, etc.

(3) **Coordination of all elements in the operation stage:** HSR operation covers mobile equipment, track maintenance, communication, signaling, traction power supply, transport organization, equipment operation and maintenance, safety control, passenger service, external environments, and other elements. Driven by model–data integration, it is necessary to integrate, share, and apply elements of multiple areas according to unified standards and comprehensively collect, intelligently diagnose, and actively give early warning of models and data, such as faults, applications, overhauls, and performance of key parts of infrastructure, through dynamic inspection, automatic monitoring, manual inspection, and other means. This is conducive to formulating more precise maintenance strategies, promoting condition-based maintenance of HSR infrastructure, and reducing operation and maintenance costs.

2.4 Key Technologies of MDD&APC Theory for IHSR

IHSR is a complex system covering the full lifecycle of HSR design, construction, and operation. A large number of models, along with dynamic and static data, are generated, starting from the design stage. In response to the needs of numerous entities, uneven granularity, massive data, and urgent sharing, it is difficult to meet the demands of axis–plane coordination for IHSR under the current situation of traditional decentralized storage of models and data, a lack of unified standards, and an inability to establish interrelations. Therefore, in order to support "model–data driven and axis–plane coordination," it is necessary to give priority to the organic interrelation of models and data of the same business object at different time sections in the full life cycle, i.e., model–data integration, to ensure the forward lossless transmission of models and data in the full life cycle and reverse iterative feedback optimization.

2.4.1 *Conceptual model of model–data integration and its connotation*

Model–data integration refers to the weaving and aggregation of the representation model, mechanism model, and dynamic and static information of the same business object based on BIM, GIS, IoT, master data, big data, AI, and other technologies in combination with the full life cycle management of railway infrastructure to achieve axis–plane coordination. The core of model–data integration is to realize the standard integration, storage integration, application integration, and transmission integration of models and data. The key is to organically integrate the representation model, mechanism model, and massive data of the same entity. The conceptual model of model–data integration is shown in Figure 2.5.

2.4.1.1 *Conceptual model of model–data integration*

(1) **Representation model:** The representation model is a visual information model of various business entities, such as railway stations/yards, tracks, subgrades, tunnels, and bridges, built based on BIM, 3D modeling, and other technologies. The representation model includes geometric and non-geometric information of entities. Geometric information refers to specific information such as shape description, length, width, and height, as well as geometric relationship information. The basic form of models is defined by these descriptions. Non-geometric information is composed of a characteristic layer, a level layer, and a category layer. The characteristic layer is composed of specific attribute sets, such as identity description, positioning description, technical information, system information, and project information, according to the hierarchical unit division characteristics of the model. Identity description includes entity name, type, and code. Positioning description refers to starting and ending kilometrage, center kilometrage, elevation, and other information in the railway linear coordinates system. Technical information includes material, standards applied, and construction process information. System information, from a professional perspective, includes areas of expertise and subsystems. Project information includes the name of the

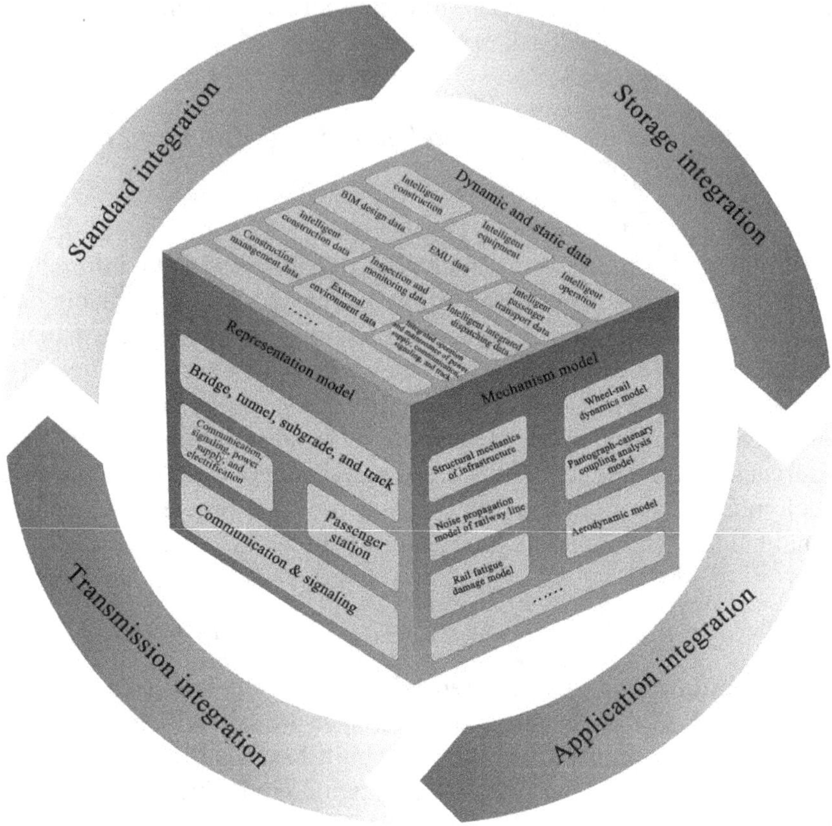

Figure 2.5. Conceptual model of model–data integration.

railway project, information of construction participants, and information of the railway line. The level layer is a management unit divided according to projects, including component level, unit level, and system level in ascending order. The category level is mainly composed of the elements of railway construction management, including personnel, machines, processes, and environments.

2. Mechanism model: A mechanism model is an analytical model that describes the interaction mechanisms between business entities. Mechanism models fall into two types. One is the deterministic model. It refers to the model of principles, theorems, and laws in the construction, manufacturing, and operation of railway infrastructure and mobile equipment, such as the structural mechanics model of infrastructure, the noise propagation model along the railway lines, rail fatigue damage model, and the electrical model of signaling equipment. The other is the non-deterministic model. It is an open and non-deterministic complex model involving many factors. It is necessary to build highly dynamic multi-factor correlation and prediction analysis models based on the big data paradigm, such as the wheel–rail dynamics model, pantograph–catenary coupling analysis model, aerodynamics model, and load analysis model, in an open environment (see Figures 2.6 and 2.7). The description of the mechanism model includes model name, model type, model function, model principle, model attribution, model input parameters, data

Figure 2.6. Schematic diagram of wheel–rail relation mechanism model.

Figure 2.7. Schematic diagram of bridge load analysis mechanism model.

model and algorithm, model output parameters, model version, operation environment, and model application effect.

3. Dynamic and static data: Data refer to all kinds of dynamic and static data generated in the full life cycle of HSR design, construction, and operation. Data generated in the design stage mainly include space–air–ground integrated engineering survey data and BIM-based engineering design data; data generated in the construction stage mainly include intelligent construction data of bridges, tunnels, subgrades, tracks, passenger stations, communication, signaling, power, and electrification works, as well as based on BIM + GIS construction management data; data generated in the operation stage mainly include the data of intelligent comprehensive inspection, infrastructure inspection and monitoring, natural disaster monitoring and warning, intelligent perimeter intrusion monitoring, and intelligent environment monitoring, as well as data of intelligent passenger transport, intelligent ticketing, intelligent integrated dispatching, intelligent traffic dispatching, integrated operation and maintenance of the track, communication, signaling, and power supply, and intelligent operation and maintenance of EMU.

2.4.1.2 *Connotations of model–data integration*

Model–data integration is the key to realizing model–data driven and axis–plane coordination of IHSRs. From the perspective of application level, model–data integration mainly includes standard integration, storage integration, transmission integration, and application integration:

(1) Standard integration is the basis and premise for realizing model–data integration. The information structure and data flow direction corresponding to the model flow should be established on the premise of interrelation, spatial relationship, and combination relationship of models, and the Industry Foundation Classes (IFC), International Framework for Dictionaries (IFD), Information Delivery Manual (IDM), master data, metadata, and other data standards should be integrated to standardize ontology, semantics, and transmission for model–data integration.
(2) Storage integration refers to the quality governance of models and data based on the integration of models and data standards and to the fusion of HSR digital twin models, mechanism models, and massive

dynamic and static big data by using model and data fabric technology and entity master data association technology, as shown in Figure 2.8. Model lakes and data lakes which associated with each other should be built to realize efficient association and centralized storage of models and data.

(3) Transmission integration is a closed-loop data transmission mode based on model and data integration in various stages and periods for the full life cycle of HSR design, construction, and operation. On the basis of standard integration and storage integration, cross-stage forward lossless transmission, efficient coordination, and reverse iterative feedback optimization based on models and data are realized, as shown in Figure 2.9.

According to relevant requirements, such as design specifications, implementation standards, delivery standards, and coding standards, the engineering entity element models, underground space facility element models, construction drawings, and geological maps in the design stage constitute the design deliverables of model–data integration. Delivery from the design stage to the construction stage is conducted based on the deliverables of model–data integration. The representation model in the construction stage carries 3D geometric information, 3D spatial coordinates information, topological relationship information, and additional attribute information in each stage of the project. All element information is stored in a unified manner to maintain the integrity and coordination of models and data. For the completion delivery, according to the concept of model–data integration, the delivery preparation, delivery process, and deliverables of 2D data and BIM models and data are specified as follows: On the basis of model–data-driven construction results and guided by the needs of operation and maintenance management, the engineering entity structure mapping and data coding chain are used as technical means to further detail the all-element information model on the BIM model and specify the modes of deliverables, such as the model information, information exchange templates, engineering drawings, construction plans, work quantities, and collision inspection based on the BIM model. With the design unit as the minimum unit, coordination is conducted with the operation and maintenance system to transmit the archives in the design and construction stages to the operation and maintenance system based on the BIM model. Models and data throughout the full lifecycle of design, construction, and operation are available in the operation and maintenance

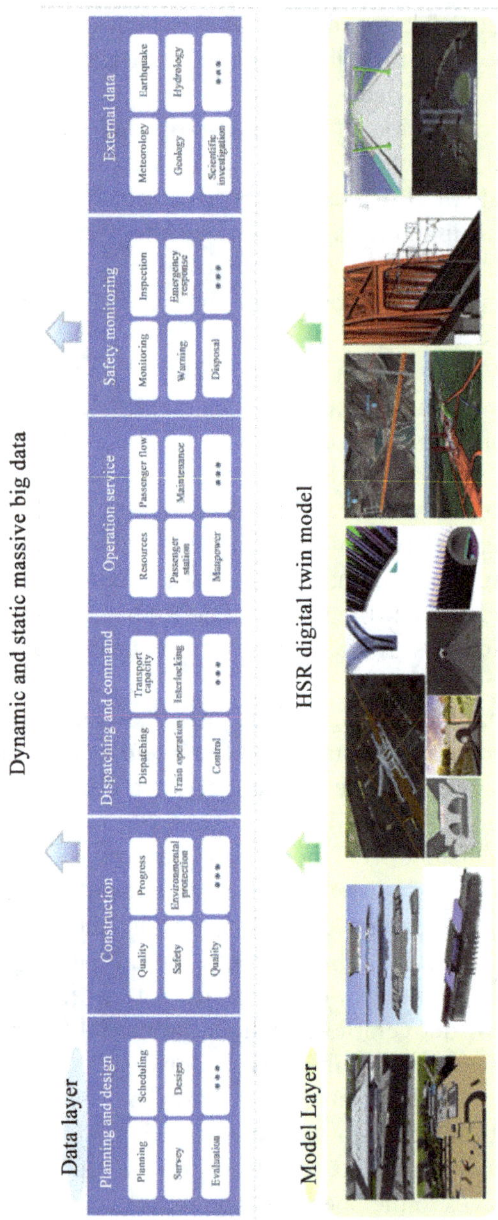

Figure 2.8. Example of model and data storage integration.

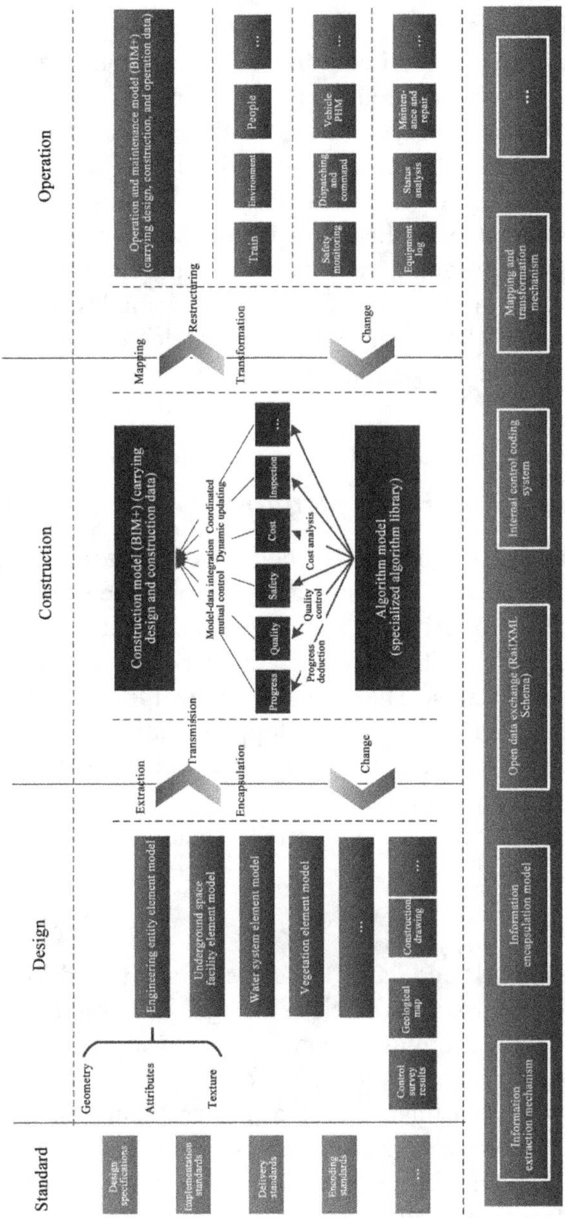

Figure 2.9. Process flow of model and data transmission integration.

system so as to provide support for the intelligent operation and maintenance of infrastructure.

Application integration is oriented toward the optimal overall efficiency of IHSR for the overall coordinated application of all elements in different stages based on model and data integration. As shown in Figure 2.10, in the design stage, it includes the coordinated application of model–data integration of such elements as bridges, tunnels, subgrades, tracks, geology, and engineering economy; in the construction stage, it includes the coordinated application of model–data integration of such elements as quality, progress, safety, investment, environmental protection, and materials; in the operation stage, it includes the coordinated application of model–data integration of such elements as track maintenance, communication, signaling, rolling stock, passenger flow, and transport.

2.4.2 *Model–data integration platform*

Following the business processes of convergence, storage, aggregation, and application of IHSR models and data, the purpose of the model–data integration platform for IHSR is to realize model–data mapping and the aggregation of entities with different granularities, such as monomers, components, and systems. The overall architecture of the platform is shown in Figure 2.11. From bottom to top, the platform consists of a data convergence layer, a model–data storage layer, a model–data aggregation layer, and an axis–plane coordination application layer. The standard system and safety assurance system for model–data integration run through the whole process:

(1) **Data converging layer**
It converges the models and data of business systems and IoT perception equipment in different stages, such as design, construction, and operation, integrates external environmental data, and implements the secure transmission, convergence, and fusion of models and data of related businesses, such as coordinated design, project management, construction management, engineering management, video surveillance, passenger transport, maintenance, and external environment, with the support of the standard system and safety assurance system for model–data

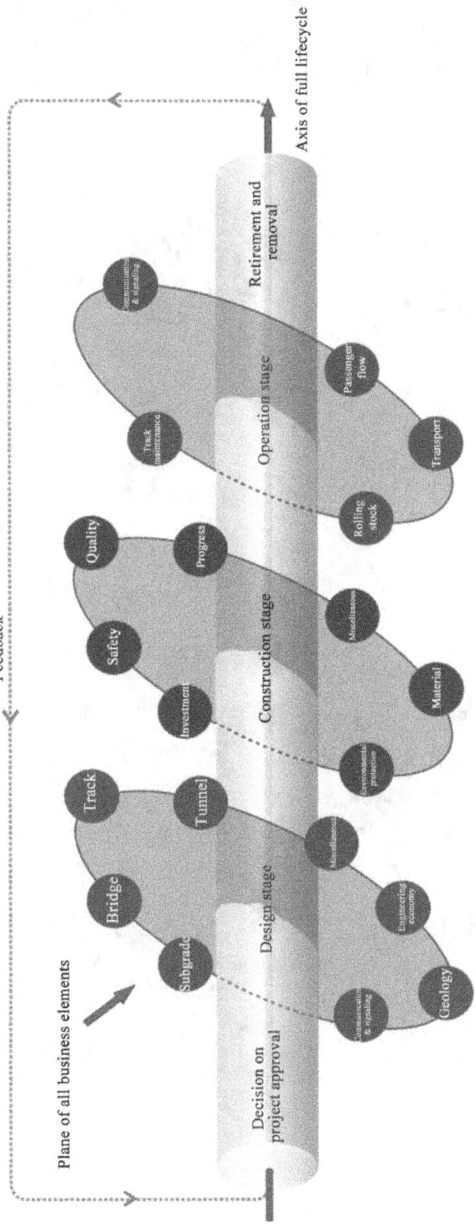

Figure 2.10. Example of application integration.

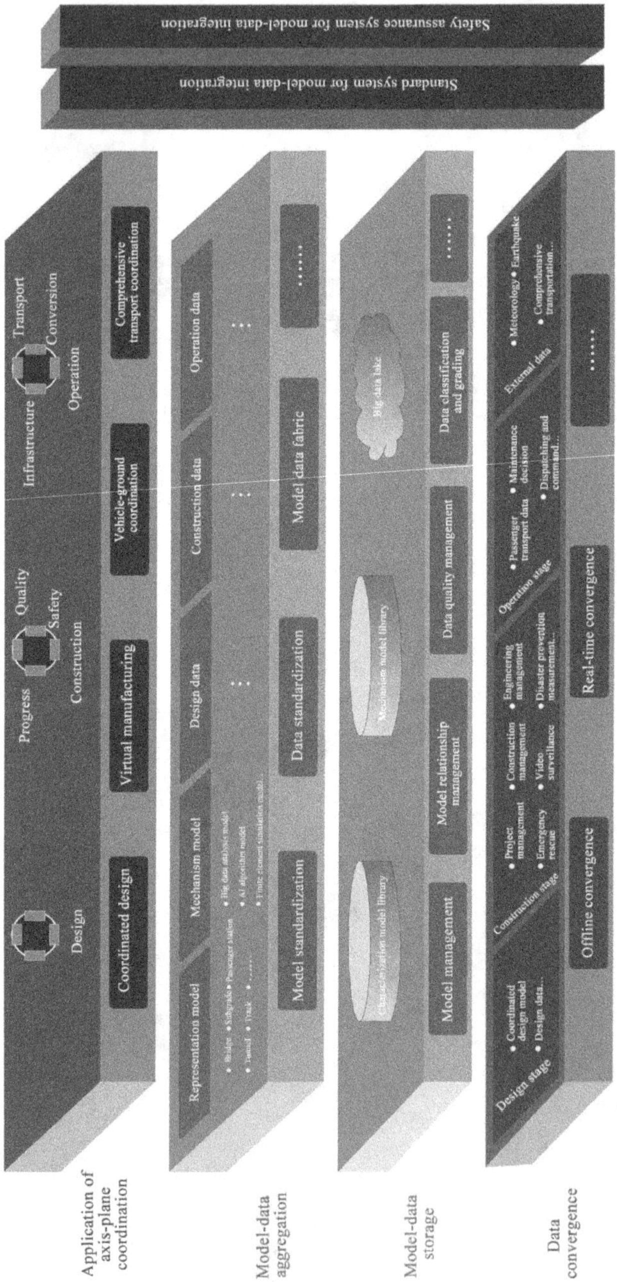

Figure 2.11. Overall architecture of model–data integration platform.

integration. The current main modes of data convergence include file exchange, data extraction, interface service, and message push:

File exchanging: It is oriented toward datasheet transmission scenarios where the real-time requirements are not high, but there is a need to transmit large amounts of data that can be easily exported as files. File exchange is divided into two modes: online SSH File Transfer Protocol (SFTP) file transmission and offline storage media copying.

Data extraction: It is oriented toward scenarios such as massive historical data and incremental data extraction from relational databases. Extract–transform–load (ETL) tools can be used for data extraction. By adapting to most types of data sources, the incremental extraction strategy is determined, and extraction tasks are set for continuous data extraction.

Interface service: It is oriented toward scenarios of datasheets or files with small daily increments and transmission volume at a single time. Rest protocols can be used for transmission.

Message pushing: It is oriented toward data transmission scenarios with high real-time requirements, large concurrency, and large data increments. Middleware such as Kafka and MQ can be used for message push. Meanwhile, Flume, Spark, and other technologies are used to implement real-time acquisition and analysis of data, such as inspection, monitoring, and operation services.

(2) **Model–data storage layer**
It fulfills the functions of model management, model relationship management, data quality management, and data classification and grading. By building a unified representation model library, a mechanism model library, and a big data lake, it performs the unified management of the converged multi-area models and data. Quality governance and storage management of massive data, representation models, and mechanism models are performed based on model management, data management, and other technologies.

(3) **Model–data aggregation layer**
It implements standardized management of models and data based on the standard system for model–data integration, extracts and semantically

analyzes the key information of models and data based on metadata technology, establishes model–data mapping and association methods, and builds model–data integration aggregates with different granularities for various business entities. In order to support the creation of the mechanism model in model–data integration analysis, it also performs structured and unstructured data analysis and mining, knowledge graph analysis, intelligent decision-making analysis, evolutionary simulation analysis, and other functions. Structured data mining includes algorithms such as data preprocessing, advanced statistical analysis, correlation analysis, and prediction analysis. Unstructured data analysis includes natural language processing, knowledge graph analysis, image analysis, voice analysis, and cross-media analysis, and a machine learning algorithm library is deployed. Knowledge graph analysis includes knowledge retrieval, knowledge extraction and encapsulation, rule reasoning, and knowledge visualization. Evolution simulation analysis includes model parametric evolution, model mechanism simulation, intelligent optimization algorithm, and finite element analysis algorithm (see Figure 2.12).

(4) **Axis–plane coordination application layer**
Various system coordination and information-sharing methods, such as Restful, File Transfer Protocol (FTP), and WebService, are adopted to open the storage, analysis, retrieval, reasoning, evolution, simulation, and other services of model–data integration to provide the business system with model–data content or analysis results and support the transmission and iterative feedback optimization for model–data integration in the full life cycle of HSR design, construction, and operation. Based on the analysis capability of model–data integration, it supports the implementation of all-element coordination and cross-industry coordination in the full life cycle, such as coordinated design, virtual manufacturing, vehicle–ground coordination, and comprehensive transport coordination.

2.4.3 *Key technologies of model–data integration*

Key technologies of model–data integration mainly include information extraction and encapsulation, data conversion and coordinated mutual control, and data association and fabric.

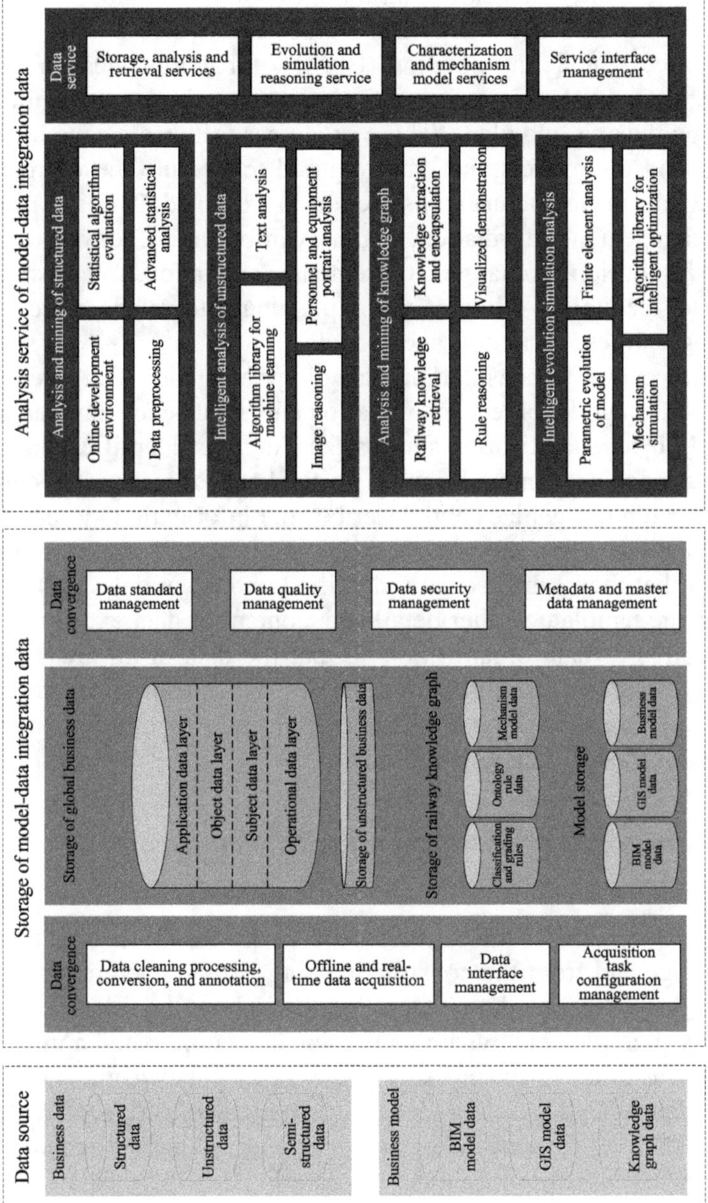

Figure 2.12. Model–data aggregation layer.

2.4.3.1 *Information extraction and encapsulation technology*

The model–data integrated information extraction technology is a complete set of information extraction processes formulated on the basis of business process modeling notation (BPMN) for the business processes and exchange requirements of specific business activities or application scenarios on the axis and plane. These processes follow after the business processes, and the models and data required to execute these processes during railway infrastructure construction are defined. The information includes management information, summary information, activity description, data description, exchange requirements description, and other information. The technology aims at accomplishing integrated extraction of engineering entities, attributes, and models.

The information encapsulation technology of model–data integration mainly includes model object specification and a data exchange template. The model object specification summarizes, defines, identifies, and sorts out the relationships of model objects on the basis of reasonable classification and coding, organizes and describes the information system structure, and establishes semantic specifications between different information systems so that they can be easily identified and accurately guided. The data exchange template further details the content of data exchange from the perspective of data organization so that the data of all areas can be identified by the system through a unified and standardized mode of exchange expression to support data exchange and transmission under the framework of model–data integration.

2.4.3.2 *Data conversion and coordinated mutual control technology*

Data conversion includes the conversion from the design stage to the construction stage and from the construction stage to the operation stage. In the application of model–data integration throughout the full life cycle, it is necessary to implement dynamic updating and coordinated mutual control of data in order to ensure strong consistency between models and data. Dynamic updating refers to the automatic acquisition of real-time status data of infrastructure using IoT and mobile internet technologies and the dynamic updating of model information driven by data elements, so as to grasp the real-time on-site situation conveniently and accurately. Coordinated mutual control refers to the two-way mutual control mode in which the model and data are kept integrated and coordinated by centering around the 3D geometric information, 3D spatial coordinates information, topological relationship information of the entity, and the additional dynamic and static engineering information in each stage of the project so that the changes in the model are driven by the data and the feedback on the data deviation is provided by the model.

2.4.3.3 *Data association and fabric technology*

In order to realize efficient association and fabric of models and data throughout the full life cycle, it is necessary to tackle key technologies such as the internal control coding system for data fabric, mapping mechanisms, and expression morphological transformation.

The internal control coding system of the data fabric automatically associates the codes of attribute data in different stages and establishes the coding mapping relationship. This aims to quickly identify entity objects and perform data interaction on the basis of defining management objects in the full life cycle and the data carried by them in accordance with coding specifications established in each stage by taking fixed attribute elements, such as entity object coordinates, as the benchmark.

The mapping and transformation mechanism of data fabric establishes the chainage correspondence table of each stage through the chainage mapping transformation method so as to quickly establish the table of correspondence between chainage positions and entity objects, such as structures and equipment. The entity objects distributed at each chainage position can be expanded according to the standard for engineering structure decomposition, and the data of each stage are mapped by taking the nodes of the engineering breakdown structure tree as the link so as to connect the data of each stage.

The expression morphological transformation of data fabric establishes data retrieval and invocation methods based on functional services by taking the engineering entity breakdown structure codes as the index. BIM model components are used in combination with entity codes to couple the data of design (construction drawings and structured design data created based on RailXMLSchema), construction (progress, quality, inspection, monitoring, engineering videos, and pictures), and operation (equipment logs, defects, and maintenance) on a single project entity to realize two-way linkage and visualization of data. After the BIM model is delivered for operation, geospatial information will be further integrated with operation, maintenance, and repair data to create the fabric and expression of data in the full life cycle of infrastructure based on model–data integration in combination with inspection and monitoring data loaded through interfaces.

Chapter 3

Intelligent Construction of High-Speed Railways

The intelligent construction of High-Speed Railways (HSR) takes BIM+GIS technology as the core, comprehensively applies new-generation information technologies, such as the Internet of Things (IoT), Cloud Computing, Mobile Internet, and Big Data, and integrates it with advanced engineering and construction technologies so as to build a closed-loop system of whole-life traceability of investigation, design, construction, acceptance, and supervision by means of comprehensive perception, ubiquitous interconnection, convergent processing, active learning, and scientific decision, and to achieve fine and intelligent management of progress, quality, safety, investment during the construction process, promoting the construction of high-speed railway from informationization to intelligence. It can realize the refined and intelligent management of progress, quality, safety, and investment in the construction process and promote the construction of HSR, from informatization and digitization to intelligence.

According to the intelligent HSR system architecture, intelligent construction is one of the three major segments, with the other two being intelligent equipment and intelligent operation, and is also the foundation of intelligent equipments and intelligent operation. Intelligent construction is horizontally divided into three fields: investigation and design, engineering construction, and construction management. The investigation and design field includes air–ground integration engineering surveys and BIM-based engineering designs. The field of engineering construction includes the intelligent construction of bridges, tunnels, roads, and rails, the intelligent construction of passenger stations, and the intelligent construction of railway communication, signaling, power supply, and electrification. The field of construction management includes the direction of BIM+GIS-based engineering construction management.

Based on the concept of "modulus-driven axis–plane synergy," this chapter takes the "whole life cycle management" of intelligent construction as the "axis," emphasizes the use of intelligent technology throughout the whole asset, industry chain, and life cycle of intelligent construction, and builds a new mode of intelligent construction management based on the whole life cycle. This promotes the seamless connection of tasks at different stages through the integration and fusion of information systems from different business fields and different phases of the entire life cycle of intelligent construction. Taking "all elements of intelligent construction" as the "plane," it emphasizes the efficient interaction of different professional elements based on information flow in the static time section of the whole life cycle and makes full use of the advantages of intelligent technologies by establishing a matching organizational structure. Effective cooperation with the relevant elements of different specialties and different fields can promote rational planning and allocation of resources. Information faults and information islands between various specialties and fields should be avoided.

The design phase coordinates and manages the design deliverables of the project at the stages of pre-feasibility study, feasibility study, preliminary design, and construction drawings, including structured delivery data, geographic maps, geological information maps, two-dimensional construction drawings, geographic information models, survey and design models, three-dimensional BIM models, and deepened BIM models, and realizes the non-destructive delivery of the design to the construction phase by means of the integration of model and data. In the construction stage, on the basis of the model and data delivered in the design stage, the following are loaded:

- construction organization, construction scheduling, project progress, construction log, and construction summary data;
- project quality inspection, parallel test, and supervision side station data;
- mixing plant, laboratory production data, advanced geological prediction, surrounding rock measurement, and settlement and deformation monitoring data;

- second lining non-destructive, CFG pile foundation, steel structure quality testing data;
- dynamic and static data, such as site photos, engineering image data, and design change data, including the professional engineering construction entity BIM model, partition, full scene BIM, completion delivery BIM model, as well as safety deduction model, quality evaluation model, precision control model, and other mechanism models.

The construction stage realizes lossless delivery from construction to the operation stage in a modular way. It provides the model and database for intelligent equipments and intelligent operation.

3.1 Intelligent Surveys for Air–Terrestrial Integration

With the development of railway engineering construction, the railway survey mode has changed from a single exploration mode dominated by drilling to a comprehensive exploration mode. In the practice of railway engineering construction, the technical framework of comprehensive geological exploration has been gradually formed. Under the guidance of remote sensing technology for geological mapping, various exploration methods such as drilling, geophysical prospecting, sounding, and digging are reasonably applied. Through close cooperation, mutual supplementation, and mutual verification, combine with the best methods to obtain multi-character geological information, and improve the accuracy of geological discrimination through comprehensive utilization and analysis in order to achieve the purpose of improving the quality and efficiency of geological exploration. This exploration method shows outstanding advantages in difficult mountain areas, large-scale key projects with complex geology, and the geological survey for railway route selection in the early stages.

In recent years, the progress of space remote sensing technology has led to the rapid development of remote sensing application technologies. Hyperspectral remote sensing, high-resolution remote sensing, airborne remote sensing, unmanned aerial remote sensing, and three-dimensional visualization technology have brought new opportunities for remote sensing geological interpretation technology and provided advanced means of remote sensing for mountainous railways. Meanwhile, with the gradual promotion of physical exploration technologies, such as geodetic electromagnetic, micromotor, and aeromagnetic, as well as the comprehensive application of technologies such as deep-hole drilling, comprehensive logging, and deep-hole *in situ* testing, the foundation has been laid for the ground exploration work. Therefore, in order to meet the major needs of railway engineering and construction and the requirements for the progress of survey technologies, it is necessary to build an integrated three-dimensional survey system for railway engineering in the space, air, and ground.

3.1.1 *Technology system*

According to the different spatial locations, railway survey technology can be divided into three categories: "air", "space", and "ground".

1. "Air": aerial remote sensing, airborne radar, unmanned aerial low altitude photography, and aerial object detection

"Empty" survey technologies include aerial high-resolution photography, aerial thermal infrared scanning, airborne laser radar, unmanned aerial tilt photography, aerial electromagnetism, aeromagnetism, semi-aerial transient electromagnetism, and other technologies based on aerial platforms and low-altitude helicopters and unmanned aerial vehicles, which are carried out in low-altitude areas within the atmosphere. Among them, for complex terrain and geological sections that need detailed investigation, drones carrying high-resolution photography and hyperspectral cameras can be fully utilized to obtain the three-dimensional spatial geometric image of the target feature by means of three-dimensional modeling of tilt photography and spectral imaging by the drone and, at the same time, obtain the spectral information of the target feature, which can realize the attribute detection of the geologic body and the quantitative analysis. It can realize the three-dimensional mapping of slopes, the survey of avalanches and rockfalls, mudslide monitoring, etc.

Aero-physical exploration is a method of physical exploration to study the distribution of underground structures and undesirable geological bodies by detecting changes in various physical fields during navigation with specialized physical exploration instruments mounted on aviation platforms, such as fixed-wing aircraft, helicopters, and drones. Aviation physical exploration methods, such as aeromagnetic and aeromagnetic surveys, are gradually gaining popularity in railway surveys. This technology is mainly used in the initial and final survey stages, which can further identify the regional geological characteristics of the survey areas and guide the selection of railway route and engineering settings.

2. "Space": satellite remote sensing, monitoring, and positioning

The "sky" survey technology mainly refers to the acquisition of various types of image information for feature extraction, classification, identification, and geological interpretation based on satellite platforms operating in space, including high-resolution satellite remote sensing, radar satellite remote sensing, satellite thermal infrared remote sensing, BeiDou satellite positioning, InSAR deformation monitoring, and multi-spectral (hyper-spectral) lithology identification.

In the process of geological investigation, high-precision remote sensing technology is first applied to obtain rich and comprehensive ground information, interpret topographic features, the distribution of fractures, lithological zoning, the distribution of landslide hazards, and other macro-regional geological distribution features along the railway. This aims to comprehensively and effectively extract the characteristics of the geological environment along the railway and to guide the geological survey and mapping by means of the interpretation of remote sensing geology. The InSAR technology is used to identify potential geological hazard sites over a wide area and to determine the distribution of risk zones with large deformation. The satellite remote sensing technology can realize the investigation of large-scale engineering geological conditions and major adverse geological conditions and also monitor the trend of changes in the geological environment. This technology is mainly used in the stage of on-site investigation and preliminary survey to guide the selection of schemes and stabilize the macroscopic route scheme.

3. "Ground": surface exploration and in-hole *in situ* testing

"Earth" survey technologies refer to exploration and testing technologies implemented on the ground and in boreholes, such as geological mapping, physical exploration, drilling, and *in situ* testing. Ground geological mapping includes geological surveys and mapping. In the air–space–ground integrated survey technology system, new survey and mapping methods, such as intelligent terminal mapping systems, fixed-point cameras, and three-dimensional laser scanning, are mainly used.

In the difficult and complex mountain surveys, the physical exploration methods used mainly include geomagnetic, transient electromagnetic, micro-motion, and resonance imaging. In the terrain of complex and dangerous areas, it is recommended to use the aviation electromagnetic method combined with ground physical exploration to carry out the work, which can improve the efficiency of the survey and the quality of the interpretation.

Drilling, as a traditional means of investigation in difficult and complex mountainous areas of railway survey, focuses on the exploration of important geological boundaries, tectonic lines, and major adverse geological points and the verification of physical exploration anomalies. In the steep slopes of the canyon sections at the tunnel entrance and exit, it is recommended to use deep horizontal drilling for the suspension bridge anchoring sections. In the arduous and complicated mountain railway survey, the role of deep hole drilling should be fully utilized. Following the principle of multi-dimensional use of a single borehole, integrated logging, hydrogeological testing, geostress testing, in-hole side pressure and shear measurements, the strength and structural parameter measurements of borehole cores, etc., are carried out in the borehole.

The "Ground" survey technology is mainly applied in the stage of finalization and supplementary finalization survey (ground geological mapping, representative physical exploration, and drilling can be selected for complex areas and important work sites in the initial survey stage). According to different geological conditions, the reasonable use of advanced physical exploration means, combined with drilling verification and comprehensive testing, can accurately and efficiently carry out various tasks such as engineering geological mapping, stratigraphic delineation, the detection of significant adverse geology, and the acquisition of geotechnical parameters, and provide real and reliable survey results.

In the process of geological survey of railways in difficult and complex areas, according to the engineering geological characteristics of various types of projects and main engineering geological problems, comprehensive engineering regional geological features, railway survey depth accuracy requirements, and different geological parameters required for engineering design, the appropriate technical means and combination modes are selected. This may range from surface to point, from macro to micro, and from qualitative to quantitative. Step-by-step, in-depth, scientific, reasonable, and efficient geological survey work is carried out in a comprehensive manner.

Deeply buried long tunnels can be surveyed through satellite remote sensing, aerial physical exploration, ground physical exploration, representative deep hole drilling, horizontal drilling and in-hole testing, and other “air–space–ground” combination of methods. Moreover, the tunnel inlet and outlet on the steep slopes of sections can be surveyed according to the needs of drone tilt photography and InSAR deformation monitoring. Satellite remote sensing, unmanned aerial photography and mapping, deep-hole drilling, and in-hole testing can be used in combination mode for deep-valley bridge crossing and steep-slope roadbed investigation. According to the applicability of various methods, investigation stage, depth, and scope requirements, the corresponding combination mode can be selected for the investigation of adverse geology, such as rupture, karst, and hollow areas, and mountain disasters, such as collapse, landslides, and mudslides. In the process of survey and evaluation, for specific survey objects and survey requirements, technical synergy and survey data fusion analysis are a very important for the railway integrated “air–space–ground” survey.

3.1.2 *Technology achievements*

The process of the integrated “air–space–ground” survey can be divided into the on-site survey stage, the geological deepening stage before the preliminary survey, the preliminary survey stage, and the finalized survey (supplementary finalized survey) stage. The main work and technical results of each stage are as follows:

1. The initial survey phase

An extensive collection of existing geological, hydrological, surveying, and mapping data, combined with medium- and low-resolution, medium- and small-scale geological interpretation maps, topographic maps, hydrogeological maps, geological hazard interpretation maps, vegetation cover maps, etc., is used to macroscopically understand the distribution range and law of geological structure and characteristics in the engineering area and obtain information about topography, geomorphology, and geological disasters. On the basis of the remote sensing interpretation data compiled in the pre-feasibility study stage (1:200000 or 1:500000) and other previous geological work, we make full use of all kinds of remote sensing data and adopt various new remote sensing techniques and methods to carry out remote sensing interpretation of geological structure, stratigraphic lithology, bad geology, special rock and soil, hydrogeology, etc.

2. Pre-primary geological deepening phase

On the basis of the pre-feasibility study of the line program, geological deepening work will be carried out in due course for the major adverse geology (mining and hollowing areas, large-scale landslides, avalanches, areas of strong karst development (dark rivers), areas with a high incidence of major active ruptures and geologic fissures, etc.) and geologically complicated sections of the long crossings of the mountain range that control the line program.

On the basis of remote sensing geological interpretation and geological large-area mapping (general scale: 1:50,000, of which 1:10,000 can be used for hollow areas and landslides), comprehensive physical exploration methods will be adopted in key areas, together with the necessary drilling and testing, so as to focus on identifying the undesirable geological problems and putting forward proposals for the selection of geological routes.

3. Initial testing phase

For the over-mountain areas, adverse geological and special geotechnical areas, complex long tunnels in Daheqiaodu, and other heavy engineering areas of the control line scheme, we should give full play to the advantages of high-end satellite remote sensing, unmanned aerial vehicle (UAV) surveying, and "air" and "space" survey technologies and carry out detailed geological mapping.

A reasonable selection of comprehensive physical exploration, *in situ* testing and other effective methods, and targeted drilling are adopted to identify the topographic and geological conditions, hydrogeological features, major adverse geological development patterns, and evaluate their impact on the line program and engineering.

A comprehensive analysis of the survey results is carried out for the selection of line design to provide more detailed data support.

4. Finalization (retrofitting) phase

The stage of finalizing the survey has the following purposes: obtaining the quantitative indexes of geotechnical, meeting the requirements of preliminary design and construction design, combining with the type of project, focusing on the "air" and "ground" types of survey methods, adopting the drone survey, geological survey, and mapping, detailed drilling, and representative physical exploration. Representative observation, testing, and other means are followed to carry out comprehensive survey work. The supplementary determination survey stage is the supplementary improvement to the work of the determination survey stage, mainly adopting the "ground" type of survey technology, further supplementing the geological mapping, drilling, and exploration data, identifying various types of tunnels, bridges, roadbeds, and other important points of site engineering geological conditions and engineering geological problems, and obtaining the physical and mechanical parameters required for the design of the construction drawings.

Bridges and roadbed sections continue to be drilled and excavated, along with corresponding *in situ* tests and trials.

For tunnels, geological mapping and physical exploration are the main methods of investigation, supplemented by drilling; for long tunnels with complex geological conditions, necessary drilling, including deep-hole drilling, must be carried out on the basis of comprehensive exploration and in accordance with the relevant provisions of the specifications to determine the geological conditions and construction methods and to ensure the quality of the tunnels and the safety of the construction.

For karst areas, manmade pits, and hollow areas, a combination of physical prospecting and drilling methods is needed for water supply sources and the identification of fracture structures, fracture zones, sliding surfaces, soft and weak layers, bedrock interfaces, rock interfaces, grading of tunnel surrounding rocks, and the thickness of the weathering layer.

3.1.3 *Technology innovations*

Intelligent technology to promote "integrated engineering survey" consists of two directions of innovation and development: the first is to innovate "air", "space", and "ground" and the development and comprehensive application of intelligent equipments at different levels; the second is to accelerate the integrated processing of internal and external data from the survey.

With the integration of new-generation information technology and traditional survey technology, a variety of intelligent survey equipments has been developed in recent years. The "air" type of survey technologies is mainly used to carry out aerial camera, airborne laser radar, thermal infrared aerial scanning, and other survey services with the help of aircraft, and the progress of UAVs has promoted the development of related technology, which is mainly applicable to the acquisition and identification of high-definition images of regional or linear projects and carrying out tectonic analysis of a larger scope, geological mapping and interpretation, geothermal and groundwater exploration, etc. In the "space" category of survey technologies, intelligent equipments such as InSAR deformation monitoring, BeiDou satellite positioning, multi-spectral/high-spectral stratigraphic lithology identification, and satellite thermal infrared remote sensing geothermal identification have emerged, which are mainly applicable to the identification of geological backgrounds, macro-geohydrological descriptions of rock formations and surface water systems, and satellite-based survey and positioning, etc. The "ground" type of survey technology is mainly aimed at smaller-scale, refined survey technology, and in recent years, there have been innovations in such technologies and equipments as three-dimensional laser scanning, hydraulic drilling, and isotropic antimagnetic flux transient electromagnetism.

On the basis of "air–space–ground" survey technology, the combination of "information" survey technology and "air–space–ground" survey technology can provide a more convenient visual management tool for the analysis and processing of multi-professional and large volume data from the railway surveys. With the support of information technologies, the data from all kinds of sources are entered into their corresponding databases through data collation and cleaning. Form the data model of the application in each stage of survey, design, construction, and operation, various application theme views on the data model are formed. Based on the integrated management platform of railway engineering geological survey, the time and space big data system is established. Surveying and mapping, engineering design, testing, and other disciplines combine to achieve data, service, and other aspects of resource sharing, which provides more comprehensive digital services for railway engineering design, construction, operation, and maintenance.

The "Information" survey technology includes video surveillance, electronic sand tables, three-dimensional BIM, and information platforms. These technologies are combined with the "air–space–ground" survey to construct an integrated management platform for the railway survey, standardize geological exploration according to the requirements of railway construction standardization, and control the survey quality and results from the source. To solve the problems of large amount of data processing in the field of exploration, high error rate in field work, and untimely information sharing in various links, establish a unified stratigraphic division standard, reduce single data processing work, and realize information processing and sharing in the whole process, from geological prospecting

Table 3.1. The integrated intelligent survey of "air–space–ground–information."

Data collection	Regional geology; hydrological data; meteorological data; existing engineering data
Comprehensive survey of air–sky–earth	Direction; verification
Space	Satellite positioning measurement technology; low- and medium-resolution optical remote sensing; high-resolution optical remote sensing technology; synthetic aperture radar interferometry technology; thermal infrared remote sensing; multi-high-spectral stratigraphic lithology recognition
Air	Low- and medium-resolution aerial photogrammetry and remote sensing technology; high-resolution aerial digital photograph measurement technology; high-resolution low-altitude unmanned aerial survey technology; high-resolution low-altitude drone survey technology; airborne LiDAR scanning technology
Ground	Geological survey and mapping; core drilling exploration technology; equivalent inverse channel transient electromagnetic technology; microseismic monitoring technology; 3D laser scanning technology; ground-based synthetic aperture radar technology; building information modeling (BIM) technology
Early stage	Reconnoite; initial test; calibration; additional calibration
Late stage	Construction; scheduled service
"Air–space–ground" three-position integrated comprehensive survey technology system	
Information processing	Survey informationization construction, 3D BIM analysis
Analyze and organize	Synthesis of data results, evaluation of synthesis papers
Compilation of documents	Compilation of survey results documents to provide design applications

scheme planning, field collection, indoor tests, indoor work arrangements, and results submission to data archiving and reuse. Related engineering practice shows that a similar information-based geological prospecting work platform can increase work efficiency by 30–50% and reduce production costs by 20–30%.

The integrated intelligent survey of "air–space–ground–information" is shown in Table 3.1.

3.2 BIM-Based Collaborative Design

Railway design is a huge and complex systematic project which has the characteristics of long lines, large volumes, complex speciality interfaces, and close relationships with topography and geology. It is difficult to coordinate with many specialties and needs close cooperation among various specialties within the design team. Under the traditional mode, the railway professional design process and the final design results all use two-dimensional drawings. In the face of huge and complex system engineering, there are some issues, such as unclear expression and understanding of planning intentions and poor communication between departments. It inevitably causes errors or omissions in the design results, which can lead to some rework or waste in the construction. With the diversified and dynamic development of refined design requirements, traditional design patterns cannot meet the design requirements of high quality and high efficiency, so it is necessary to introduce collaborative design.

Collaborative design based on BIM has the characteristics of digitalization, visualization, multi-depersonalization, cooperation, and simulation, which can run through the whole life cycle of design, construction, operation, and maintenance. Based on a unified building information model, all relevant personnel can share engineering construction design drawings, design materials, design manuals, and other materials and achieve multi-professional cooperation and multi-stage seamless connection. This can reduce the cost waste caused by the design change in the construction stage, improve the ability of project management, and make the management more refined and the decision-making more scientific.

3.2.1 *Collaborative design platform based on BIM*

3D collaborative design is the process of reengineering traditional design patterns, that is, based on the building information model, all the design data and information in the model are interrelated and interdependent. Once a parameter in the model changes, all the contents related to it will change accordingly. During the design process, the unique central file on the server side keeps the data synchronized with the model of each client.

The collaborative work based on BIM should be based on the technical standards and management system of the railway industry, establish a collaborative work platform, determine a standardized work flow, establish a standardized project breakdown structure and work breakdown structure, and finally create an information model in collaborative mode.

The content, format, and structure of collaborative process delivery and stage outcomes should meet the requirements of the standard.

1. BIM collaborative platform

Modeling software is in the core position in the application process of BIM technology and plays a fundamental role in realizing the value of BIM. At present, the core modeling software of BIM and its corresponding collaborative platform are mainly foreign "ABC" platform and domestic BIM platform.

"A" is Autodesk's BIM series software and collaborative platform, such as Revit, Civil3D, Inventor, AutoCAD, InfraWorks, and Navisworks. Its advantage is that various departments can directly or secondarily develop complete railway BIM modeling, can manage the design process through collaborative management software, model ancillary information to add convenience, flexibility, and strong parameterization ability, with a wide range of application basis and usage habits.

"B" is Bentley's Bentley series software products, such as MicroStation CE version, Open Rail Designer, Substation, ProjectWise, and Restructures. Its advantages are that all softwares are developed based on MicroStation, the model format is unified, it is convenient for professionals to refer others' work, and it has a strong ability to support a large number of large and linear models, in line with railway design habits, and provides design collaborative management software with strong collaborative management ability.

"C" is Dassault's CATIA software, or 3D experience platform. Its advantage is that it provides a complete solution for modeling-construction simulation performance and three-dimensional results display. Each department uses the same software for modeling, and it is convenient for each department to refer each other's model. The software has strong parameterization abilities. Positioning assembly engineering with "skeleton lines" is similar to railway line positioning. It also includes specific collaborative management software and has strong collaborative management abilities.

In recent years, the domestic BIM platform has continued to strengthen research and development and has made remarkable progress, forming domestic leading products such as Zhongwang 3D, Shanda Huatian SINOVATION, structural BIMBase, and railway BIM platform software for linear engineering, as shown in Figure 3.1. The domestic BIM platform can support a variety of modeling methods, can quickly and flexibly complete design expression, and provides a solution for the whole process of product design and development, which has been used in machinery, molds, auto parts, and other design and manufacturing fields.

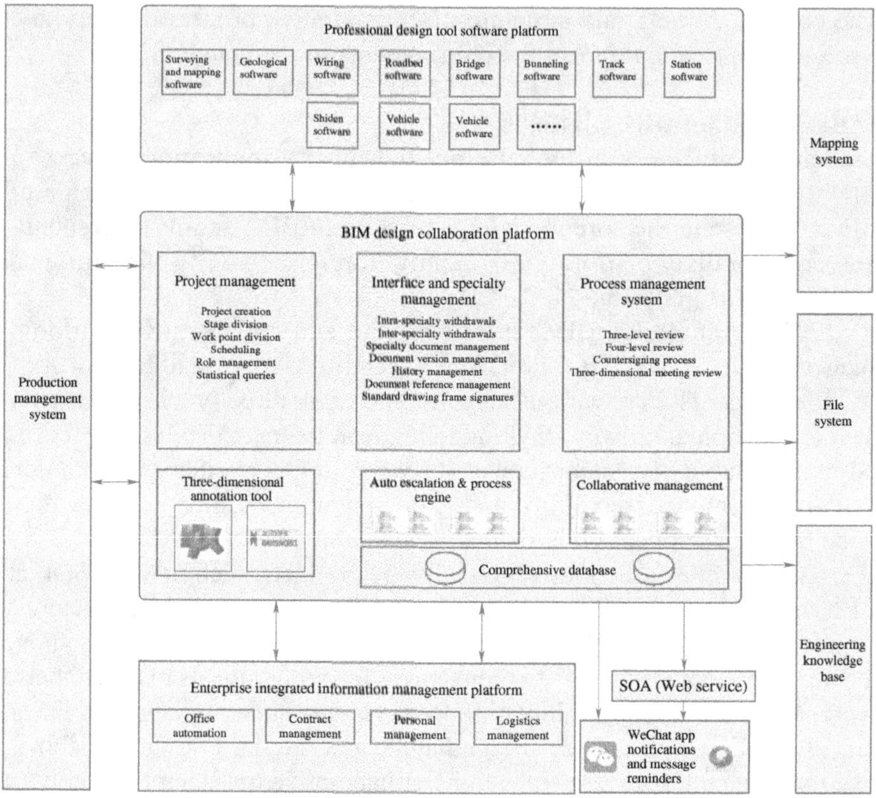

Figure 3.1. Collaborative design platform architecture example.

2. Key links of collaborative design

(1) *Standard system formulation*

The BIM standard is the premise for ensuring the effective transmission of the information model throughout the whole life cycle of the project. In the process of collaborative design, special attention should be paid to the implementation of BIM standards, including industry-level, enterprise-level, and project-level BIM standards.

Industry-level BIM standards: At present, the collaborative design of railway projects mainly adopts the "Unified Standard of Railway Engineering Information Model" issued by the State Railway Administration, "Railway Engineering Information Model Classification and Coding Standard," "Railway Engineering Information Model Expression Standard," "Railway Engineering Information Model Delivery Accuracy Standard," "Geographic Information Delivery

Standard for Railway Engineering Information Model," "Railway Engineering Construction Drawing Design Document Compilation Method Based on Information Model," and other standards issued by the Railway BIM Alliance.

Enterprise BIM standards: Under the guidance of industry body standards, enterprises combine their own selected collaborative design platform, internal speciality division of labor and workflow formulation, BIM collaborative design workflow, modeling standards, etc., and stipulate the content and form of information transmission, involving data coding and storage standards, information attachment and delivery standards, etc. The operation steps of each stage of planning and design, preliminary design, technical design, construction drawing, and deepened design are explained; the principles of collaborative design based on BIM are formed to guide designers in carrying out standardized design work.

Project-level BIM standards: The project BIM application implementation standard for a specific project BIM application, which can be standardized in three aspects: design resources, behavior, and delivery. It clarifies the responsibilities and permissions of overall and speciality leaders and designers and guides the landing application of project-level BIM design.

(2) *Design process reengineering*

Collaborative design takes topography and geology as the basic design data and carries out route design, pre-station engineering design, and post-station engineering design according to the design process. The design process is similar to the traditional two-dimensional design process, but the collaborative design workflow and BIM data flow need to be organized in an orderly manner.

First of all, according to the results of the railway survey, route selection, and alignment, the centerline of the line is fitted based on the horizontal and vertical sections of the line, and the civil engineering in front of the stations, tunnels, bridges, subgrades, and other stations is segmented, combined with topographic and geological strips, and then the sites are subdivided step by step according to the structural relationship of the project and the needs of management requirements. A parameterized component library is established according to surveying and mapping, geology, lines, station yards, tracks, subgrades, bridges, tunnels, and other specialties, so that it can be reused, and each specialty carries out design under the leadership of a engineering breakdown structure. According to the engineering structure tree, the tasks are decomposed, the tasks are

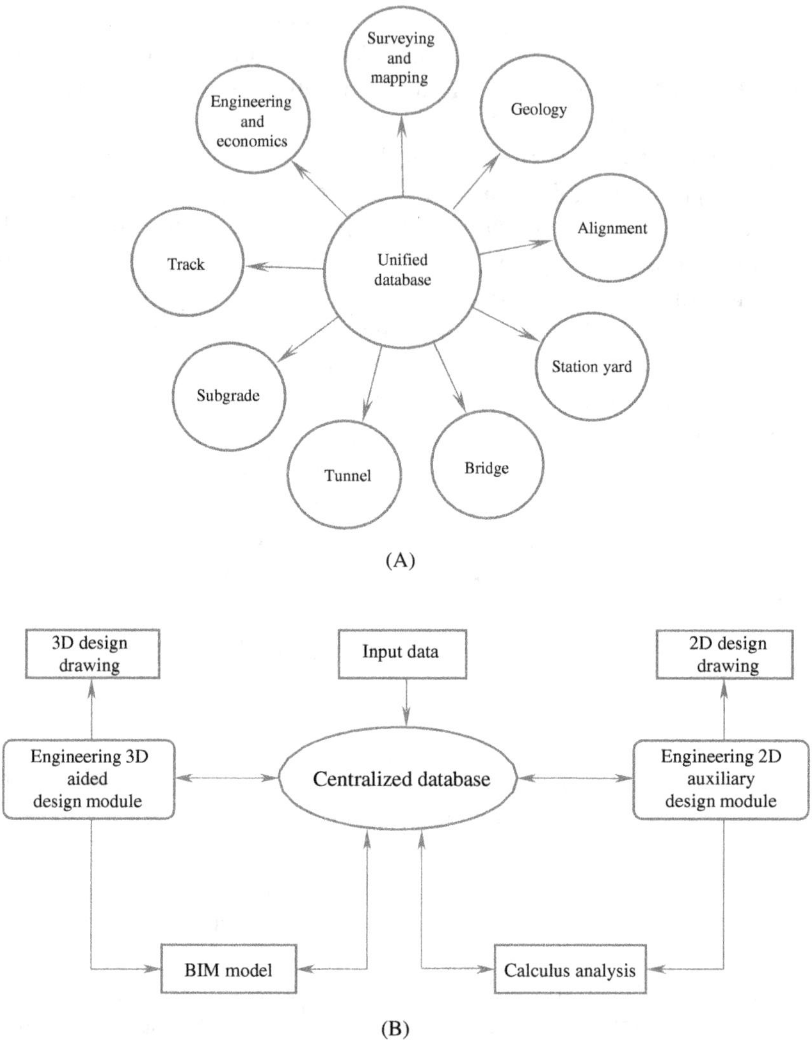

Figure 3.2. Collaborative design pattern with data as the core.

distributed and graded on the collaborative design platform, and the design process is managed.

When the design scheme is adjusted within a certain range, the BIM model can be updated quickly. All the above work is carried out in a unified database, which ensures the real-time cooperation design of multi-speciality, design, as shown in Figure 3.2.

The characteristics of collaborative design are as follows:

(1) Each department completes the design process in the 3D integrated application environment.
(2) Multiple departments complete design collaboration in a unified integrated application environment through the network.
(3) The form of data transmission between departments has changed, from the traditional provision of documents to the release of a shared information model through the BIM model, document transmission, circulation, and sharing, to achieve the coordination of the model and drawing integration. The collaboration within specialties is reflected through the design, verification, and review of models and documents, and the collaboration between specialties is reflected by the state change of models and documents between departments.
(4) The collaborative design platform realizes multi-speciality collaborative design, which ensures the standardization of the design process, the consistency of design standards, the sharing of real design resources, the timely discovery of speciality conflicts, and the effective avoidance of rework.
(5) Digital delivery: Limited by the development level of BIM technology, the external environment, and supporting laws and regulations, two-dimensional design drawings are still legal deliverables and the main basis for paying design fees. It still need a certain amount of transition time to take the information model as the legal deliverable, so it is necessary to deliver two-dimensional drawings based on BIM collaborative design.

The application of information model drawing refers to the process of generating two-dimensional section drawings or design drawings through projection, cutting, or information extraction of the information model, which generally requires software development according to specific drawing requirements. At present, domestic design institutes have realized the arbitrary cutting function of the BIM model based on spatial lines, solid boundary lines, and routes, so that some two-dimensional drawings of sections, such as bridges and tunnels, can be obtained by customizing the template and using the cutting projection of the model. According to speciality needs, the profile of the section is extracted after the plane model is cut, and the section is dimensioned automatically by customizing the speciality templates, which improves the efficiency of two-dimensional drawing.

Taking the information model as the final deliverable is the development direction of BIM technology. There is an essential difference between the delivery information model and the delivery two-dimensional drawing: the delivery information model takes the information model as the core, and the information model is visual, computable, and machine-readable; the core of the two-dimensional drawing is the drawing, which is mainly recognizable to people. At this stage, individual projects are recommended to accurately express the design intention in the way of "two-dimensional drawing + three-dimensional view," as shown in Figure 3.3.

Considering the division of railway design stages, the digital delivery of BIM design includes a pre-feasibility study model, a preliminary design model, a construction drawing design model, a construction drawing deepened design model, and a design change model. These models are the general name of the deliverables of information models in all stages of the whole life cycle of railway engineering projects. When the information model of each stage is applied for a specific purpose, it is generally necessary to carry out the adaptive delivery of the information model according to the application requirements.

3.2.2 *Application of collaborative design of HSR*

The collaborative design of HSR is carried out according to the process of "planning first, then implementation, and then evaluation." Before carrying out BIM design, the overall design principles and BIM implementation standards of the project are compiled, the engineering organization breakdown structure tree is established, and the collaborative design platform, with the help of BIM, is used to realize the digitization of design results, the standardization of design information, the integration of results delivery, the virtualization of key projects, and the visualization of engineering decisions. A full-line and full-speciality BIM collaborative design is developed. The application process for project collaborative design is shown in Figure 3.4.

In the BIM collaborative design process, standards such as "Railway Engineering Information Model Data Storage Standard" and "Railway

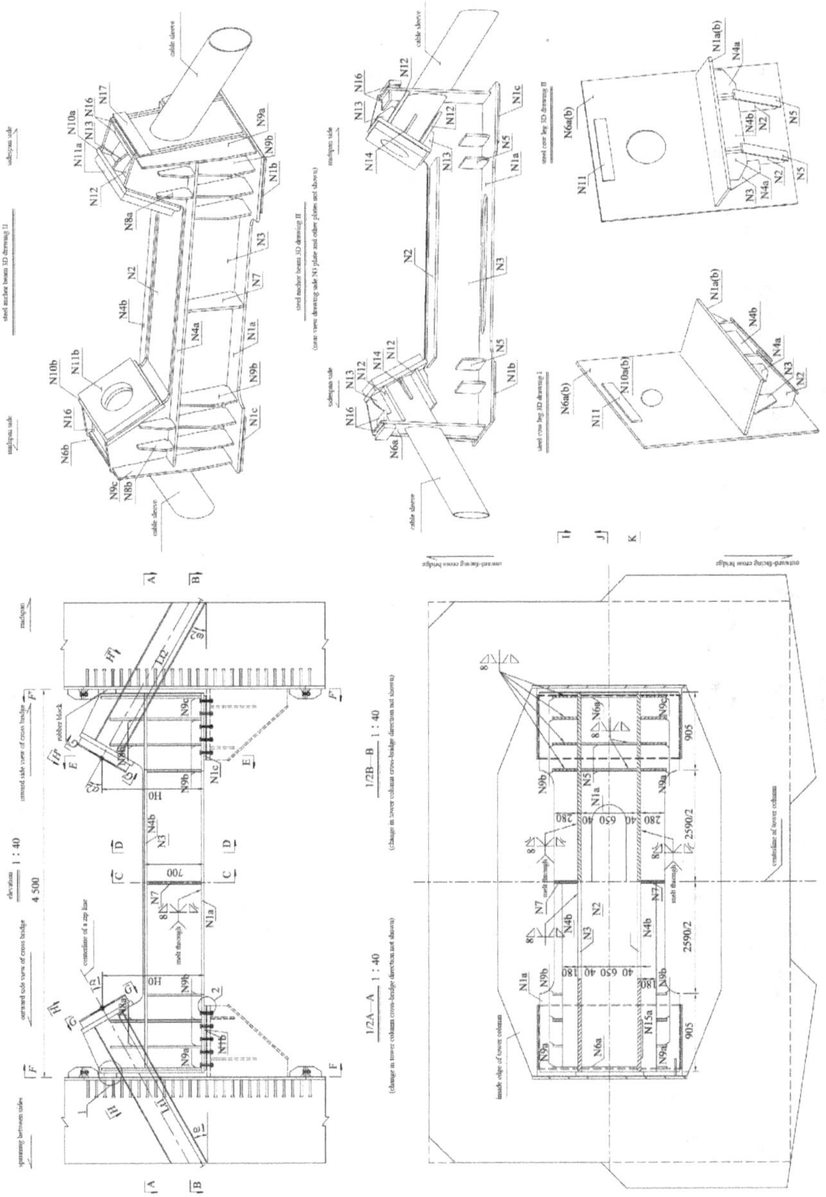

Figure 3.3. Reinforcement model and drawings of a 32m simply supported box girder of a railway.

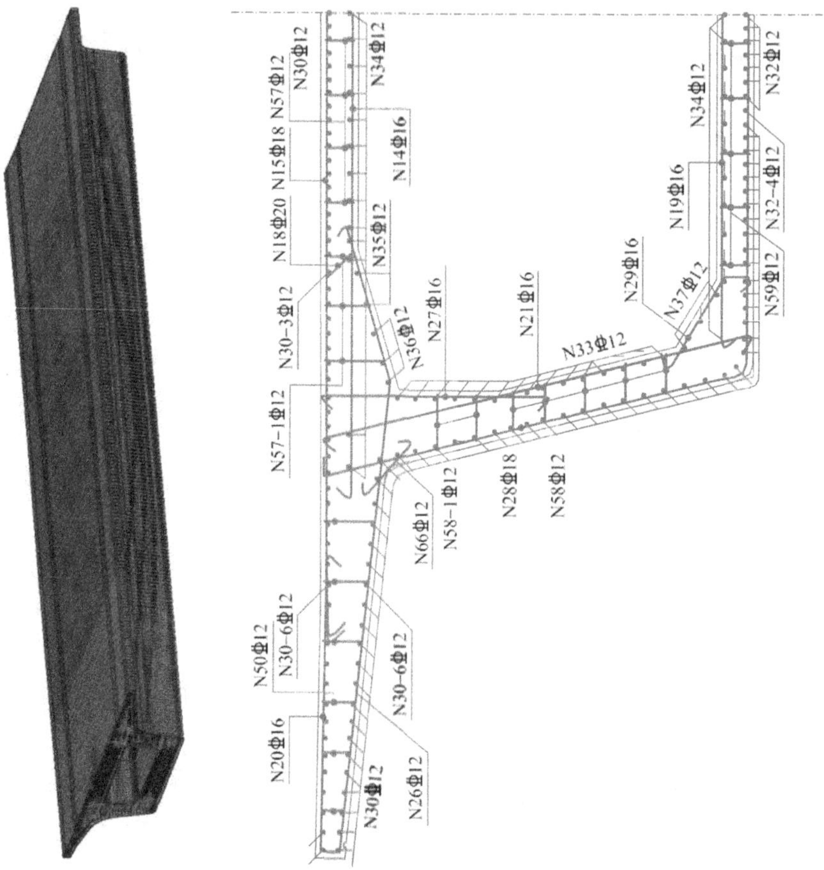

Figure 3.3. *(Continued)*

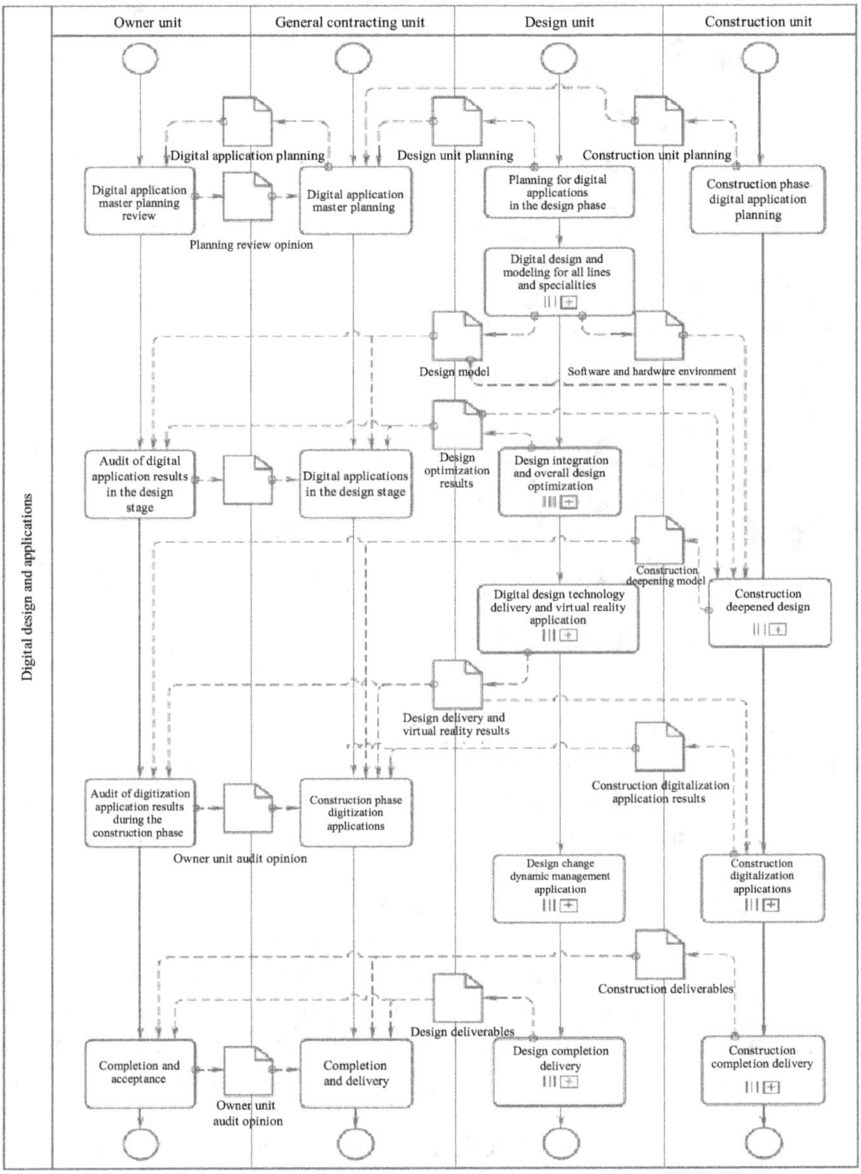

Figure 3.4. Project co-design application flowchart.

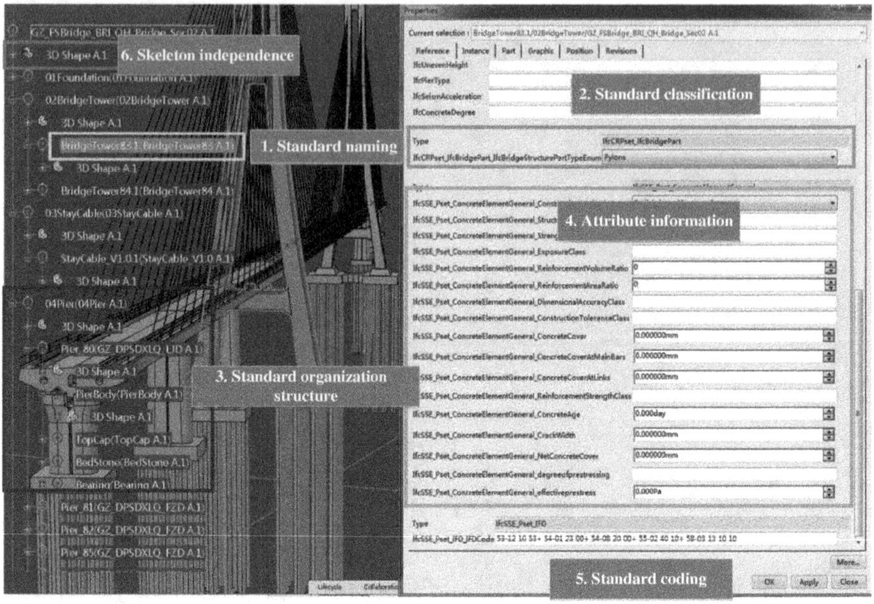

Figure 3.5. Standardized model example.

Engineering Information Model Classification and Coding Standard" are followed, and project-level BIM design implementation standards are compiled to ensure the standardization of the BIM design model by making agreements in terms of model information, model organization, and design behavior. An example of a standardized model is shown in Figure 3.5.

The purpose of full-professional BIM design model integration is to separate the railway engineering information model from the "heavy customer end" of BIM modeling software and to realize efficient browsing and viewing of large amounts of railway engineering BIM data with the help of the multi-LOD hierarchical resource scheduling strategy of the lightweight graphics engine. The effect of data integration is illustrated in Figure 3.6.

Based on the above BIM design model integration, omni-directional design inspection and optimization are carried out.

(1) **Coordination check between roads, bridges, and tunnels and the environment:** Carry out the coordination inspection of subgrades,

Figure 3.6. BIM design model integration example.

Figure 3.7. Checking the coordination between the tunnel opening and the topography of the mountain.

bridges, tunnel entrances, and other projects with the environment, focusing on the relationship between subgrade slope protection, tunnel entrances, and natural terrain. A check of the coordination between the tunnel entrance and the mountain topography is shown in Figure 3.7.

Figure 3.8. Road and bridge transition section interface inspection and optimization.

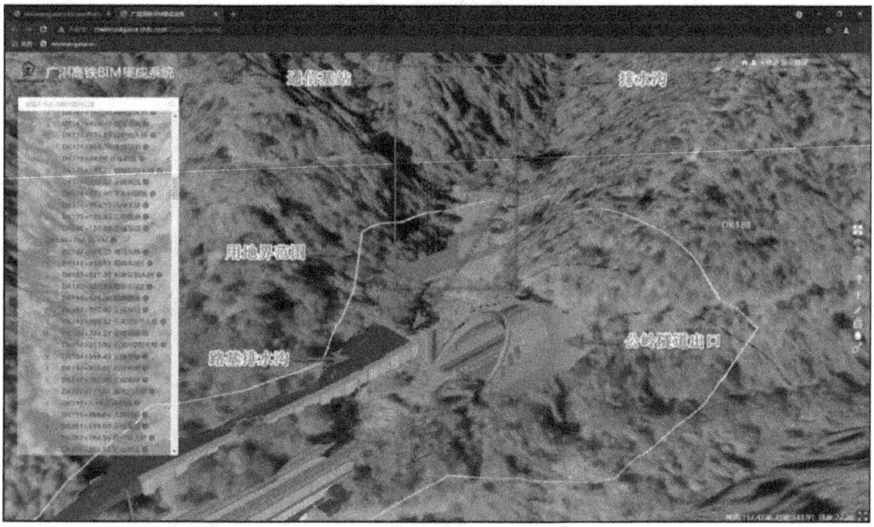

Figure 3.9. Drainage system inspection.

(2) **Interface check of road, bridge, and tunnel engineering:** Carry out the inspection of the interface projects of roads and bridges as well as roads and tunnels, with emphasis on checking and optimizing the connection of drainage ditches, the transition of cable trenches, and the connection of slopes. The interface inspection and optimization of the road–bridge transition section are shown in Figure 3.8.

(3) **Design inspection of drainage system:** Check the rationality of the drainage engineering setup of the whole line, focusing on the rationalness and consistency of the design of drainage facilities, such as tunnel openings, road riffles, the periphery of the site apron of the section office, natural washout ditches, side ditches of the local low-lying areas, drainage

Figure 3.10. Refined design of pre- and post-station interface works.

ditches, interceptor ditches, and culverts, under the conditions of complex terrain. The drainage system inspection is shown in Figure 3.9.

(4) **Engineering inspection of the interface between the front and back of the station:** By integrating the post-station interface engineering models of cable crossing, cable upper and lower bridges, comprehensive grounding interfaces, etc., the interface position distribution, space occupation, and other speciality requirements are clearly shown, so as to realize one-time checking of the post-station interface engineering of the whole line within a short period of time and to optimize the design of the interface engineering. The refined design of interface engineering pre and post the station is shown in Figure 3.10.

(5) **Inspection of the location of the section and ancillary works:** Through the integration of high-precision GIS terrain and the four-electrical field model, the four-electrical site selection and supporting drainage project verification are carried out, and the location scheme of the field is optimized. The selected site and drainage facilities in the section are checked, as shown in Figure 3.11.

(6) **Verification of land use design:** Carry out the verification of the flat land of the four-electrical plants on the whole line, check the location and land use of the field site by site, and minimize the land replenishment and secondary land expropriation on the premise of meeting the needs of usage. The verification of the flat land requisition of the four-electrical fields is shown in Figure 3.12.

Figure 3.11. Sites and drainage facility installation calibration for interval sections.

Figure 3.12. Four-electrical field pad acquisition site verification.

(7) **Traffic space inspection:** Carry out collision inspection of spatial boundaries, such as expressways, provincial highway crossings, and navigable waterways, and check the rationality of the setting of spatial limits for all-line traffic. The verification of traffic space limits is shown in Figure 3.13.

Figure 3.13. Verification of traffic space limits.

Figure 3.14. Comprehensive pipeline inspection and optimization of HSR station.

(8) **Comprehensive pipeline inspection and optimization:** Based on the railway BIM collaborative design platform, the integration of a multi-speciality comprehensive pipeline 3D model and collision inspection are completed, and the comprehensive optimization design of the pipeline is carried out to achieve the purpose of a reasonable layout, a smooth process, and zero conflict. The comprehensive pipeline inspection and optimization of the HSR station are shown in Figure 3.14.

3.3 Digital Factory

Typical representatives of a railway digital factory include an intelligent girder yard, a track panel yard, and a sleeper yard, whose intelligent construction is represented by two aspects of technology: one is information collection and comprehensive management for the whole production process, and the other is the intelligent upgrading of key production equipment. Information collection and comprehensive management rely on production management systems, including production plan management, production process data collection, production logs, progress/safety/quality warning, intelligent equipment data access, and other functions. Intelligent equipments are applied to key aspects of factory production, forming an automated operation assembly line and gradually realizing quality diagnosis, intelligent control, equipment health management, and other functional enhancements.

3.3.1 *Intelligent girder yard*

One of the characteristics of China's HSR is the large number of bridges, and the length of the standard design simply supported box girder bridge accounts for more than 50% of the mileage of the operating line. The construction method of the simply supported girder of HSR is mainly prefabricated in the girder yard, with a high degree of mechanization, standardization, and industrialization, and has the basis for intelligent development. In the process of intelligent HSR construction, the intelligent girder field is built by means of information technology, robot technology, and intelligent control technology, which comprehensively improves the production efficiency and management level and ensures the construction quality of HSR simply supported beam bridges.

The overall construction idea of the intelligent girder yard is to take the unified information platform as the traction, break the "islands of information", and comprehensively coordinate the project management elements such as "man, machine, material, method, and environment." It takes intelligent equipments as the basis to transform the key production process and reduce the labor intensity of the workers. The overall construction goal is to realize the intelligent, unmanned, self-perception, and scientific decision-making of production management in the girder yard.

3.3.1.1 *Function and composition*

The intelligent girder yard specifically includes a girder yard comprehensive information management platform, key process intelligent equipment, and a quality control information system.

(1) **Girder yard integrated information management platform**
The integrated information management platform of the HSR girder yard is supported by BIM technology, combined with the management requirements of construction, and revolves around five modules: three-dimensional navigation, progress management, safety and environmental protection management, comprehensive management, and system management. Taking quality management as the main line, it closely combines material management, test management, concrete mixing plants, process management, warehouse management, and statistical analysis. At the same time, the beam yard production progress, safety, environmental protection and other aspects of management and control are considered to achieve the construction process, safety, quality, operation, and maintenance of the whole life cycle management. The integrated management platform of the railway girder yard is shown in Figure 3.15.

(2) **Intelligent equipment for key processes**
The key production processes of a simply supported girder in HSR include steel bar processing, formwork installation, concrete maintenance,

Figure 3.15. Railway girder yard integrated management platform.

prestressed tensioning, and prestressed pipe grouting, among which formwork installation and adjustment and the steel bar binding process use the most manual labor. The tensioning of prestressed tendons and prestressed pipe grouting are the most important to ensure the quality of prestressed construction, and girder maintenance has a great influence on the quality of concrete construction. At present, these key processes are made intelligent:

(a) The CNC processing system for steel bars is shown in Figure 3.16. With the application of BIM technology, the semi-finished model of rebar is imported into the CNC rebar processing equipment, such as the intelligent automatic rebar shear, the intelligent inclined rebar bender, and intelligent rebar bender. The equipment automatically analyzes data, such as the part of rebar used, processing quantity, processing size, and angle, and carries out rebar processing according to the imported data, which has the advantages of high efficiency, precision, safety, and savings.
(b) A self-driven hydraulic inner mold is used for the box girder. Wireless remote control technology, telescopic control, and electric drive walking electromechanical equipment are adopted to replace the traditional winch dragging process, realizing the rapid installation and dismantling of box girder hydraulic internal mold.
(c) The automatic maintenance system is shown in Figure 3.17. The automatic sprinkler system with a high pressure water pump, induction temperature measurement, and an automatic controller is adopted to

Figure 3.16. Rebar CNC machining system.

Figure 3.17. Automated maintenance system.

Figure 3.18. Automatic prestressing tensioning system.

automatically carry out natural sprinkler maintenance according to the actual temperature change and the demand for maintenance water according to the development law of concrete maturity.

(d) The prestressed automatic tensioning system is shown in Figure 3.18. Using prestressed automatic synchronous tensioning equipment and its computer-controlled system, full-scale, one-button operation and a tension qualified rate of 100% are achieved.

(e) The automatic grouting system for prestressed pipes is shown in Figure 3.19. It is composed of pulping, vacuum, grouting, auxiliary, and other hardware components and a control system, which can automatically mix slurry according to the specifications of the prestressed pipeline and ensure the compactness of the pipeline.

Figure 3.19. Automatic grouting system for prestressed pipelines.

(3) **Quality control information system**

(a) *Information management system of girder yard mixing station*: The IoT technology is used to monitor the raw material consumption and material proportion of the beam mixing plant to realize real-time on-line analysis of mixing errors and control the mixing quality of concrete.

(b) *Informatization management system of girder yard testing room*: It utilizes IoT technology to collect, transmit, and analyze real-time data from the presses in the testing room of the girder yard to enhance the reliability of test data, such as concrete test blocks and raw material inspection.

(c) The automatic loading system for static load tests is shown in Figure 3.20. Automatic control technology, sensor technology, mechanical technology, and a computer control system are used to realize automatic synchronous loading, dynamic balance, and precise control of load in static load tests of HSR simply supported girders, so as to improve the work efficiency of static load tests of box girders.

3.3.1.2 *Key technology*

(1) **Information integration technology based on BIM**

In the girder yard integrated information management platform, BIM technology is used to show the layout of the girder field and the effect of personnel management and control. By using the lightweight BIM model

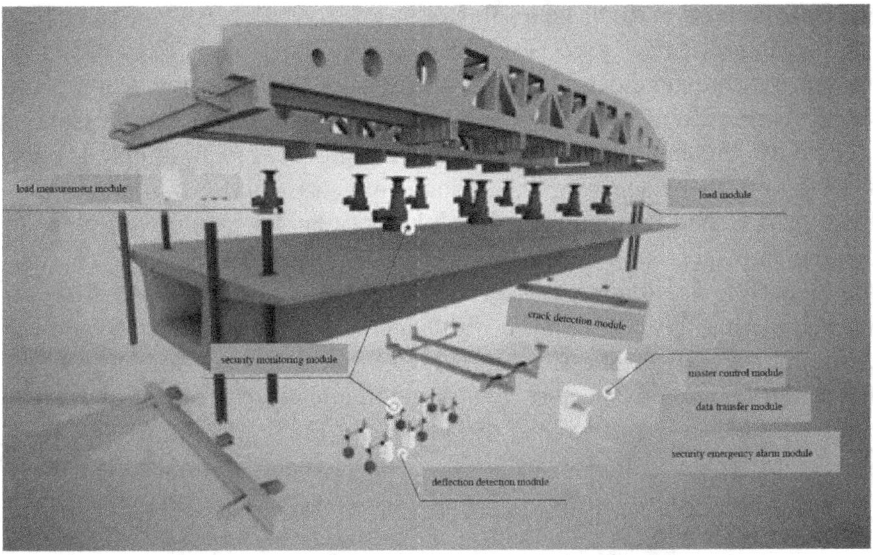

Figure 3.20. Static load test automatic loading system.

of simply supported box girders, the large samples of steel bars, prestressed tendons, and other design information can be displayed in detail, and the CNC processing system for steel bars, the prestressed automatic tensioning system, and the automatic grouting system for prestressed pipes can be directly connected. The detailed information of each process of each hole box girder is recorded in a separate simply supported beam BIM, including concrete pouring start and stop times, concrete use quantity, concrete block strength, tension, elongation, tension time, grouting time, and grouting volume.

(2) **Advanced sensor technology**

A large number of advanced sensors are used in the intelligent beam field system, which is the perception end of each subsystem and the main source of production data. It includes the weighing sensor of the mixing station, the temperature sensor of the automatic maintenance system, the pressure and displacement sensors of the prestressed automatic tension system, the weighing and pressure sensors of the automatic grouting system for prestressed pipes, and the pressure and displacement sensors of the automatic loading system for static load tests.

(3) **Data-driven control technology**
The intelligent girder yard is realized as a key process control based on design and statistical data. It utilizes the design reinforcement sample and the CNC processing system for steel bars to realize accurate processing of steel reinforcement, design tension, and the prestressed automatic tension system. It realizes accurate tensioning of prestressing, design pipe specification size, and the automatic grouting system for prestressed pipes to achieve high-precision and high-compactness prestressing pipe grouting, the automatic calculation of test loads, and accurate application of test loads by using the information of the age of box girder concrete and the design of the second-phase constant load in the automatic loading system for static load tests. The test load is automatically calculated and accurately applied in the static load test automatic loading system.

At present, the intelligent girder field has been applied in the 32 m simply supported box girder fields of Zheng Wan, Jingzhang, Jingxiong, and other HSR and the 40 m simply supported box girder field of Jizheng HSR, which significantly improves the production efficiency of precast beams and reduces the amount of labor. The construction quality of the simply supported box girders is guaranteed.

3.3.2 *Intelligent track panel yard*

CRTS III slab track is a ballastless track system with completely independent intellectual property rights which was developed on the basis of summarizing the previous application practice of ballastless track in our country. At present, it has become the standard form of ballastless track of HSR in China. The CRTS III track panel is the core product of the ballastless track system. With the rapid development of China's HSR and the transformation and upgrading of industrial structures, the track panel manufacturing industry has put forward the goal of realizing intelligent manufacturing. Intelligent manufacturing technology research and development has been carried out at three levels: intelligent manufacturing equipment research and development, information fusion and application technology research, and intelligent management system.

At present, there are two main production techniques of the CRTS III pre-tensioned track panel of domestic HSR: the fixed pedestal method and the flow unit method. This section focuses on the domestic CRTS III ballastless track panel intelligent production line with international first-class level and independent intellectual property rights, which is intelligently upgraded on the basis of the original production line. The intelligent control system is adopted to realize the automation and informationization of track panel production so that the track panel production process is intelligent and efficient and the quality of track panels is significantly improved.

1. Function and composition

The intelligent production of CRTS III slab tracks is mainly embodied in three aspects: the informationization of production management, the automation of production process and the intellectualization of key production stations.

(1) *Informationization of track panel production management*:

The track panel manufacturing information integrated production line system is adopted, the automation equipment is configured, and centralized intelligent control is carried out in the production control center so that track panel production can realize information management, production whole process video monitoring, and production network scheduling management. Through the implementation of track panel manufacturing process information, visualization, and intelligent management, the track panel, from raw materials to factory inspection, production process control, product inspection, inventory, shipping, and other management elements, can be managed and controlled, which effectively improves the quality of track panel products. Figures 3.21 and 3.22 show the production control center and information management system of the intelligent track panel factory.

Figure 3.21. Production control center.

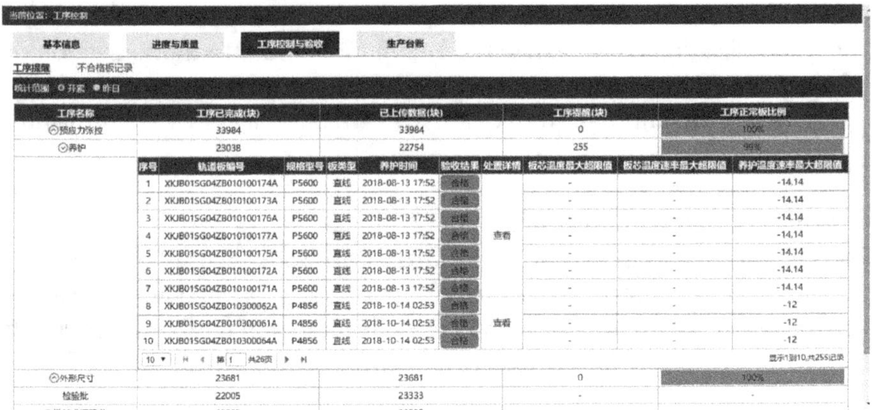

Figure 3.22. Information technology management system.

(2) *Automation of track panel production process*

The production key technology station is equipped with automatic control, with each working procedure standardizing the operation and achieving control over the whole production process. The whole production process of track panels is divided into several stations according to the requirements of production technology. During production, the products are moved from one station to the next along the industrial assembly line with the aid of lifting and transportation equipment, and each station completes the prescribed process that should be completed, respectively. According to the production process of track panels, the automatic production process mainly includes automatic mold cleaning, automatic spraying of release agent, automatic installation of pre-buried casing, automatic binding of RFID tags, automatic detection of steel bar skeleton, automatic pouring and maintenance of concrete, and automatic release and demoulding. Figures 3.23 and 3.24 show the application of an automatic installation system for embedded pipes in a track panel and the prestressed automatic tensioning system in an intelligent track panel factory.

(3) *Intelligence of key production stations*

CRTSIII track panels usually adopt a two-way pretensioned prestressed structure, which has the characteristics of multiple production processes and high precision. In order to ensure the production efficiency and finished product quality of track panels, track panel production has realized

Figure 3.23. Automatic installation of track panel embedded sleeves.

Figure 3.24. Automatic tensioning of prestressing for track slabs.

Figure 3.25. 3D scanning inspection of the external dimensions of a track panel.

intelligent control at key stations, including automatic prestress tensioning of track panels, automatic maintenance of concrete, and 3D scanning inspection of shape dimensions. Figure 3.25 shows the application of the track panel 3D scanning detection system in an intelligent board factory.

2. Key technology

The intelligent production line of CRTS III pre-stressed concrete track slabs is implemented through optimizing the configuration of the original production line, developing a new structural model, accurately locating steel bar binding tools, upgrading model automatic cleaning, automatic brushing release agent, automatic installation of pre-embedded casing, automatic prestress tension and release, automatic concrete maintenance, 3D scanning inspection of track panels, etc. The integration of automatic and continuous operation of the main process of track panel production is greatly improved, and the information management of key technical station, and product quality control of track panels are realized so that the whole process of production can be controlled, and the quality of the engineering entity is guaranteed. Automation, information, and intelligent management of production lines are realized. The key technologies of an intelligent track panel factory mainly include:

(1) *Intelligent manufacturing technology of two-way pretensioned prestressed track panels under complex process conditions*

According to the key production process and quality control points of track panels, the design technology of prestressed integral dies under the conditions of complex mechanical behavior, high frequency vibration, high temperature, and high humidity maintenance is put forward. It is based on the tension control technology of single root control, synchronous tensioning and automatic locking of prestressed tendons, and the flatness control technology of track panels based on deformation compensation and water film maintenance, a complete set of modular equipment and tools is developed. Production automation and the reuse of equipment and tooling are realized.

(2) *Intelligent detection technology and equipment for track panels based on 3D laser image measurement*

Based on the 3D laser image measurement technology and the optimization algorithm of data processing techniques, such as point cloud fusion, surface fitting, and convex silhouette, an intelligent detection system of track panel shape dimensions and appearance quality based on machine vision and depth learning is developed. The full parameters and unmanned detection of track panels are realized.

(3) *Integrated management platform of track panel manufacturing information based on IoT technology*
Based on the technology of track panel manufacturing process simulation and workshop layout optimization, the central control system of the production line is constructed, and a comprehensive management platform for track panel manufacturing information is built. It realizes the optimization of workshop layout, centralized control of key equipment, automatic collection of process information, process data interconnection, collaborative management of "man, machine, material," and product quality traceability. It can also provide systematic data support for production management decision-making and optimization of industrial parameters.

At present, the complete set of technologies in intelligent track panel factory has been widely used in the manufacture of track panels for dozens of HSR lines, such as Jingxiong, Zhengji, Chaoling, and Weilai. Its continuous operation has a high degree of integration and automation, and the core technology and equipment are the first of their kind in the world and rank at the international leading level. The application of intelligent production line has significantly improved the manufacturing efficiency and product quality of track panels and achieved good economic and social benefits.

3.3.3 *Intelligent double-block pillow field*

Double-block ballastless track is the longest ballastless track structure in China, in which double-block sleeper is a field prefabricated reinforced concrete structure and is the core component of the double-block ballastless track. At present, double-block sleepers have been laid for more than 20,000 km in China.

Since the introduction of double-block sleeper manufacturing technology in 2005, China has carried out the work of digestion, absorption, and re-innovation, and it has re-innovated many technological aspects of structural design, raw materials, processing of reinforcement bars and embedded parts, production process, work equipment, inspection and testing, etc., and formed a complete set of double-block sleeper manufacturing technology. This technology has been popularized and applied on a large scale in the construction of China's HSR. However, the traditional railway sleeper manufacturing technology has a low level of intelligence, and it is difficult to adapt it to the large-scale, high-precision, and high-efficiency construction needs of China's HSR. In recent years, the IoT, big data, machine vision, and other emerging technologies have emerged rapidly, providing good technical support for the intelligence of double-block sleeper manufacturing.

1. Function and composition

Through the integration of intelligent information technology, such as robots, the internet, and the IoT, in the production of HSR sleepers, the intelligent management of the whole process of sleeper production is realized, from steel bar processing, transportation, placement, concrete pouring, maintenance, and product inspection to storage and other links.

(1) *Automation of sleeper production process*

According to the production process of double-block sleepers, through automatic and intelligent upgrading, optimization, and transformation of each process of the double-block sleeper production line, such as mold cleaning, mold release agent spraying, pre-buried casing installation, truss steel bar processing, concrete pouring, and maintenance, the traditional manual operation mode is completely replaced by intelligent information technologies, such as robots, automation equipment, the IoT, and big data. It makes the prefabrication of double-block sleepers more accurate, safe, efficient, standard, intelligent, energy-saving, and environmentally friendly. Figures 3.26 and 3.27 show the application of steel bar automatic processing technology and concrete automatic precision distribution technology in the intelligent double-block pillow field.

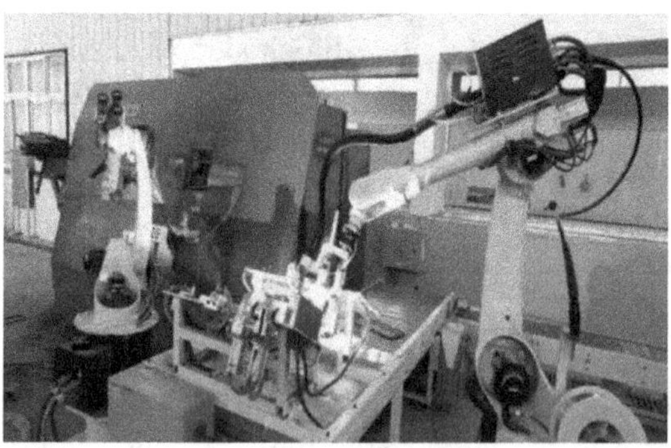

Figure 3.26. Automated rebar processing technology.

Figure 3.27. Automatic precision concrete placing technology.

Figure 3.28. Rail sleeper intelligence.

(2) *Intelligent testing of sleeper*

In view of the faults in double-block sleepers on each operation line in recent years, machine learning technology has been applied to realize the rapid detection of all sleeper shape dimension parameters and shoulder cracks. Figure 3.28 shows the application of the sleeper intelligent detection system in the double-block pillow field.

(3) *Informationization of sleeper production management*

The intelligent production management system of double-block sleepers realizes the unified and centralized online management of personnel, materials, equipment, production processes, and finished products in the sleeper field. Among them, the radio frequency identification (RFID) card is embedded in the construction of the on-line bed board as the electronic tag of the block sleeper in the intelligent detection system of double-block sleepers, which is reserved with the data interface of double-block sleepers and rail fine adjustments. The whole process of double-block sleeper production is standardized and managed using information and database technologies.

2. Key technology

The intelligent double-block sleeper yard adopts an advanced cluster processing architecture using advanced technologies, such as the internet, IoT, automation, data warehousing, digital modeling, data mining, data analysis, and laser image recognition. By integrating production data of the whole manufacturing process of double-block sleepers, automation of manufacturing processes, standardization of management, graphical production progress, digitization of data storage, and flow of business reports are realized. At the same time, through a combination of advanced QR code and electronic label management technology, the whole process of sleeper production, recording, and tracing to sleeper laying, operation, and maintenance stages can be traced back to the whole process, which can effectively improve the efficiency of sleeper production management and provide effective data support for later operation. The key technologies of intelligent pillow yards mainly include the following:

(a) The process machinery automation of the double-block sleeper production line, based on industrial robot technology, effectively reduces the input of personnel, reduces the labor intensity of construction personnel, and eliminates man-made errors. This achieves staff reduction and efficiency, improved material utilization, and reduced material waste and environmental pollution.

(b) The production management information system oriented toward the Industrial Internet is adopted to realize the online unified and centralized management of all elements of the sleeper yard, and the efficiency of data statistics is significantly improved.

    With the application of mobile internet, IoT, cloud computing, and other technologies, taking the production batch as a unit, a unique "digital identity card" of the sleeper is created, which organically combines the construction process of the sleeper yard. Thus, a digital quality tracing mechanism of the whole construction process centered on sleepers is established, which enables the integration of sensing, computing, and networking, opens up isolated islands of information, and realizes data sharing. The Industrial Internet-oriented double-block sleeper production management information system can meet the needs of unified and centralized online management of personnel, materials, equipment, process links, finished products, and other information in the sleeper field. The integration of human, information, and physical systems in double-block sleeper production is realized.

(c) The intelligent and rapid detection technology of double-block sleepers based on laser measurement and machine learning has realized the transformation from batch sampling inspection to full inspection one by one, and the detection accuracy and efficiency have been significantly improved.

According to the inspection requirements of double-block sleepers, an intelligent detection technology for double-block sleepers based on the principles of laser measurement and machine vision depth learning is developed, which can realize automatic, intelligent, unmanned operation, and information management of inspection data. The detection parameters cover the requirements of the field inspection of the double-block sleepers, and the root-by-root test of the double-block sleeper is realized, and the uniqueness and traceability number of each sleeper are constructed. The automatic spray printing on the surface of the sleeper facilitates the data management of the sleeper during the laying cycle, realizes the transformation from product sampling inspection to full inspection, and improves the detection precision and efficiency.

Double-block sleeper automation and intelligent prefabrication technology have been applied in the prefabrication tasks of double-block sleepers on the Beijing–Zhangzhou, Huanghuang, Guinan, and other HSR lines, leading to the industrialization of railway manufacturing by replacing the traditional decentralized, low-level, and inefficient handicraft production mode with a modernized production mode. This is marked by the factory production of railway components, mechanization of construction, and the organization and management of the scientific process, gradually adopting the synergy of the whole process to improve labor productivity, speed up construction, reduce project costs, and improve project quality.

3.3.4 *Intelligent mechanized sand processing plant*

With the wide span of railway areas, the uneven distribution of river sand resources, and the promulgation of the limited mining order of river sand in China, railway construction is facing a serious shortage of river sand resources. Mechanized sand is made by mechanical crushing, sieving, and shaping with qualified parent rock as raw material after soil removal treatment. The particle size is less than 4.75 mm, and the particle shape and gradation meet the requirements. Compared with the seasonal mining of natural river sand, the great fluctuation of quality in different regions, and the high cost of long-distance transportation, mechanized sand has the advantages of high economic benefit, strong controllability of quality, and easy adjustment of output due to the characteristics of local materials, industrialization, and large-scale production.

1. Function and composition

Mechanized sand production has gradually transformed from a simple and scattered artificial or semi-mechanical production mode to a standardized and large-scale factory. An intelligent mechanized sand processing plant uses multidisciplinary technologies, such as automation control, electrical design, mechanical design, computer arithmetic analysis, and network transmission, and develops an automated mechanism sand production quality online monitoring system, an intelligent mechanized sand production line control system, and an informationized and mechanized sand production management system.

(1) *Mechanized sand production quality online monitoring*

Stone powder content, fineness modulus, particle gradation, and particle morphology indirectly reflect the nature of the parent rock and mechanical processing level of mechanized sand, which are the key parameters to characterize the quality of mechanized sand production. The online monitoring system for the production quality of mechanized sand is based on the dynamic particle image analysis method of microscopic imaging and monitors the stone powder content, fineness modulus, particle gradation, and particle morphology of mechanized sand during the production process. The testing principle is illustrated in Figure 3.29.

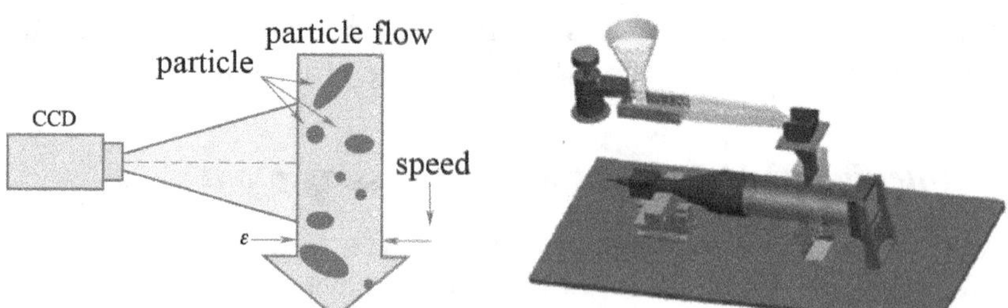

Figure 3.29. Test principle.

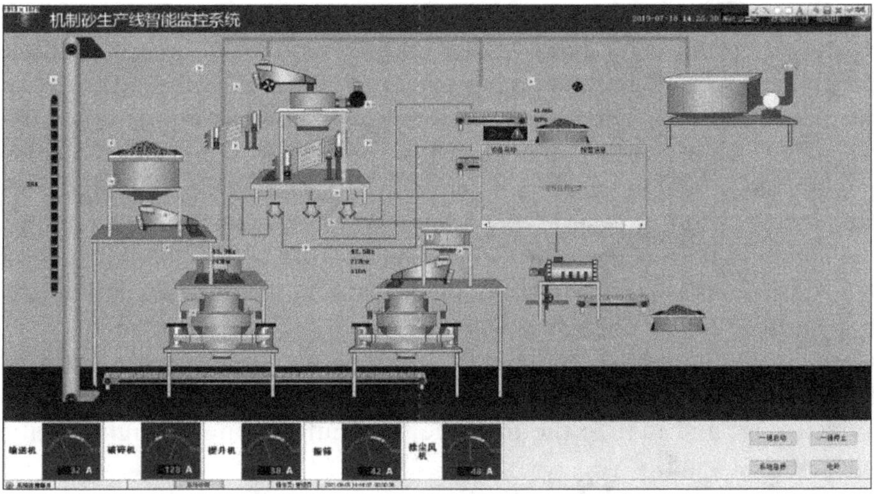

Figure 3.30. Operation interface of sand production control system.

(2) *Control of mechanized sand production line*

The mechanized sand production control system (Figure 3.30) has the functions of equipment visualization, intelligent operation, equipment maintenance and monitoring, etc.

Equipment visualization: Through real-time monitoring of the amplitude of the sand-making machine, the oil return temperature, and the load rate of other key hosts, the operator can fully understand the operation of the equipment and alarm the abnormal situation at the same time.

Intelligent operation: Through one-click start-stop buttons, the equipment can start or stop according to the preset program, or the designated equipment can be started or stopped separately. At the same time, the abnormal state of some hosts can be warned in advance, and the feed quantity can be adjusted actively.

Equipment maintenance and monitoring: Display and record the operation status of the equipment and generate the history curve of the key parameters of the equipment, such as current, voltage, frequency, and wind pressure, which makes it convenient to view and analyze the load of the equipment and give alarm prompts for replacement of vulnerable parts or equipment maintenance.

The linkage between the online monitoring system of mechanized sand production quality and the control system of mechanized sand production line can further realize the intelligent control of machine-made sand production quality. To establish the correlation between the production quality evaluation parameters of mechanized sand (such as fineness modulus, particle gradation, stone powder content, roundness, and aspect ratio), the electrical parameters (current, voltage, power, etc.), and mechanical parameters (rotation speed, vibration frequency, wind pressure, etc.) in crushing equipment, screening equipment, and powder removal equipment, the quality parameters of real-time monitoring of machine-made sand on the production line are transmitted to the control system. When the quality parameters are offset from the preset parameters, the control system automatically adjusts the equipment parameters to correct the quality offset of mechanized sand and realize the intelligent manufacturing of high-quality mechanized sand.

(3) *Production management of machine-made sand*
The mechanized sand production management system is used to meet the needs of the managers of the mechanized sand processing yard for the supervision of the production process. Information such as the status of key parts of the production line, operation parameters of production equipment, quality of finished products, incoming materials in the yard, types and stocks of finished products, output of finished products, etc., are obtained in real time to ensure the safe and stable operation of the production line and the quality and quantity of products up to the standard. When there are problems such as equipment failure, abnormal quality parameters, insufficient stock, etc., an alarm will be issued.

2. Key technology
The intelligent mechanized sand processing yard realizes the standardization of finished product quality of the mechanized sand, the automation of regulation and control of the mechanized sand production equipment, and the informationization of production data through modules such as online monitoring of the production quality of the mechanized sand, intelligent control of the mechanized sand production line, and information management. This promotes the intelligent manufacturing process of the mechanized sand processing plant. The key technologies of intelligent mechanized sand processing plant mainly include following:

(1) *Particle morphology analysis technique for large-scale particle size range from μm to cm*
The online monitoring system for the production quality of mechanized sand reflects the key parameters such as powder content, fineness modulus, particle gradation, and particle morphology in the large-scale range from "particle + powder" μm to cm level and automatically sends out a production quality warning by setting a threshold, which realizes the real-time monitoring of finished product quality in the large-scale production of mechanized sand.

(2) *Automatic control technology of key parameters of production equipment*
The control system of the mechanized sand production line is a field network computer control system composed of an industrial computer, programmable logic controllers (PLCs), and a motor starting cabinet, which adopts a DCS structure control network. The mechanized sand production control system effectively improves the level of the visual setting state, intelligent operation, and online maintenance of equipment and significantly reduces the complex and heavy manpower control and maintenance work. It provides a stable, efficient, and convenient operation and control method for mechanized sand production personnel.

(3) *Integrated management and control technology of production data*
The mechanized sand production management information system adopts comprehensive control technology, sensing technology, network technology, and big data analysis to manage the mechanized sand parent rock, test results, production process, material flow direction, etc. It alleviates the current situation of rough management, low automation, and low traceability of production process data in railway engineering mechanized sand production and improves the degree of automation and monitoring of the mechanized sand production process. The traceability of production data for mechanized sand is realized.

At present, the technology of intelligent mechanized sand processing yards has been applied in Sichuan–Tibet, Chihuang, Xuanji, Hangwen, and other railway engineering sand processing yards, which improves the automation level of the production process, promotes the purpose of scientific management, quality control, and information sharing of mechanized sand production, and speeds up the process of railway engineering information and intelligent construction.

3.4 Intelligent Construction

Using intelligent construction technology to promote railway construction with high level, high quality, and high efficiency and build a railway network with reasonable layout, extensive coverage, high efficiency, convenience, safety, and economy, it is of great significance to improve the national comprehensive transportation system and support and lead the economic and social development of China. Based on the practice of large-scale railway construction projects since the 13th Five-Year Plan, China has formed the world's leading intelligent construction technology for railway engineering in bridges, tunnels, roadbeds, tracks, passenger stations, power stations, etc.

3.4.1 *Intelligent bridge construction*

HSR bridges are mainly concrete girder bridges, including commonly used span simply supported girder bridges, continuous beam bridges, and continuous steel structure bridges. The largest number of simply supported girders have realized intelligent prefabrication in the girder yard, and the on-site construction is mainly the transportation of girder vehicles and the erection of bridge-erecting machines. At present, intelligent management has been realized for large-scale equipment. The prefabricated assembly of concrete bridge structures is one of the important achievements of HSR bridge construction, including the segmental assembly of girders, the segmental assembly of bridge piers, the application of precast pipe piles and the prefabricated assembly of bridge deck facilities, with good intelligent construction conditions. In order to ensure the operation safety of the existing line, the rotation construction continuous girder or continuous rigid frame is adopted in the HSR bridge, and the rotation tonnage has repeatedly broken the world record. The intelligent control of the rotation process is one of the key means to improve its construction quality.

Among the HSR bridges, there are long-span arch bridges, cable-stayed bridges, and suspension bridges, and their span, weight, height, and other construction indicators have reached the leading level in the world. Represented by Shanghai–Sutong Yangtze River Railway Bridge, Wufengshan Yangtze River Railway Bridge, and Shanghai–Kunming HSR Beipanjiang Bridge, intelligent construction technology has been adopted in steel structure processing, BIM technology application, foundation construction, etc.

Figure 3.31. 1,000 t class bridge erectors used in Jizheng HSR.

3.4.1.1 *Intelligent transportation of simply supported girders for HSR*

The weight of the 32 m simply supported girder of China's HSR is about 800 t, and the transportation and erection equipment adopted is 900 tons, including a beam transport vehicle, a walking bridge erection machine, and an integrated transport and erection machine. In 2018, Jizheng HSR applied a prefabricated 40 m simply supported box girder for the first time, the beam weighs 925 t, and 1,000 t transport equipment was used to realize intelligent transportation and frame construction, as shown in Figure 3.31. The intellectualization of transportation and shelf construction is embodied in: the intelligent management of large-scale transportation equipment is realized by using the IoT technology; the technical data, maintenance history, and working status of the equipment are recorded in real time; and the intelligent management of simply supported box girder construction is realized by using GIS technology and Beidou technology, including transportation track recording, construction progress, and key construction quality control information management.

3.4.1.2 *Intelligent construction of HSR concrete bridge precast assembly*

The construction of Jingxiong Intercity Railway, Heluo Railway, and Lianxu HSR has realized the fully assembled structural design and construction from foundation, pier, beam to bridge deck system and completed the forward design of prefabricated structures based on BIM and production information management of prefabricated yards, which has laid a solid foundation for the transformation and upgrading of China's railway bridge construction in the direction of industrialization and intelligence, as shown in Figure 3.32.

Figure 3.32. Intelligent construction of precast concrete bridges for HSR. (a) Beijing–Hungary Intercity Railway Assembly Bridge Deck Facilities, (b) Lianxu HSR Donghai Special Bridge Section Assembly Bridge, (c) Jingxiong Intercity Railway Assembled Bridge Pier Construction, (d) Zhengzhou-Jinan HSR Assembled Bridge Pier Construction, (e) Wako Railway Assembled Piers, and (f) Jingxiong Intercity Railway Large Diameter Pipe Pile.

(1) **Prefabricated, assembled bridge deck facilities:** For the first time, the Jingxiong Intercity Railway realizes prefabricated assembly of bridge deck facilities, turning the original need for cast-in-place bridge The vertical and protective walls, which need to be cast-in-place on the bridge deck, are made into prefabricated components with the capacity of safety railings and sound barriers, which are connected with the bridge deck using high-strength bolts, and the assembly construction is mechanized. The assembled bridge deck facilities have also been applied in many HSR, such as Yantong, Fuxia, and Guinan.

(2) **Segmental assembly of bridges:** The concrete beam is divided into parts and several segments for prefabrication to reduce the area of the precast yard and the difficulty of transportation. The bridge is assembled by a bridge-building machine and connected into an integral structure by prestress.

(3) **Segmental assembly of bridge piers:** The pier body is precast in segments to reduce the on-site concrete pouring work and improve the construction quality, and the precast sections are connected by means of a grouting sleeve, a concrete wet joint, a prestress, etc.

(4) **Large-diameter pipe pile:** The prestressed concrete pipe pile with a diameter of 1.0 m is adopted in Jingxiong Intercity Railway. The segments of prefabricated pipe piles are manufactured, the segments are welded by flanges, and the pipe piles are driven into the foundation soil using a diesel hammer or hydraulic hammer.

3.4.1.3 *Intelligent turning construction of HSR concrete bridge*

The span and weight of concrete bridge transits of China's HSR have reached the leading level in the world, and Xixi Special Bridge of Fuxia HSR is one of the representative bridges (Figure 3.33), with a transiting weight of 19,000 t and a transiting length of 168 m. The intelligent construction system of concrete bridge transits consists of three parts. The intelligent construction system for concrete bridge transitions consists of three parts: first, intelligent weighing system, which quickly calculates the eccentricity and required counterweight values according to the on-site weighing test data and uses the principle of unbalanced moment calculation to guide the on-site construction; and second, intelligent

Figure 3.33. Construction of Xixi Special Bridge of Fuxia HSR.

monitoring system for transitions, which monitors the wind speed of the transition process in real time, provides early warning of the transitions' speeds, transitions' accelerations, and attitude deviations, and monitors the ball hinges and provides a warning of the transitions' deviations. The second one is the intelligent monitoring system for rotating construction, which monitors the wind speed in the process of rotating in real time, provides the rotating speed, rotating acceleration, and attitude deviation warning, monitors the stress state of the ball hinge, girder, and bridge, and guarantees the quality and safety of the construction. The third is the intelligent turning system, according to the feedback information on turning speed and acceleration from the monitoring system, it can finely control the starting state of the traction jack and the traction state, as shown in Figure 3.34.

3.4.1.4 *Intelligent welding of steel structures for HSR bridges*

The difficulty of welding the steel structure of a HSR bridge is mainly reflected in the processing of orthogonal anisotropic steel bridge panels, and the construction quality and efficiency are effectively improved through the application of intelligent welding technology. The application of intelligent welding technology effectively improves the construction quality and efficiency, specifically including: U-shaped rib bevel

Figure 3.34. Intelligent monitoring system for transformer bridge.

(a) (b)

Figure 3.35. Intelligent welding of orthogonal anisotropic deck plates for the Beijing–Zhanghai HSR Guanting Reservoir Bridge. (a) Automatic assembly machine and (b) Gantry robot automatic welding system.

processing CNC machine tools, automatic assembly machine tools for U-shaped ribs and panels, a gantry robot automatic welding system based on welding robots, and a welding data monitoring system, which realizes the workshop assembly and manufacturing of large-scale orthogonal anisotropic steel bridge panels. This technology has been successfully applied in projects such as the Guanting Reservoir Bridge of Beijing–Zhangzhou HSR (Figure 3.35).

3.4.1.5 *Construction technology for kilometer-scale mega bridges*

The Shanghai–Suzhou–Tongshan Yangtze River Public Railway Bridge (Figure 3.36) adopts a steel truss girder cable-stayed bridge structure with a main span of 1,092 m. It is the first dual-purpose cable-stayed bridge with a span of more than 1,000 m in the world that was designed and constructed independently by China. The design and construction technology realizes five "world's firsts," namely, the design and construction technology of kilometer-level dual-purpose cable-stayed bridge, the manufacturing technology of 2,000 MPa strength cable-stayed cable, the realization of 1,800 t steel girder erection complete set of equipment technology, the realization of 15,000 t huge sinkhole accurate positioning construction technology, and the realization of the bridge pier collision avoidance technology based on the real ship — real bridge *in situ* impact test.

The Wufengshan Yangtze River Bridge (Figure 3.37) is a suspension bridge with a main span of 1,092 m. Upon completion, it will fill the gaps

Figure 3.36. Hu-Su-Tong Yangtze River Railway Bridge.

Figure 3.37. Wufengshan Yangtze River Railway Bridge.

of the world's HSR suspension bridges, China's dual-use suspension bridges, and China's railway suspension bridges and take the lead in the international arena to establish China's HSR suspension bridge design methodologies, calculation theories, and related technical standards.

The construction of Shanghai–Sutong Yangtze River Public Railway Bridge and Wufengshan Yangtze River Public Railway Bridge shows that China's HSR mega-bridge construction technology has reached an internationally advanced level, in which intelligent technology has also been successfully applied, which is mainly embodied in the BIM-based whole-life cycle management system, intelligent construction monitoring and control, and intelligent construction of steel girders and other aspects.

1. BIM-based full life cycle management system: Based on the management concept of the whole life cycle, it has carried out all-round exploration and formed the project BIM application concept led by the owner, insisting on "the same model, the same platform, and one data center." A perfect project-level bridge BIM standard has been compiled, and based on the application mode of "three ends and one cloud," a BIM management system for sharing information among all parties involved in the construction has been set up, which solves the problem of lightweight BIM models, realizes multi-user and multi-terminal access, and ensures the smoothness of model display and operation. The BIM management system of Hsu-Su-Tong Yangtze River Public Railway Bridge and the BIM model of Wufengshan Yangtze River Public Railway Bridge are shown in Table 3.2 and Figure 3.38.

2. Intelligent construction monitoring: The intelligent construction monitoring system was utilized to guide the whole construction process of the Hsu-Su-Tong Yangtze River Public Railway Bridge and the Wufengshan Yangtze River Public-Railway Bridge, and fine control was carried out in the aspects of the deep-water foundation construction, the construction of the super-high bridge towers, the construction of the cable-stayed system, and the construction of the main girders, etc., as shown in Figures 3.39 and 3.40.

3. Intelligent construction of steel girders: The construction of steel girders for kilometer-class mega bridges has realized "factorization and specialization," adopting the integration of three-dimensional drawing design and processing detail design, digital nesting system, measurement and control system, welding group control technology, virtual pre-assembly technology,

Table 3.2. Hsu-Su-Tong Yangtze River Railway Bridge BIM management system.

Project management system outlook for the full BIM lifecycle		
Engineering management center railway engineering management platform	Digital manufacturing; material management; geographic information system; electronic construction logs; visualization of delivery; prestressing tensioning; mixing plant; electronic supervisory log; laboratory; construction monitoring progress; safety management construction; implementation standard; Intelligent building system	
Shanghai-Suzhou-Tongshan Yangtze River Bridge BIM Management System; technology standard talent cultivation	Building unit; construction unit; manufacturing unit; monitoring unit; supervision unit; Design stage; construction stage	Modeling process 3D Information modeling; component base; design implementation standard; collaborative design system
Operation and maintenance stage	Health monitoring; emergency management; equipment management; electronic inspection; resource base, knowledge base; condition assessment conservation; staff training; operations and maintenance implementation standards	Digital custodial system

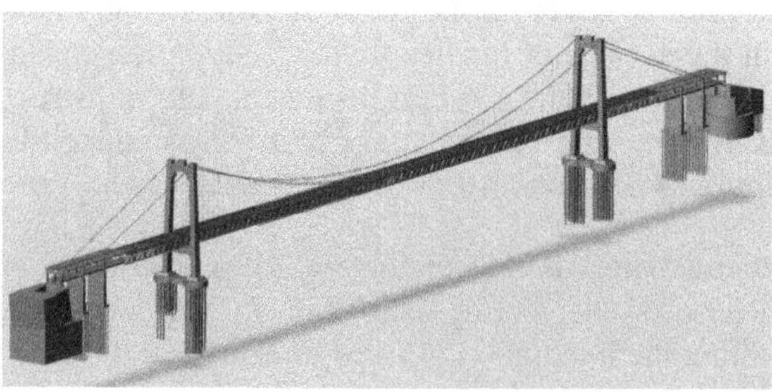

Figure 3.38. BIM model of Wufengshan Yangtze River Railway Bridge.

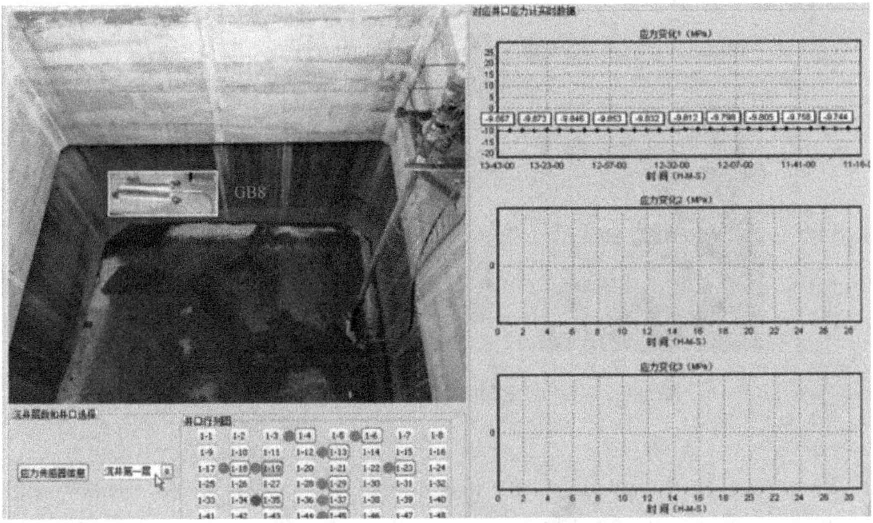

Figure 3.39. Large sinkhole construction monitoring system.

Figure 3.40. Application of underwater 3D sonar technology.

and IoT quality tracking system, which promotes the innovation of steel girder manufacturing technology.

3.4.1.6 *Intelligent construction technology for large-span concrete arch bridges*

Beipanjiang Special Bridge of Hukun HSR (Figure 3.41), with a main span of 445 m, is a top-bearing reinforced concrete arch bridge with strong skeleton, which is the largest span reinforced concrete arch bridge for HSR

Figure 3.41. Shanghai-Kunming high speed railway beipanjiang special bridge.

in the world. In the process of construction, it realizes three-dimensional parametric modeling based on BIM technology, develops a nonlinear stability assessment system, solves the problem of stability control of steel pipe skeleton construction, puts forward the six-working surface construction method of high-performance concrete, and solves the problem of outsourcing concrete construction. During the construction of the bridge, the three-dimensional parameter modeling based on BIM technology was realized, the nonlinear stability assessment system was developed, the stability control problem of steel pipe skeleton construction was solved, and the six-working face construction method of high-performance concrete was proposed.

3.4.2 *Intelligent tunnel construction*

The main construction methods of railway tunnels include the drilling and blasting method and the boring machine method (shield, TBM). During the construction stage of Beijing–Zhangzhou HSR, the overall architecture of intelligent construction for China railway tunnels (as shown in Table 3.3) was established by using technologies such as mechanized supporting, sensing, IoT, three-dimensional visualization and the comprehensive management of geological, structural, mechanical, personnel, and material information, intensive use of time, space, and human resources. The tunnel construction technology has made great progress, and the overall construction level is already in the advanced

Table 3.3. Overall architecture of intelligent construction of railway tunnels.

Railway tunnel intelligent construction overall architecture	
Intelligent control	Intelligent survey and design; intelligent construction; intelligent management
Intelligent decision making	Big data analytics; machine learning; deep learning; interactive analytics; advanced intelligence
Data resources	Ground investigation data; design drawings; construction data; supervision files; material information; quality evaluation; management files; data service platform
Intelligent sensing	Sensors; internet of things; capture chips; RFID; GIS; NFC; 3D scanning
Intelligent equipment	Rock drilling cart; anchor injection trolley; multifunctional drilling rig; arch frame trolley; wet spraying trolley; paving trolley; second lining truck

ranks in the world. Tunnel construction technology has made great progress, and the overall construction level has been in the advanced ranks in the world.

China's railway tunnel drilling and blasting construction operation process mainly includes advance geological forecasting, excavation, slagging, monitoring and measurement, support, and quality inspection. Since 2013, China has carried out intelligent technology research in the railway tunnel over-the-horizon geological forecasting, monitoring and measurement, support, and quality inspection, forming an intelligent command and scheduling "people, machines, materials, and methods" of the tunnel construction system through machine-group coordination and intelligent control, realizing less manpower operation in tunnel construction. Through group coordination and intelligent control, tunnel construction has realized less manned operation, increased construction safety, improved construction efficiency, and strengthened construction quality.

In recent years, the number of railway tunnels excavated by tunnel boring machines (shield, TBM) has gradually increased, and the construction of China's railway tunnel boring machine method mainly includes the start of tunnel boring machines, advanced geological prediction, tunneling, monitoring and measurement, tube piece assembly, grouting, slagging, and construction of under-rail structures. At present, for the digging process, the informationized monitoring and intelligent judgment of the construction process have been realized; for the under-rail structure in holes, the intelligent construction of all prefabricated under-rail structures has been realized.

3.4.2.1 *Advanced geological forecasting information system*

Based on the technical characteristics of the geological survey, physical exploration, and overdrilling methods, the overdrilling geological forecast information system has formed a rapid response mechanism for the overdrilling geological forecast of railway tunnels. It has established a closed loop of information of the overdrilling geological forecast of railway tunnels, realizing early warning of geological risks and timely feedback on risk control measures, and guaranteeing the authenticity of the forecasting operation and the completeness of the data. Further, it has realized the standardization of the paperless format of the overdrilling geological forecast of railway tunnels, upgrading from the traditional language expression to a comprehensive forecast display combining language and text with intuitive visualization. It has established an informatization system of "real-time uploading, platform reporting, and timely control," realizing the whole process of forecasting and providing timely and accurate decision-making basis for the managers. The system has been popularized and applied to national railways, providing strong technical support for risk control in tunnel excavation and construction safety. The interface of the system is shown in Figure 3.42.

3.4.2.2 *Tunnel excavation and support intelligent construction*

Based on the mechanization and informationization construction technology and the deep integration of the IoT technology, intelligent construction equipment and systems for tunnel excavation and support have been formed, which have improved construction efficiency, construction quality, and management level. Intelligent tunnel construction equipment includes intelligent grouting carts (Figure 3.43), intelligent rock

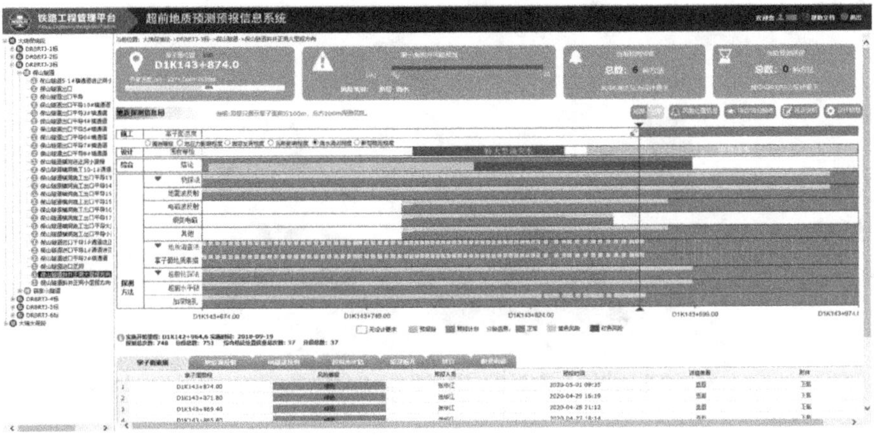

Figure 3.42. Interface of the tunnel advance geologic prediction information system.

Figure 3.43. Intelligent grouting trolley.

drilling carts (Figure 3.44), intelligent arch installation carts (Figure 3.45), intelligent injection carts (Figure 3.46), intelligent anchor carts (Figure 3.47), intelligent lining carts (Figure 3.48), and intelligent maintenance carts (Figure 3.49). The intelligent equipment integrates all kinds of sensing technologies, realizes construction information transmission to the platform through IoT technology, and the platform analyzes and processes the working data, status data, real-time video, and alarm information of the intelligent equipment, and displays them in a classified manner, so as to realize remote and real-time monitoring of all kinds of working information of the intelligent equipment. At the same time, according to the design parameter data, the intelligent construction

Figure 3.44. Intelligent rock drill trolley.

Figure 3.45. Intelligent arch installation trolley.

Figure 3.46. Intelligent jet trolley.

Figure 3.47. Intelligent anchor trolley.

Figure 3.48. Intelligent lining trolley.

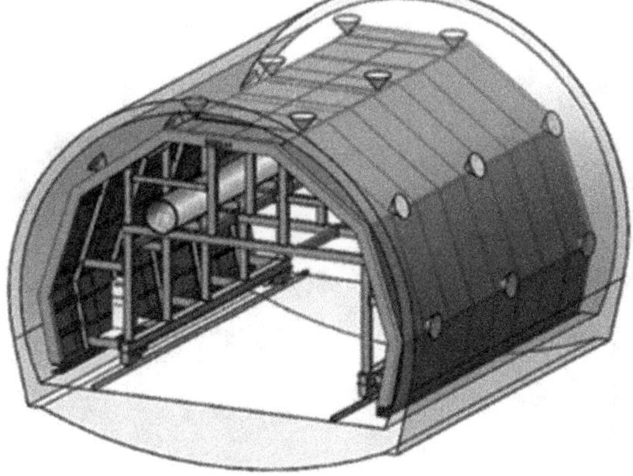

Figure 3.49. Intelligent maintenance trolley.

equipment is used: to realize the intelligent construction of tunnel excavation and support and real-time monitoring of the construction status of intelligent equipment, intelligent command, and scheduling of the "man, machine, material, and method" tunnel construction system and through the synergy of the machine group and the intelligent management and control; to realize the construction of tunnels with fewer people (unmanned) operations; to enhance the safety of the construction; and to speed up the construction progress and shorten the construction time. At present, the tunnel excavation and support intelligent construction equipment and systems have been applied in many routes, such as Beijing–Zhangzhou and Zhengzhou–Wanzhou.

3.4.2.3 *Intelligent construction of tunnel boring machines*

The intelligent construction of tunnel boring machines, through sensor technology, BIM + GIS, and IoT technology, sets up a tunnel construction visualization and cooperative management platform integrating boring parameters, monitoring data, and construction management information. It realized the whole process of shield construction visualization and dynamic monitoring and management, with real-time prediction and forecasting of the dangers of the adjacent buildings (structures), and laid a solid foundation for the eventual realization of unmanned construction of tunnels. The tunnel construction progress and system prediction and warning platform display interface are shown in Figures 3.50 and 3.51, respectively. This platform has been successfully applied to the construction of

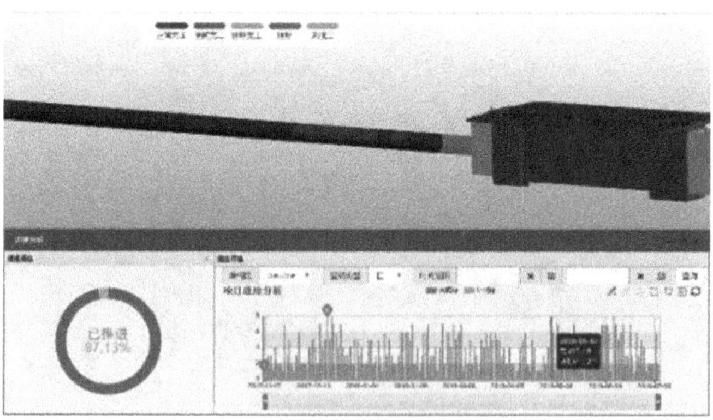

Figure 3.50. Tunnel construction progress analysis and presentation.

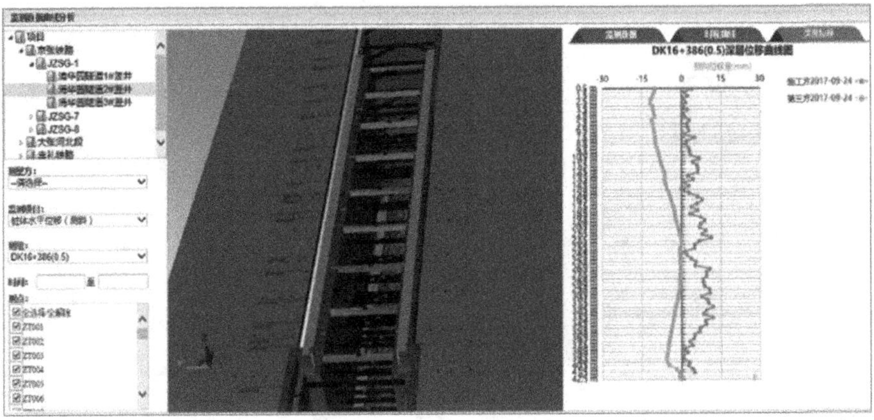

Figure 3.51. System forecast early warning real-time forecast.

the Qinghuayuan Tunnel of the Beijing–Zhangzhou HSR, which has played a positive role in guiding the integrated and coordinated management of the tunnel construction of the tunnel boring machine method and, at the same time, strengthened the ability to control the construction quality and construction risks and improved the efficiency of the tunnel construction of the tunnel boring machine method, thus laying a technical foundation and an example of the application of the platform in order to realize the intelligent construction of railway tunnels.

3.4.2.4 *Tunnel monitoring and measurement information system*

The tunnel monitoring and measurement information system utilizes Bluetooth technology to control an ordinary total station on site, realizing all-round collection of data, such as cross-section, measuring points, headroom change, and surface settlement, during tunnel construction. The working composition of the tunnel monitoring and measurement information system is shown in Figure 3.52. For the first time, the system is designed to collect data from the terminal app via Bluetooth and transmit the data to the back-end platform through the network, which solves the problems of data authenticity and generalization. The system first designs the terminal app to collect data through Bluetooth and transmit data to the back-end platform through a network, which solves the problems of data authenticity and universality. Through the real-time alarm function of the

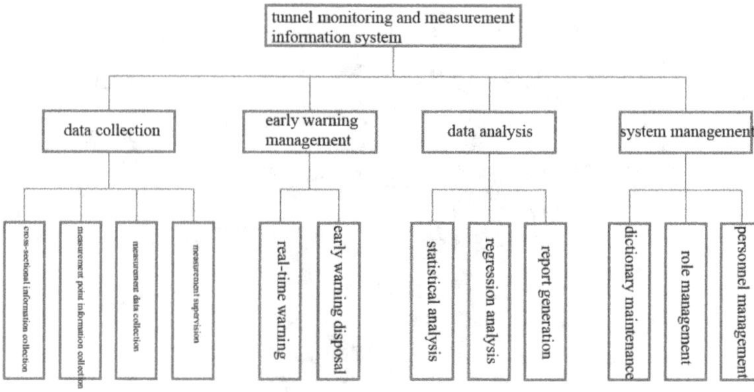

Figure 3.52. Tunnel monitoring and measurement information system components.

cumulative deformation amount and deformation rate dual-control cave, timely warning of deformation measurement is realized. The system changes the traditional management mode of reporting level by level into a common supervision mode at all levels, which significantly shortens the response time of alarm disposal. Eventually, a large amount of monitoring data will accumulate, laying a solid foundation for tunnel safety and big data analysis. Currently, this system has been popularized and applied to railway tunnels nationwide.

3.4.2.5 *Intelligent construction of fully prefabricated under-rail structures for shield tunnels*

As a new type of tunnel substructure construction method, the fully prefabricated intelligent construction technology of under-rail structure solves the interference of cast-in-place under-rail structure construction to the boring machine and realizes the parallel and synchronous construction of under-rail structures and boring operation, which not only improves the efficacy and quality of the project but also greatly improves the operation environment. The tunnel side box culvert prefabricated parts assembling machine is shown in Figure 3.53. The successful application of full prefabrication mechanized construction technology to the under-rail structure in the Qinghuayuan Tunnel of Beijing–Zhangzhou HSR has created a precedent for the prefabrication assembly method of tunnel base structure construction using the tunnel boring machine method in China.

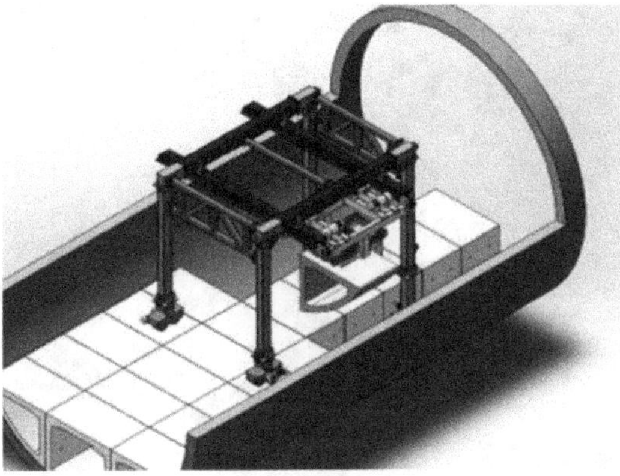

Figure 3.53. Tunnel side box culvert precast assembly machine.

The application of intelligent construction technology accelerates construction progress, saves construction time, avoids a large amount of reinforcement damage to the shield tube sheet, improves the durability and reliability of the structure, reduces the interference of the external environment on the construction operation, improves the operation environment, and reduces the impact of vibration on the surrounding environment. This has laid a solid foundation for realizing the factorization, mechanization, and specialization of tunnel construction by the tunnel boring machine method.

3.4.2.6 *Tire-type tunnel lining quality inspection technology and equipment*

Aiming at the shortcomings of low efficiency, poor accuracy, and low standardization such as visual inspection of tunnel lining quality, knocking, and hand-held radar detection, and focusing on the quality inspection demand during the construction of HSR tunnels, we have solved the problem of rapid inspection of lining quality during the construction period of tunnels by starting from the aspects of inspection equipment, method principles, key defects, data processing, informatization, and comprehensive inspection technology. The tire-type tunnel lining quality inspection technology and equipment are shown in Figure 3.54. In order to achieve standardization, informationization, intelligentization, and fast and accurate detection, the tire-type tunnel lining quality inspection vehicle adopts the robot automatic tracking

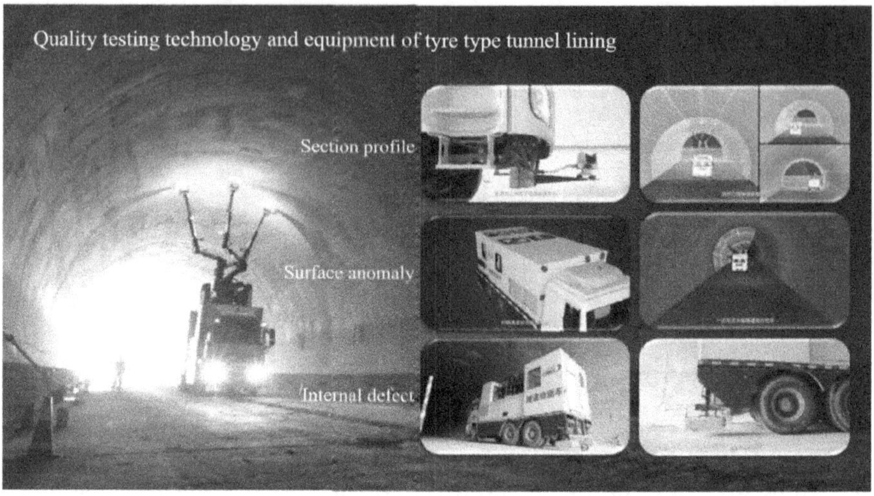

Figure 3.54. Tunnel side box culvert precast assembly machine.

detection technology, which realizes the spacing maintenance and position tracking between the robot lifting radar antenna and the tunnel lining inspection point, so as to ensure the continuity and accuracy of the tunnel lining quality inspection data, the carrying radar antenna and the tunnel lining quality inspection data, and their accuracy. Equipped with a radar antenna, it can realize simultaneous detection of three lining measuring lines and two up-arch measuring lines and can identify the lining behind the hollow, insufficient thickness or density, steel abnormalities, etc. Laser scanning technology is adopted to achieve rapid detection of the tunnel profile and ensure that the limit is normal. The image recognition system adopts an integrated line array imaging module, which overcomes the technical difficulties of fast and clear imaging of large sections, realizes rapid acquisition of high-definition images of rail tunnel lining, and is able to carry out full-section inspection at a speed of 50–60 km/h. It effectively identifies apparent defects, such as cracks, water seepage, spalling, and chunking. The large-scale mechanical equipment enables the inspection efficiency to be more than twice that of traditional manual inspection. This technology and equipment have been successfully applied in the construction of many HSR tunnels, such as Zhengwan, Yinxi, and Taijiao.

3.4.3 *Intelligent subgrade construction*

Intelligent construction of subgrade is based on a comprehensive perception of the parameters of subgrade investigation, design, and construction work and combines the results of new generation information technology to realize the intelligent mechanical construction of foundation treatment, subgrade filling, slope support, protection structure, subgrade deformation monitoring, etc. The technologies and equipment researched and developed include the management of piling construction, informatization of piling inspection, continuous compaction and vibration inspection, intelligent filling of subgrade, and automatic monitoring of subgrade deformation. Initially, it has solved the quality control problems of foundation treatment, subgrade layered filling, settlement, deformation monitoring, and other key links. Meanwhile, the railroad subgrade project is vigorously developing intelligent construction technologies, such as slope assembled support structures and BIM-based construction organization and management. It is further improving the precision of automatic detection of subgrade construction quality and enhancing the research, development, and application of subgrade intelligent construction machinery, with the goal of realizing construction organization and technology. The goal is to realize intelligent analysis and decision-making in construction organization and process optimization, quality control, etc., and provide support for the whole life cycle management of subgrade projects.

The architecture of intelligent construction of subgrade includes four levels: intelligent perception, data resources, intelligent decision-making, and technology development and upgrading, as shown in Figure 3.55.

Intelligent perception aims at safety, quality, and progress supervision, and it collects and uploads diversified and ubiquitous information of subgrade construction through various types of sensors, informatization systems, and intelligent equipment. Data resources are the important foundation for realizing intelligent construction of subgrades, and the data collected by intelligent perception enters the decision-making layer after screening and categorization to provide support for intelligent decision-making. Intelligent decision-making includes a site decision-making center and a railroad engineering management platform. The site decision-making center is oriented toward the organizational and management needs of the construction side and makes decisions on construction planning, process organization, personnel, machinery and material scheduling, intelligent equipment control, etc.

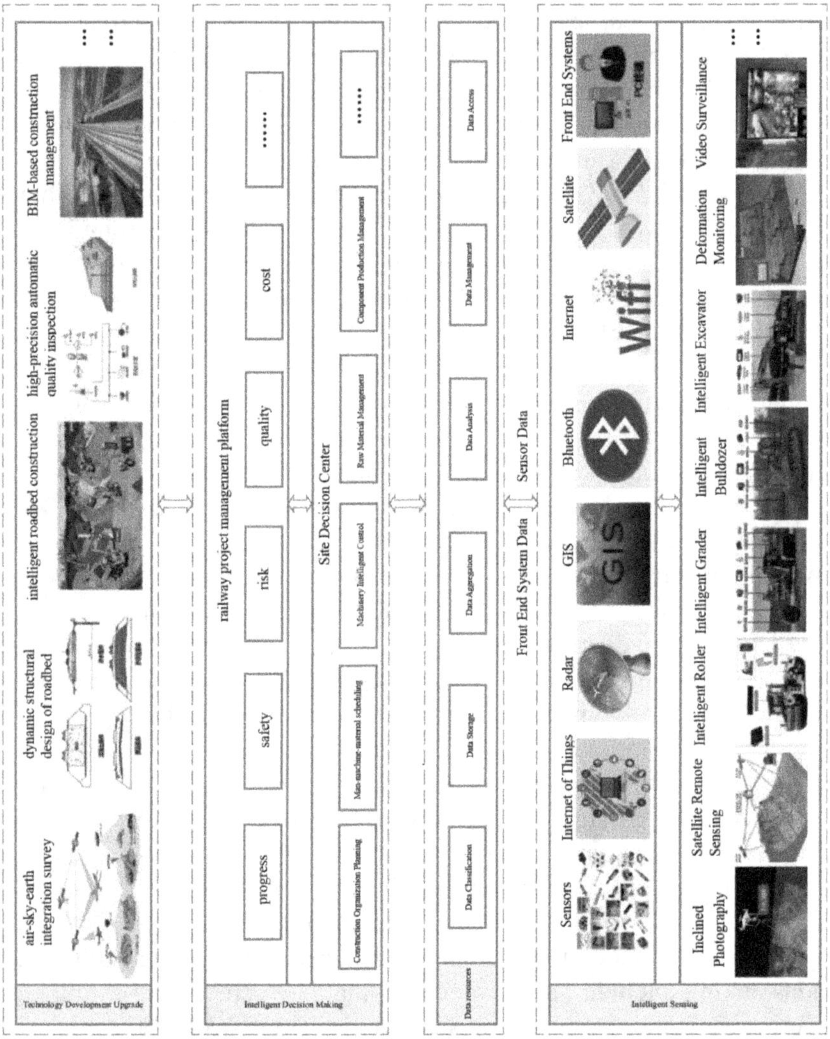

Figure 3.55. Intelligent construction architecture for subgrades.

Key data on safety, quality, and progress are collected, analyzed in depth, and displayed in a comprehensive manner for overall decision-making and project control in response to the needs of the State Railway Group and construction companies. Using the whole-process data provided by the railroad project management platform, with the support of the big data center, the relevant parties involved in the construction will continue to research and develop intelligent construction technology, BIM-based construction organization and management, and intelligent construction standards of subgrade, so as to continuously promote the progress of intelligent construction technology of subgrade and provide support for the health management of the whole life cycle of the subgrade project.

3.4.3.1 *Intelligent construction for foundation treatment*

Pile foundation construction requires strict control of process parameters such as pile length, pile depth, verticality, etc. By integrating BeiDou positioning, multi-source information sensing, and IoT, the intelligent construction machinery for pile foundations can realize automatic monitoring of key construction parameters, provide real-time feedback, and guide the operator to drill into the work. At present, we have developed an automatic monitoring system for the construction management of CFG piles, screw piles, rotary spray piles, mixing piles, and other common pile types, with the following main functions: (i) realize drilling rig attitude adjustment and drilling rod positioning guidance, instead of traditional manual sampling; (ii) adopt automated monitoring means to realize real-time monitoring of key process indicators, such as drilling depth, pile verticality, drilling rate, drilling current, and other key process indicators during the construction process, and provide on-site prompts for cases of insufficient drilling depth and insufficient drilling current that do not meet the design requirements; (iii) data are uploaded to the back-end management platform, a monitoring database is established, image progress display and quality management functions are provided, and querying, analysis, and statistics of specified data are realized.

On the basis of the popularization and application of the railway project management platform, the uploading and management of raw data and the test results of pile foundation testing are realized by using information management technology, combining pile foundation

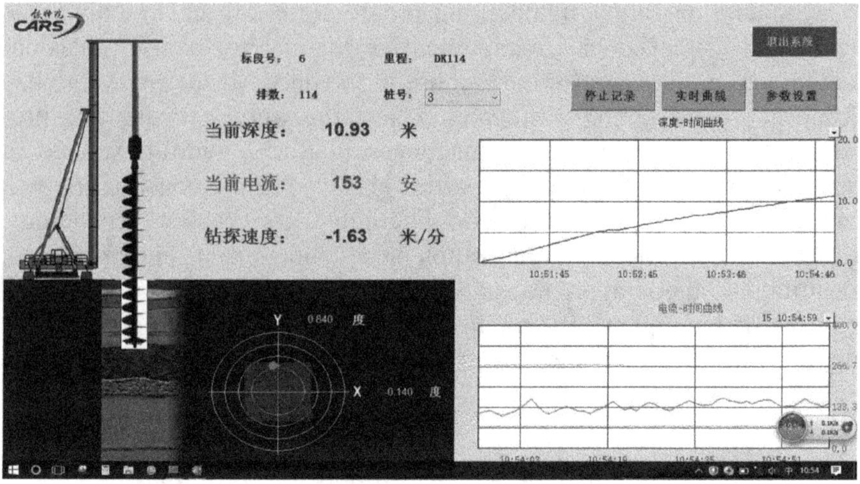

Figure 3.56. Pile foundation construction site operation management.

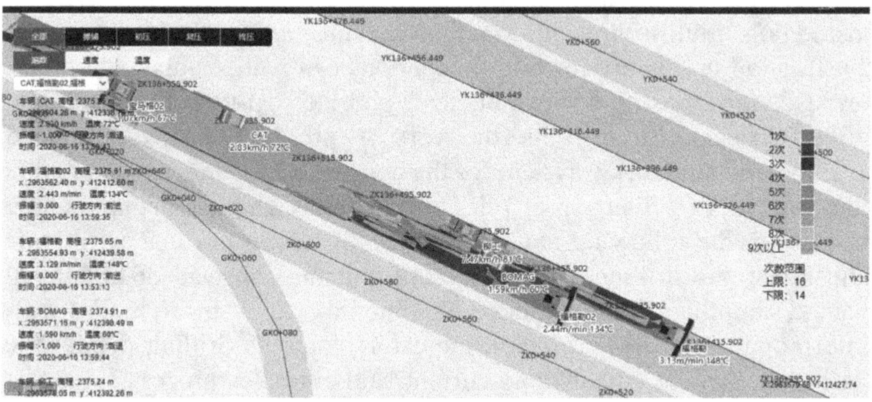

Figure 3.57. Pile testing data management.

construction (Figure 3.56), pile foundation testing data pooling, categorization, and synthesis.

The analysis (Figure 3.57) can realize the linkage analysis between the piling construction parameters and the piling quality to improve the management of the site.

The development direction of foundation treatment intelligent construction is to further develop integrated and intelligent piling equipment and technology on the basis of automatic monitoring of key parameters of foundation treatment piling construction: establish regional geological stratification database according to geological drilling and static touching data; judge the soil stratum demarcation line and soft and weak conditions by comparing and verifying piling machine drilling dynamics parameters and geological parameters, and dynamically adjust the piling machine operation parameters of different strata in a timely manner; judge holding layer drilling depth according to piling machine operation parameters, and automatically judge final pile conditions with reference to different pile construction requirements by combining deformation monitoring, laser ranging, and other methods; judge the drilling depth of the holding layer according to the drilling parameters; refer to the construction requirements of different pile types, and automatically judge the final pile conditions by combining deformation monitoring and laser distance measurement. Through the accumulation of big data, a regional database of pile foundation design, construction, and testing parameters will be constructed, and a big data-based pile foundation quality assessment method will be formed to guide the design of foundation treatment programs and automatically judge the quality of pile foundation construction.

3.4.3.2 *Intelligent filling of subgrade*

The quality of subgrade filling is crucial to the long-term operational safety of HSR. At present, the compaction quality of railway subgrades in China is mainly controlled by compaction coefficient K, dynamic deformation modulus E_{vd}, and foundation coefficient $K30$, and this kind of point-sampling inspection method is unable to comprehensively master the compaction quality of subgrades, which leads to the phenomenon of under- and over-pressures. The compaction vibration continuous testing technology was released in 2011 as a railway standard, and after continuous research in recent years, three main types of time domain frequency domain indexes, mechanical type indexes, and compaction energy indexes have been developed. Among them, the compaction energy index is based on vibratory roller-coarse-grained soil filler power coupling analysis to reveal the energy transfer mechanism of the compaction process and feedback the degree of compaction, which has proven to have good applicability for coarse-grained soil filler of HSR. It provides a new way of high-precision detection of the compaction quality of subgrade based on machine–soil power coupling analysis and intelligent construction. In addition to the continuous compaction vibration detection, the continuous subgrade compaction quality detection also covers the automated continuous detection of water content, filler gradation, etc. The goal is to comprehensively analyze the water content, filler gradation, roller operating parameters, and geotechnical dynamic characteristics and adopt intelligent algorithms to evaluate the state of subgrade compaction, so as to improve the accuracy of automated detection of subgrade quality.

On the basis of continuous quality inspection of road subgrades, a complete set of intelligent filling technology and equipment for road subgrades has been developed by synthesizing the theories and technologies of soil dynamics, BIM, BeiDou positioning, automatic control, and image recognition. Intelligent filling of subgrade (Figure 3.58) includes several main components, including a BIM-based command system, a BeiDou positioning system, a continuous quality inspection system, and a mechanical intelligent control system. The command system (Figure 3.59) contains modules of image progress, construction organization plan, real-time working condition, filler management, process quality

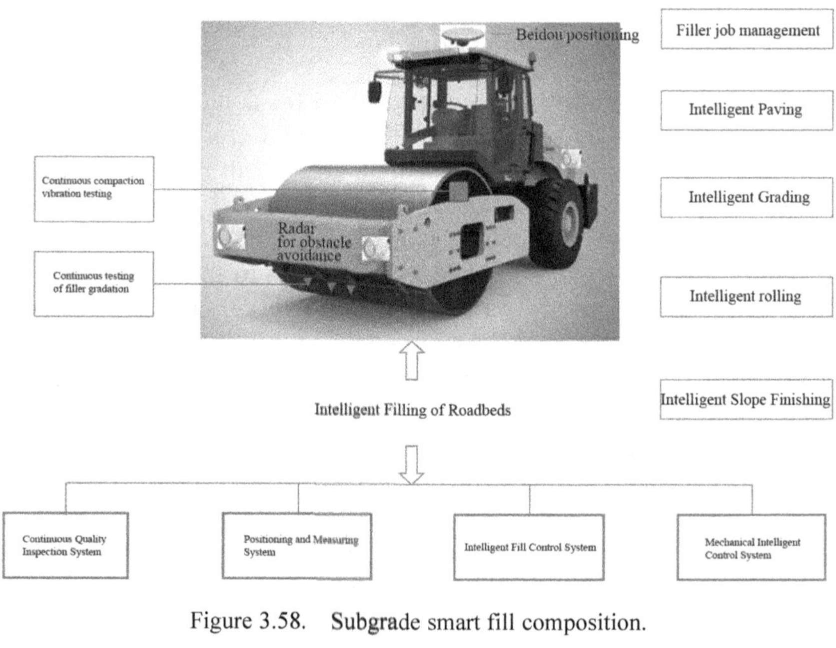

Figure 3.58. Subgrade smart fill composition.

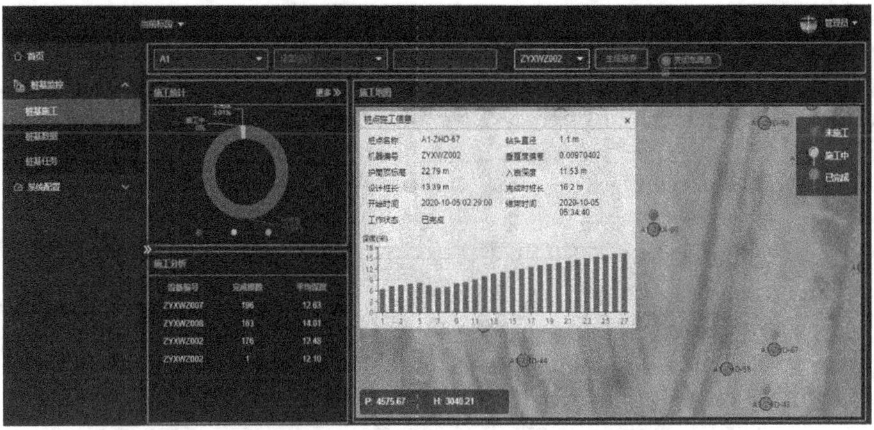

Figure 3.59. Intelligent filling command system.

management, etc. It uses BIM models to realize the progress and quality display of the subgrade filling project, as well as on-site task assignment and scheduling management. It promotes the factory production of filler, carries out the informatization management of the production, quality inspection, and transportation processes, and realizes the dynamic balance between the production of filler and the demand. The Beidou positioning system provides high-precision positioning information for on-site construction and guides the machinery to carry out construction according to the design. The continuous quality inspection system includes continuous inspection of subgrade compaction quality and automatic inspection of filler gradation, which realizes comprehensive control of subgrade filler and crushing quality. The intelligent control system of machinery includes the three-dimensional guidance system of bulldozers, the automatic control system of graders, the unmanned rolling system of rollers, and the three-dimensional guidance system of excavators. The three-dimensional guidance system of bulldozers and excavators compares the current position of shovel (bucket) with the design position and accurately guides the attitude regulation of shovel (bucket). The automatic control system of motor graders can automatically control the attitude of shovel according to the design requirements, and the driver only needs to operate the motor grader. The unmanned roller system guides the roller to carry out automatic rolling according to the requirements of the process parameters and automatically compensates for the unqualified areas according to the results of the continuous compaction test. The unmanned roller system guides the roller to roll automatically according to the requirements of process parameters and, at the same time, automatically compensates the unqualified area according to the continuous compaction results. The above technology realizes the operation standardization, management informatization, and construction digitization of subgrade filling, realizes quality assurance in the whole process of construction, and promotes the development of HSR subgrade filling from mechanization to intelligence.

Anchor/cable frame beams account for more than 60% of the railway slope protection works, and the assembled frame beam structure has the advantages of high mechanization, a short construction period, easy control of construction quality, etc. It is especially suitable for the application of slope protection structure under complex geological conditions, and the attempts of assembled soil nail wall, anchor cable frame structure, anchor frame structure, and skeleton slope protection structure have been carried out in the railways at present. The force pattern of assembled frame beams is different from that of cast-in-place frame beams, and the future development direction is to consider different slope conditions, put forward the selection principles of assembled anchor/cable frame beams under different topographic and geological conditions, establish the internal force calculation model of assembled frame beams considering prestressing and reinforcement of the beams, and put forward the method of calculation of cross-section setup and reinforcement of frame beams based on the design requirements of prestressing members. Finally, the design method of the assembled anchor/cable frame girder structure and the complete set of intelligent construction equipment and technology methods are formed.

3.4.3.3 *Informationization of subgrade deformation observation*

There are two types of roadbed deformation observation systems: automatic monitoring system for subgrade deformation and informationized management system for roadbed deformation observation. The automatic subgrade deformation monitoring system adopts automated monitoring sensors and instruments to realize 24 h real-time observation of subgrade deformation and data uploading (as shown in Figure 3.60), with the goal of centralized supervision of the observation process and data quality.

To improve the management efficiency of settlement and deformation observation and the quality of settlement and deformation observation,

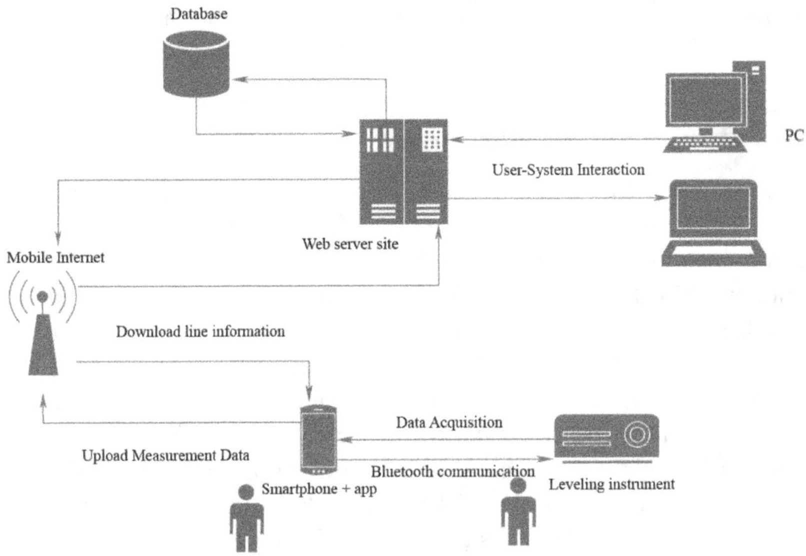

Figure 3.60. General architecture of automatic monitoring of settlement and deformation.

the informationized management system of subgrade deformation observation adopts the mode of "terminal app + railway engineering management platform." The terminal app collects operation data, uploads the operation data to the settlement and deformation observation management module of the railway engineering management platform, and then realizes the supervision of the operation of settlement observation and the tracking of the settlement overrun on the platform after processing. After processing, the platform can realize the supervision of settlement observation business, tracking of settlement exceeding limit, analysis of deformation development trend, etc. Meanwhile, the accumulated data can provide support for the big data analysis of subgrade specialty. The informatization of subgrade deformation observation has achieved breakthroughs in the following aspects: the use of automatic monitoring technology realizes accurate perception of subgrade deformation; the authenticity and validity of observation data are controlled through informatization means; supervision and management of observation frequency, settlement exceeding the limit of measurement points, and retesting of working base points are realized; the unified digital management of internal and external data of observation operation is realized; and the sharing and full life cycle application of the data are supported. It also realizes unified digital management of internal and external observation data, supports data sharing and full life cycle application, and improves the efficiency of deformation observation work and management efficiency.

3.4.4 *Intelligent track construction*

Ballastless track has become the preferred form of track structure for China's HSR due to its advantages of good stability, high smoothness, long-term maintenance of track condition, substantial reduction of maintenance workload, etc. A ballastless track structure is adopted in all HSR with a design speed of more than 300 km/h in China. In order to meet the initial development demand of HSR construction, China introduced different types of ballastless track structures from Germany, Japan, and other countries and formed a ballastless track construction technology that is adapted to China's national conditions through digestion, absorption, and re-innovation. After more than 10 years of development and application practice, track construction has basically realized mechanized flow operation, but the key links in the construction process still generally have such problems as a large manual workload, a low automation degree, and difficulty taking into account the construction precision and efficiency. The ballast track construction technology has not yet reached the level of intelligent and industrialized construction.

With the launch of the "Made in China 2025" plan and the development of HSR construction technology, intelligent railway and green railway have become the key directions of railway construction, which puts forward higher requirements on the process monitoring of construction quality and the management of measurement data. In this context, given the demand for intelligent construction of HSR track engineering, all parties involved in HSR construction have researched intelligent construction equipment and key technologies in ballast track construction. They have developed several types of intelligent construction technologies for the key aspects of ballast track bed construction, track slab/pillow fine adjustment, rail construction, etc. This section mainly introduces ballasted track intelligent laying technology and seamless line intelligent stress relief locking technology.

3.4.4.1 *Intelligent laying of ballasted track*

At present, China's new HSR mainly adopts two structural forms of CRTS III plate: ballastless track and double-block ballastless track. The CRTS III plate ballastless track paving process mainly includes base and limit groove construction, isolation layer, elastic bedding layer construction, track slab paving, and self-compacting concrete construction. The process of double-block ballasted track-laying mainly includes the construction of a supporting layer/base, an isolation layer, an elastic bedding layer, the assembly and fine adjustment of rail row, and the concrete construction of roadbed plates. At present, ballasted track laying has basically realized intelligent operation in key processes such as concrete construction of the base and fine adjustment of the track slab/rail sleeper. Further optimization of related equipment and processes to enhance its adaptability to complex line conditions is the main development direction of ballasted track intelligent laying in the future.

According to the structural characteristics and construction process of two main types of ballast tracks: CRTS III slab ballast track and double block ballast track, intelligent paving construction mainly includes key technologies of intelligent paving, such as integrated automatic leveling of the base, intelligent fine adjustment of track slabs and sleepers, and automatic maintenance of ballast track concrete.

1. Base-integrated automatic leveling technology

A base-integrated automatic leveling machine has its own power supply, remote control operation, automatic control of elevation, push flat fabric, vibration compacting, pulp, leveling and shaping, and other functions. In addition, it can efficiently pave high strength concrete and low slump concrete during operation at the same time. Its powerful vibrator plate vibration frequency of up to 3,000 times/min ensures that the concrete vibration compacts, so that the entire paving of the concrete substrate is homogeneous and dense. An integrated automated leveling machine consists of a frame, generator, walking mechanism, lifting mechanism, front push plate, rear shaping mechanism, vibration system, high-pressure cleaning system, electronic control system, etc. An automatic leveling machine field application is shown in Figure 3.61.

Based on the digital design, the base-integrated automatic leveling technology can realize automatic vibration, automatic smoothing, and automatic tracking construction, effectively improve the construction quality and precision, improve construction efficiency, reduce the labor intensity of the operator and the investment of the operator, and realize the mechanization and automation of the concrete pouring process.

Figure 3.61. Integrated automatic leveling of the base.

2. Intelligent fine-tuning of track plates/sleepers

An intelligent fine-tuning machine for rail plate/rail sleeper is a kind of special equipment used to replace the manual fine-tuning operation of CRTS III rail plate or double-block rail sleeper rows. This equipment mainly consists of a fine adjustment bracket, a fine adjustment machine execution system, and a fine adjustment software system. The fine-tuning machine execution system mainly consists of a fine-tuning machine frame, walking mechanism, guiding mechanism, lifting and lowering mechanism, traversing mechanism, alignment mechanism, fine-tuning system, and so on. The fine-tuning machine software system consists of fine-tuning data reading, fine-tuning data conversion, and fine-tuning control system. The software system adopts OCR data recognition technology to read and accurately transmit adjustment data. The fine-tuning operation is not interfered by human factors, which effectively improves the fine-tuning efficiency, reduces the number of fine-tuning operators, and ensures the precision of fine-tuning data and data stability.

Intelligent fine-tuning machine has 24 h all-weather working ability. Through automatic measurement of the standard frame prism, industrial control machine calculation, and a servo motor precision reducer through a universal telescopic drive shaft to transfer precision rotation adjustment, the total station can completely replace the manual adjustment. The system utilizes wireless network technology to control the on-site total station through the terminal software installed in the laptop or tablet PC, and it collects the key data of the line construction, including the track

Figure 3.62. Intelligent fine-tuning machine on-site operation.

elevation and rail direction data, etc., and implements dual-control alarms for the adjustment accuracy and the smoothness of the line. The on-site operation of intelligent fine adjustment equipment is shown in Figure 3.62.

Based on floating point coordinate system modeling and least squares system identification techniques, intelligent fine-tuning of rail plates / rail sleepers are **realized, which can enable the automatic transmission and real-time adjustment of fine-tuning data**; multi-point linkage adjustment is adopted to meet the requirement of adjusting as and when it is measured without repeated adjustments, reduce interference, and improve efficiency. Intelligent fine-tuning machines can realize automatic measurement, automatic fine-tuning, data tracking, data storage, report query, and other functions without human interference, realizing the traceability of construction quality data.

3. Automatic ballast concrete curing

The ballasted track concrete automatic curing system consists of an intelligent curing membrane, a temperature and humidity intelligent monitoring system, and a water supply system. It is used for curing the concrete of base and roadbed slabs after the completion of concrete casting in the process of ballasted track construction.

An intelligent maintenance membrane uses a high water absorption polymer resin material, realizing the conservation goal of "one time water replenishment, seven days of moisturizing." It had been verified by practical application that the water retention rate of the intelligent maintenance membrane can be reduced from 90% to 60% in seven days. The temperature and humidity intelligent monitoring system consists of temperature and humidity sensors, temperature and humidity data acquisition, analysis, processing control program, and an electronic water-saving control system. It has a humidity monitoring accuracy of up to 0.1%, a temperature monitoring accuracy of up to 0.1 °C, and the shortest monitoring beat of up to 10 ms.

The use of intelligent concrete curing system can effectively solve the current problems of water waste, manual work, and difficulty in maintaining curing quality in the construction of roadbed concrete, thus simplifying the curing process and improving the quality of curing. By adopting a sprinkling cover, water-saving moisturizing, and intelligent curing technology of curing film, one-time water absorption and saturation can be moisturized for more than seven days. It saves the amount of water used for curing, reduces the labor intensity, and avoids the problems of concrete cracking caused by untimely curing. Intelligent curing equipment for roadbed can monitor the humidity of the concrete curing environment in real time, automatically control the curing system in real time through the monitoring data of the monitoring system, and the monitoring data can be transmitted to the cell phones of the site management personnel in real time, which is convenient for monitoring at any time.

Intelligent ballast track-laying technology has been applied in a large number of HSR projects, such as Beijing-Zhangjiakou high speed railway, which realizes efficient ballast track construction under the situation of tight schedule and large amounts of work, effectively improves the quality of ballast track construction, and reduces the inputs of personnel and machinery, as well as secondary reworking. In the key operation link of ballast track construction, through the application of a series of intelligent equipment and technologies such as intelligent leveling of the base, intelligent fine adjustment of the track slab/pillow, and intelligent maintenance of concrete, the intelligent construction level of the ballast track is effectively improved, the construction quality, progress, and cost of key processes are ensured, and good economic and social benefits are achieved.

3.4.4.2 *Intelligent track laying*

Track-laying operation is an important part of HSR track engineering construction, and its construction quality and efficiency directly affect the smoothness of HSR lines and the overall project progress. The traditional track laying operation is labor-intensive, with many auxiliary staff in the operation process, and has a low degree of automation.

Intelligent rail-laying technology takes unmanned or less human-assisted operation as the principle and carries out a series of operations before, during, and after rail-laying through intelligent and automated equipment, the core of which is the application of intelligent equipment and information integration. Using the IoT, cloud computing, big data, machine vision, and other new-generation information technologies, it builds an intelligent, efficient, energy-saving, and environmentally friendly rail-laying operating system, improves the controllability of the operation process and the level of scheduling management, and realizes the intelligent optimization of rail-laying operation, management, and decision-making.

According to the HSR track-laying construction process, intelligent track-laying mainly includes rail intelligent group lifting, long rail intelligent laying, rail line area intelligent scheduling command, and other key technologies.

1. Intelligent group lifting of rails

Traditional long rail lifting mostly adopts full manual control mode, requiring 32 operators to operate the gantry crane at the same time for hooking, hook picking related loading and unloading operations, high labor costs, and low operating efficiency. For long rail lifting synchronization problems, the rail intelligent group lifting system applies PLC, gray bus positioning, yard information technology, automatic grabbing, placing of rails, and other technologies, with synchronous action, accurate positioning, centralized control, automatic lifting, fault alarm, and other functions. In the lifting process, the central control room can be realized on the gantry crane with multiple linkage control or single control switching, and real-time monitoring of lifting capacity, lifting height, traverse stroke, and other parameters of the work status, saving labor costs. At the same time, rail loading and unloading efficiency increased by more than 20%. It in fact realizes the intelligence and automation of long rail loading and unloading operations. A rail intelligent group lifting system application as shown in Figure 3.63.

2. Smart laying of long rails

The ballastless rail long rail intelligent track-laying unit integrates tires, wheel rails, and crawler walking into one, which consists of a long rail intelligent sorting car, a long rail intelligent pushing car, long rail intelligent roller recycling car, a long rail intelligent traction car, and other equipment units. The track-laying unit makes full use of cutting-edge and

Figure 3.63. Rail intelligent group crane.

Figure 3.64. Intelligent track-laying units.

new technologies, such as intelligent robots, IoT, big data, laser scanning, precise positioning, and automatic navigation equipment (AGV) trajectory control. It can automatically complete a series of processes, such as precise traction and pushing of long rails, precise placement of rollers and recovery, stacking, and inverted transportation of rollers. Compared with the traditional way of track-laying, the intelligent track laying unit can save more than 60% of auxiliary workers, effectively reduce the number of workers, and reduce labor intensity. In high-cold, high-altitude, and other special areas with labor shortages, the promotion of this system is more significant. An intelligent track-laying unit is shown in Figure 3.64.

3. Intelligent dispatching command in the rail area

The level of railway engineering line (rail line area) scheduling management is related to the integration and balance of construction and track laying, the safety of people, vehicles, and construction, and whether the track laying progress affects the joint commissioning and operation, which is the key link in the process of railway engineering construction and also an important part of the construction of railway engineering informatization. Through the application of BeiDou positioning, wireless radio frequency, big data analysis, and other technologies, the intelligent scheduling command system in track areas realizes the real-time monitoring and unified scheduling command of the running status and operation position of track-laying units, long rail carriers, welded rail cars, and other equipment in all-weather conditions and carries out the whole-process management and control of the progress of track-laying operation, construction plans, and personnel by means of information technology so as to ensure that the track-laying construction is in a safe and orderly manner.

Intelligent track-laying technology has been widely used in HSR projects such as Beijing–Zhangzhou, Zhengzhou–Wanzhou, Zhangji–Huai, and Guinan, which effectively improves the operational efficiency and quality of track-laying construction of HSR.

3.4.4.3 *Intelligent stress relief and locking of seamless lines*

Seamless line stress relief and locking ensures the smoothness and stability of the railway track key operations and seamless line intelligent stress relief and locking system using information technology and high-precision sensing technology. It can realize the seamless line construction process of information management to improve the quality of construction operations and management levels.

The seamless line intelligent stress relief and locking system includes a management platform and terminal system, and its system architecture is shown in Figure 3.65.

The management platform is mainly oriented toward the industrial management center, construction unit, operation unit, construction unit, and supervision unit, and the main functions of the platform include basic data management, construction organization management, query management, report management, and statistical analysis.

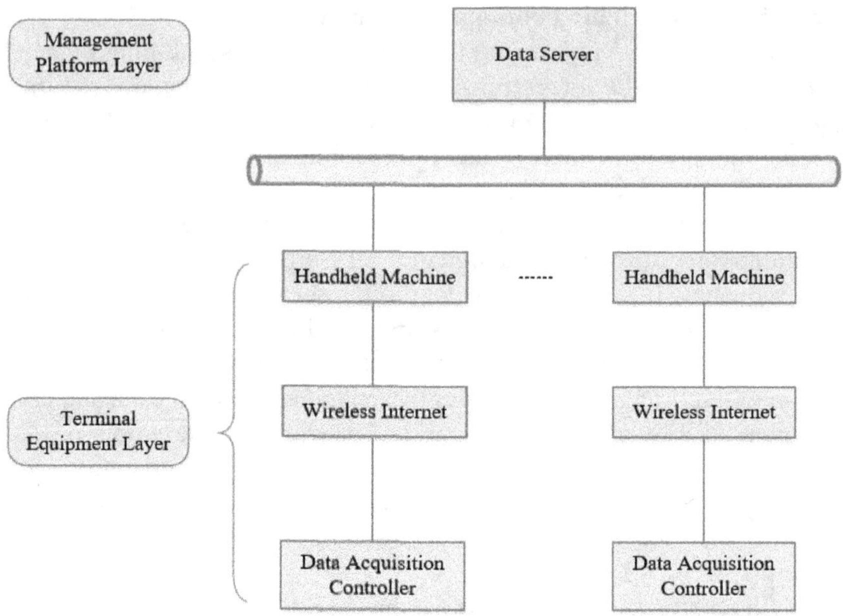

Figure 3.65. Seamless line intelligent stress discharge and locking system architecture.

The terminal system is oriented toward the disbanding and locking operation team, including hardware units (power supply unit, sensing unit, distance measuring unit, collecting unit, communication unit, electronic label, industrial plate, etc.) and app. The terminal system receives the rail temperature and displacement data of all collecting equipment in real time, and according to the received rail temperature, it automatically detects whether it conforms to the rail temperature specification and prompts that the construction conditions don't have the conditions of disbanding when it doesn't conform to the specification. When it meets the specification, it automatically determines the construction method, and at the same time, according to the displacement requirement in the specification of the construction method, it automatically determines the displacement in real time until the displacement of all collecting devices meets the requirement.

The system adopts high-precision sensing technology, intelligently senses the on-site rail temperature and stress dispersal displacement, and uploads, summarizes, and calculates the real-time rail displacement and operation rail temperature data of each measuring point through zigbee wireless self-organizing network and transmission technology, providing real-time and accurate operation data for the process of stress dispersal and guidance for the standardization of on-site construction.

Figure 3.66 shows the application of a rail displacement sensor based on surface array laser scanning in the system of automatic measurement.

The key technological innovations in the development of the seamless line intelligent discharge and locking system are as follows:

(i) Realizes the whole-process information management of seamless line stress relief and locking construction methods (tensioner roller method or natural roller method): rail temperature, air temperature, and rail displacement can be monitored automatically;

real-time monitoring of the impact process, automatically determine whether the impact of the rail to reach zero stress state; real-time monitoring of the tensile process, the whole process of monitoring the longitudinal tensile displacement value of the rail, and uniformity.

(ii) The long-distance self-organized network transmission technology is proposed to collect on-site data quickly and in real time, which breaks through the difficulty of real-time transmission of long-distance self-organized networks without network coverage along the railway. The system realizes "source control, process control, and detail control," reduces the random and non-standard operation of seamless line stress relief, and provides important support for realizing the concept of

Figure 3.66. Application of surface-array laser scanning rail displacement sensors in systematic automatic measurement.

all-round and whole-process quality monitoring of seamless line locking operation.

Since 2016, the seamless line intelligent stress relief and locking system has been applied on a pilot basis in Baoji-Lanzhou, Beijing–Shenyang, and other HSR lines. The application effect shows that the system has realized the digital management of the whole process of stress relief and welding-linking locking operation of seamless lines by means of informatization and can collect and analyze in real time the temperature of the rails, rail displacement, and ambient temperature in the process of unit-rail bar stress relief and locking operation, accurately guide the construction unit to comply with strict construction specifications, and effectively improve the quality of the construction of seamless lines. It can accurately guide the construction unit to strictly comply with the construction specification and effectively improve the quality of seamless line construction. The intelligent construction technology has been fully popularized and applied to all HSR projects.

3.4.5 *Intelligent station construction*

The main body of the HSR station building is related to the railway service level and affects the passengers' railway transportation experience, which has the characteristics of large span, large space, special form, varied changes, components of different shapes, difficult construction, etc. The application of intelligent construction in the construction of the HSR station building has changed the traditional construction mode and improved construction efficiency and quality. BIM, artificial intelligence, IoT, virtual reality, three-dimensional scanning, intelligent equipment, construction robots, and other technologies are integrated and applied to the intelligent construction of passenger stations.

3.4.5.1 *Mechatronics BIM applications*

The station building project involves many electromechanical specialties, pipeline layout is complex, and the cross-operation of various specialties is substantial. In the construction process, there were often situations where a professional pipeline installation did not take into account other specialties, resulting in the inability to complete the installation of other professional pipelines, and the situation had to be reworked.

Figure 3.67. Mechatronics BIM applications.

Applying BIM technology to make comprehensive arrangements in the pipeline model can ensure reasonable arrangements among pipelines of various specialties with decoration works and other specialties. An integrated bracket is designed according to the arrangement of pipelines, and after the completion of the construction of the main structure, it is retested and fine-tuned for the arrangement of pipelines, and the prefabricated processing bracket is unified after the final plan is determined. Each specialty organizes prefabricated pipeline construction according to the fixed position of the bracket in sequence. It realizes factory prefabrication and processing of electromechanical pipelines, reduces on-site cutting and welding, improves construction quality, and accelerates construction progress. The application of BIM for electromechanical integration is shown in Figure 3.67.

3.4.5.2 *BIM applications for assembly rooms*

Through the application of BIM technology in the assembly machine room, the construction efficiency and quality are effectively improved, and the specific BIM model design effect and installation effect are compared, as shown in Figure 3.68.

3.4.5.3 *Intensive design BIM applications*

Intensive design BIM application refers to modeling through software, analyzing and checking the model, discovering clashes between various disciplines and problems of reserved holes in advance, and correcting them through collision checking and other functions. A scene is shown in Figure 3.69.

Figure 3.68. Assembly room design and installation.

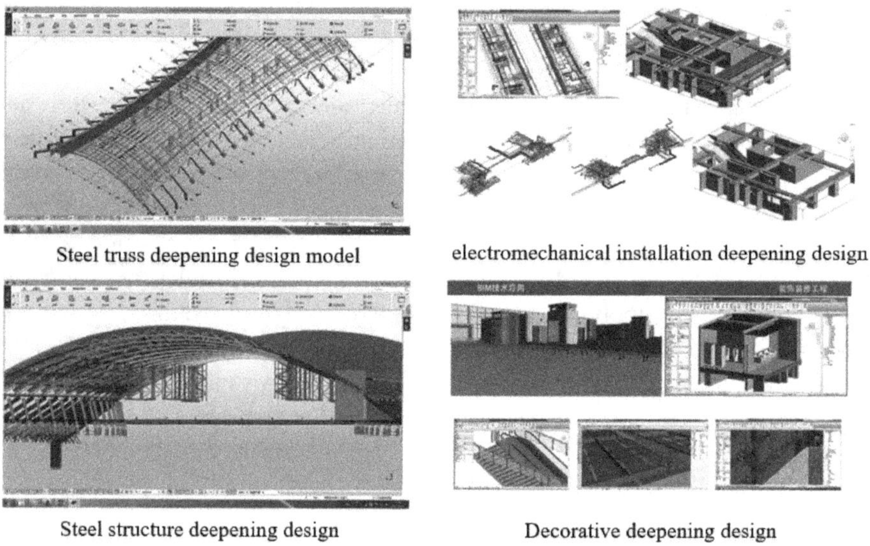

Figure 3.69. BIM intensive design applications.

(1) **Civil engineering**
The collision check of the model verifies that the main structure is not in an unreasonable position and that the working surface is undersized.

(2) **Electromechanical specialties**
Identify spatial collisions in the design and construction process through the collision checking function, analyze the collision points through the collision checking function of the BIM software, exclude reasonable

collisions, and then discuss the collision points to solve the problem before construction. This enhances design quality, saves man-hours, and reduces the generation of design changes and abandoned works due to collisions.

(3) **Decoration specialization**

Using the model in advance to grasp the decorative material modulus, effect, an accurate grasp of the type of material selection, and procurement of quantities, we can avoid rework after construction, insufficient material purchases, or purchases beyond the actual use. It also ensures accurate control of the details of the decoration in advance of the development of the construction of small nodes of the construction method to avoid program lag affecting the construction period.

3.4.5.4 *Collision detection BIM applications*

Collision detection BIM application means that based on the collision checking function, the collision points are quickly and accurately found and dealt with, and the problematic conflicts are front-loaded, as shown in Figure 3.70. The results of professional coordination are comprehensively checked, and the conflicts between E&M, architecture, and structure and the collisions between E&M specialties are the key points of checking. The checked problems are fed back to the designers of each specialty, and after adjustment, theoretically, "zero collision" can be achieved. Through software modeling, model analysis, inspection, collision checking, and other functions in advance to find out the conflict between the various disciplines, the problem of reserved holes is

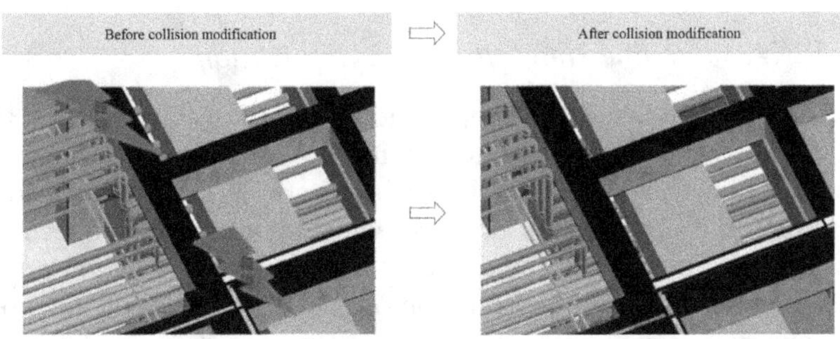

Figure 3.70. Collision detection BIM applications.

Figure 3.71. Construction program simulation BIM application.

addressed, and we can effectively make up for the mistakes in the front end of the design. For the project effect, schedule and cost control will be very helpful.

3.4.5.5 *Construction program simulation BIM application*

Construction program simulation BIM application refers to the simulation of major construction programs, such as steel mesh frame installation and electromechanical installation, based on the BIM model to verify the feasibility of the program and improve its expressiveness. Virtual simulation installation simulation (as shown in Figure 3.71) can be carried out in advance to preview and optimize the position of construction machinery, material transportation channels, and personnel deployment scheme during the construction process, so as to formulate a more perfect construction scheme.

3.4.5.6 *Key node fine simulation BIM application*

Key node fine simulation BIM application means to establish a fine BIM model for key nodes and apply it to briefing, coordination, program preparation, program thesis, etc., as shown in Figure 3.72. According to the BIM model, the construction organization design can be deduced, and construction simulation can be carried out for each construction process and construction sequence so as to find out the difficulties and key points of the construction and reasonably arrange the input of resources, such as construction personnel, materials, and machinery. We can also understand

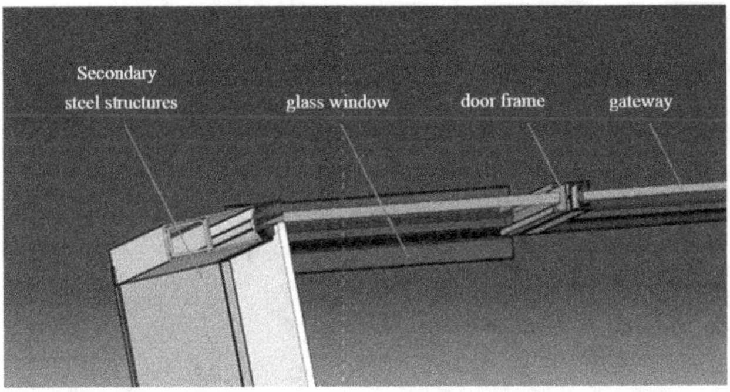

Figure 3.72. Key node fine simulation BIM application.

the hidden dangers of the construction process in advance so as to take precautions and prompt the attention of the construction personnel.

3.4.5.7 *Visualization of handover BIM application*

Visualization BIM application refers to the visual display of the content of the delivery in various forms, such as video and pictures, based on the BIM model for the key work processes. The visualization BIM application (Figure 3.73) allows construction and management personnel to understand the construction technology and construction process more intuitively, reduces errors in the construction process, and provides a strong guarantee for the improvement of project quality.

3.4.5.8 *IoT tracking BIM applications*

IoT tracking BIM application means that by attaching BIM data information to a QR code or an RFID chip, it can accurately locate steel components, concrete components, curtain wall components, etc., monitor the transportation and installation of components in real time, and track and trace the information during the construction process, as shown in Figure 3.74.

3.4.5.9 *Process method library BIM application*

Based on the BIM model, process simulation can be carried out to guide the on-site construction and gradually form the BIM process and method library, as shown in Figure 3.75. After engineering practice, a

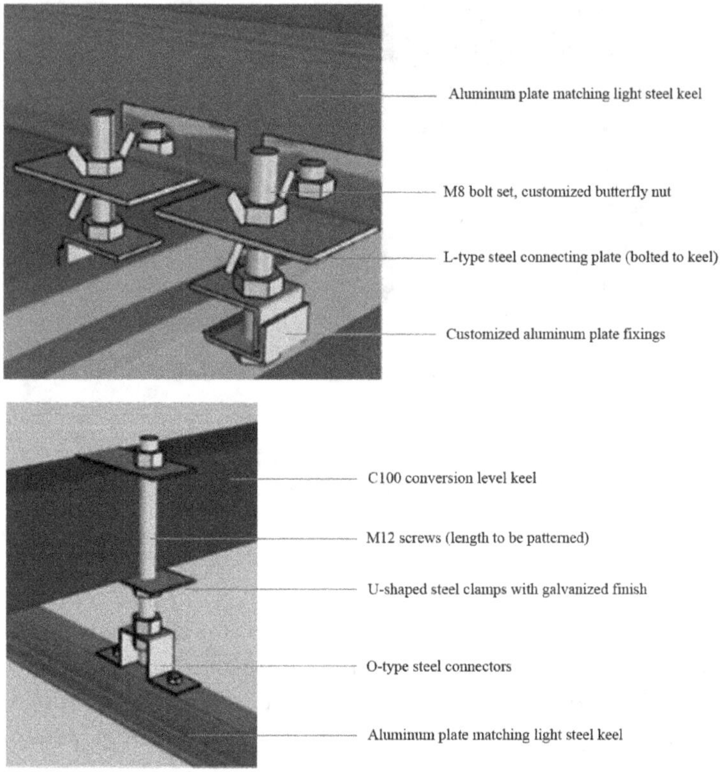

Figure 3.73. Visualization handover BIM application.

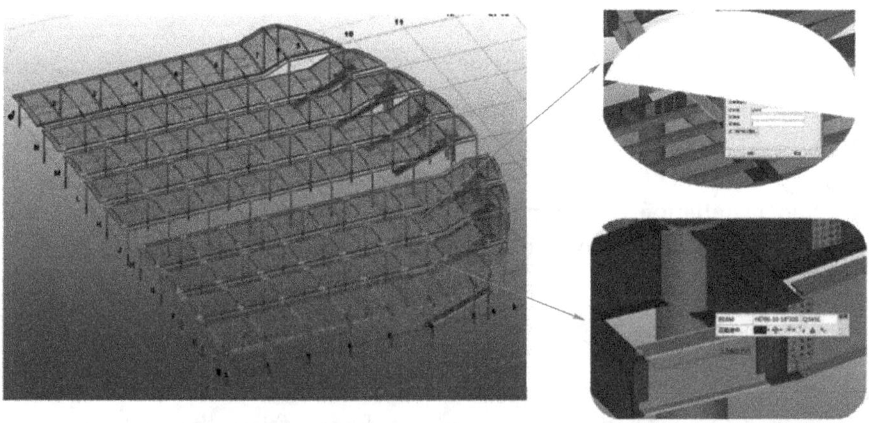

Figure 3.74. IoT tracking BIM applications.

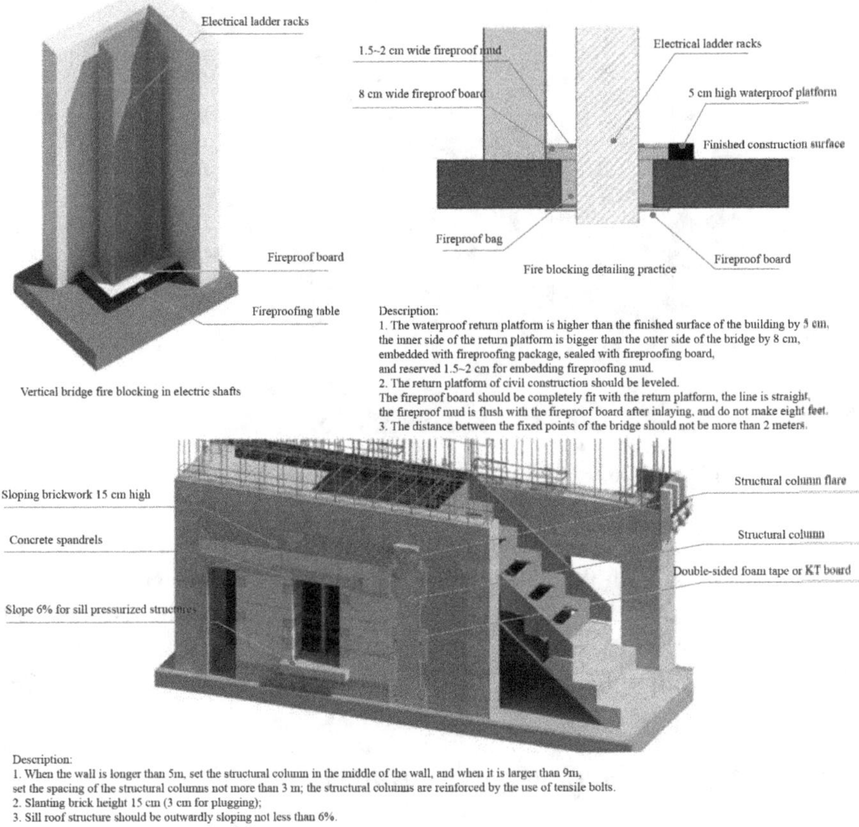

Figure 3.75. Process method library BIM application.

comprehensive supporting construction method query application is formed so that construction personnel can query the standards of technology and the construction process in real time, providing a basis and guidance for on-site construction safety and quality acceptance.

3.4.5.10 *Curtain wall optimization and control BIM application*

Curtain wall optimization and control BIM application refers to the establishment of a BIM model for the characteristics of curtain wall projects, such as strong professionalism, complex nodes, and numerous components, and adjusting and optimizing the model. For the waterproofing nodes and other complex detailed processes, BIM technology is used to

Figure 3.76. Curtain wall optimization BIM application.

establish an animation guidebook for visual delivery. The use of BIM technology, combined with the material management system of RFID technology, can carry out the whole-process management of the component information. The BIM software can be used to count material consumption. A curtain wall optimization BIM application is shown in Figure 3.76.

3.4.6 *Intelligent electrical engineering construction*

The four-aspects of railway power engineering are communication engineering, signal engineering, power engineering, and traction power supply engineering, specifically including communication, signaling, information, disaster monitoring, power, contact network, substation, power housing, and other engineering content.

The four power projects have the characteristics of a short construction period, multiple specialties, complex interfaces, wide and scattered work sites, etc., and are subject to more constraints in the actual construction. The intelligent construction of the railway four-electricity project follows the principle of "informatization, mechanization, automation, and intelligence." It aims to realize the fine and intelligent management of progress, quality, safety, and investment in the construction process of railway four-electricity projects, promote the construction of railway four-electricity projects from informatization and automation to intelligence, and improve the construction efficiency and quality of the projects.

3.4.6.1 *Contact network intelligent pre-wiring and automation installation*

The intelligent construction of a contact network project of HSR realizes the whole process of automatic operation and management and realizes the automation and intelligence of contact network assembly, which effectively improves the assembly precision and installation efficiency and enhances the construction quality, safety, and reliability. The contact network intelligent pre-assembly and automatic installation mainly includes contact network engineering factory processing, contact network pillar assembly, contact network wrist arm installation, contact network parameter detection and hanging string calibration, automatic installation of contact network lifting pillar, etc., which forms a whole set of manufacturing factories, modularization, informatization, and intelligent construction technology.

1. Contact network engineering factory processing

The factory processing of contact network engineering mainly includes the automated pre-assembly of hanging chords and contact network wrist arms. Through contact network engineering factory processing, it can improve the mechanization rate and automation rate of contact network construction, especially reduce manual participation in the high-altitude heavy work, improve efficiency and precision, enhance the informatization and artificial intelligence of the contact network construction, ensure the safety of the project, improve the efficiency and quality of the construction, and improve the service life of the contact network.

(1) *Automated pre-wiring of contact wire strings*

The smoothness of the contact line is a direct factor affecting the relationship of the bow network. The automated pre-assembly production line for the contact wire suspension line of HSR can complete the production of the whole suspension line body and the automatic assembly of the related accessories, such as wire clips at both ends, hexagonal bolts, nuts, and stop spacers. Intelligent preassembly can improve the quality, progress, and precision of preassembly.

The nominal length of the automated pre-assembled chord can be made with an accuracy of ± 1 mm, which improves the quality of the product and the stability of the batch operation. It also reduces the waste of raw materials, as shown in Figure 3.77. At the same time, the self-test function

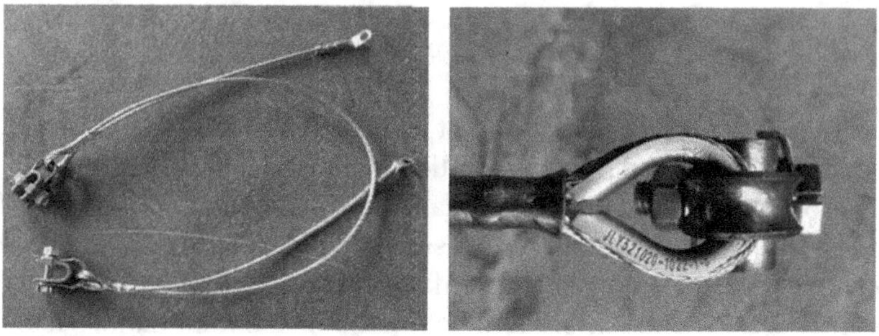

Figure 3.77. Pre-assembled chords from automated chord pre-assembly lines.

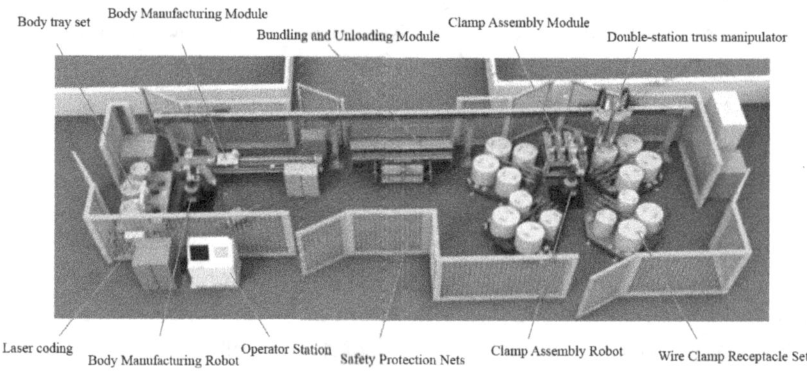

Figure 3.78. Schematic diagram of automated pre-assembly production line for hanging strings.

of the power-on sensor effectively improves the reliability of the mechanical equipment and improves the quality of preassembly.

The HSR contact network hanging string automated preassembly production line mainly consists of the operation desk, body manufacturing module, wire clip assembly module, double-station truss module, bundling and discharging module, etc., as shown in Figures 3.78 and 3.79.

Operation desk: The data are directly imported and intelligently extracted by adopting communication control of the upper computer with a six-axis robot system, a multi-axis servo motor control system, a hydraulic system, a pneumatic system, and a large-scale double-station truss system. It realizes intelligent pre-assembly, such as online inspection of chord length, intelligent assembly of wire clip components, laser coding of chord

Figure 3.79. HSR contact network string automated preassembly production line.

numbers, automatic bundling, etc. It also has the function of outputting relevant inspection and construction data.

Body manufacturing module: The body part is manufactured based on the string data given by the host computer. The production accuracy of the nominal length of the chord can reach ±1 mm, which improves the quality of the product and the stability of the batch operation.

Clip assembly module: According to the pre-matching data of hanging strings given by the host computer, using special assembly molds, together with a six-axis robot system, a pneumatic system, and an intelligent feeding system, it realizes the automatic selection and pre-matching of different specifications of hanging string clips and, at the same time, meets the different needs of straight line, in-curve, and out-of-curve uses of the line. The robot pickup ensures the accuracy and completeness of the matching of each wire clip part, avoiding the uncertainty of manual work and indirectly improving the construction quality of the line.

Dual-station truss module: A servo system, which can realize synchronous operation of a dual-station and independent operation of a single-station, meets the size variability of hanging string products.

Bundling and discharging module: The defective products are detected online, grabbed by a double-station truss manipulator, and moved directly to the discharging slot of the defective products in the bundling. The qualified strings after the assembly of wire clips are grabbed by the truss manipulator and placed directly in the waiting area of bundling in the bundling module. It realizes 2 min/pc string preassembly speed, which improves the efficiency of preassembly and reduces the investment of construction personnel; it improves the accuracy of string preassembly, with a dimensional accuracy of up to ±1 mm and realizes the remote control of the equipment through the BIM platform, which improves the level of project management.

(2) *Contact network wrist arm automation pre-provisioning*:
The quality, accuracy, and progress of the wrist arm preassembly are the key factors affecting the application of the wrist arm system. Based on the customized characteristics of contact network system components for HSR, the wrist arm pre-fitting platform needs to realize intelligent pre-fitting and flexible production.

The contact network wrist arm pre-matching platform is mainly used for the automation and intelligentization of the 350 km/h HSR contact network wrist arm pre-matching. It is shown in Figures 3.80–3.82.

The contact network wrist arm pre-assembly platform mainly includes the functions of reserve, loading rack, sawing, drilling and grinding, spraying and marking, assembling station, and output of finished products.

Stocking and loading rack: The dual-station modular assembly structure design supports a single 6 m or a single 12 m raw material length. It has the functions of automatic material turning, material organizing, material

Figure 3.80. Wrist arm pre-assigned for automated wrist arm pre-assignment line.

Figure 3.81. Schematic diagram of wrist arm automated pre-assembly production line.

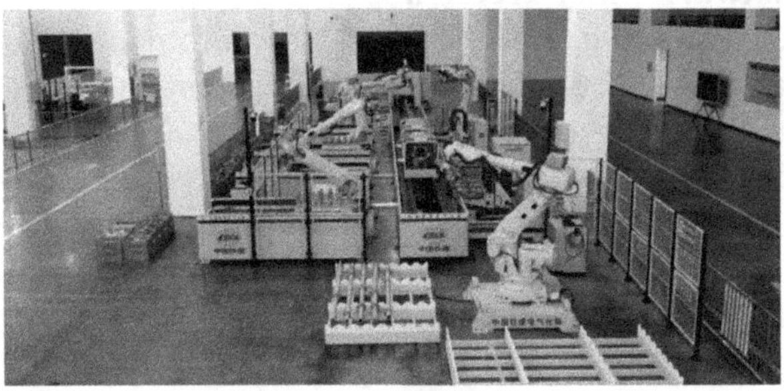

Figure 3.82. Intelligent production line for contact network wrist arm pre-assembly for HSR.

discharging, material loading, material pushing, material shortage detection, and safety protection.

Sawing and drilling trimming: According to the pre-set data of the wrist arm mechanical automatic judgment, fixed-length sawing is performed, which has a dimensional error of ±1 mm.

Spraying code and marking: A customized Chinese marking code, strut number, wrist arm parts positioning triangle marking, 2D code or bar code, and other information are used. According to the wrist arm pre-fitting data and user settings, we automatically locate the position of sprayed markers and information codes with an error of ±1 mm.

Fitting station: It has automatic pipe threading and positioning, alternating truss and manipulator transfer, and automatic identification and rotation of part mounting angles.

Finished product output: Pre-assembled flat and inclined wrist arms are automatically unloaded by the truss manipulator and placed on the buffer receiving plate. The conveyor belt rotates synchronously and can buffer and store four wrist arms at the same time.

It realizes 2–4 min/group of wrist arm pre-fitting speed, which improves the pre-fitting efficiency and reduces the investment of construction personnel; it improves the pre-fitting accuracy, with a pre-fitting dimensional accuracy of ±1 mm and a torque error of ±3 N m; and it improves the level of project management.

2. Contact network pillar erection

Based on the existing integrated process data information, the HSR contact network pillar erection intelligent equipment realizes reliable and accurate intelligent construction control and management, reduces the physical labor intensity of construction personnel, and improves construction efficiency and quality. In the process of pillar erection, the intelligent equipment detects and measures the user's foundation construction at the same time and interfaces with the BIM management system to share the installation process data and measurement data information, as shown in Figure 3.83.

The contact network pillar erection intelligent equipment has the functions of adapting to multiple working conditions (railway, highway, or track plate) and walking by itself, realizing automatic positioning

Figure 3.83. Intelligent equipment for HSR pillar erection operation.

installation of robot arms based on field vision guidance systems, multiple safety measures guarantee, high-precision multi-degree-of-freedom control of heavy-duty robot arm, collaborative operation with BIM platforms, etc. It is mainly composed of a chassis system, a robot arms, a pole erection robot, an electric control system, a hydraulic system, a pipeline system, and other parts. Through the new technologies of high-precision control of heavy-duty robotic arms, automatic guiding of machine vision, and spatial multi-dimensional path planning, it can realize the automated operation of pillar forming and erecting.

The working principle is to use the dual camera vision system as a guide in the appropriate position of the pillar foundation to place the guide target plate. When the pillar arrives near the foundation, through the vision system, the operator completes the position of the location calibration, automatically completes the position of the pillar, the ground bolt docking, and the pillar fixation.

Key technologies include:

(1) *Precise adjustment of strut slope based on intelligent measuring base*
The intelligent measuring base accurately detects and displays the pre-installed tilt rate of the pillar in real time and adjusts the bottoming nut to make the tilt rate of the pillar vertical line and smooth line meet the standard inspection requirements, shorten the time of the pillar straightening operation, and improve the one-time construction rate of meeting the standard, and at the same time, it can realize the collection of coordinates of the pillar position to be installed and the real-time transmission so as to provide accurate data for the locomotive interconnection and control.

(2) *Robotic arm automatic alignment mounting based on vision guidance system*
Based on a high-precision control work robot arm equipped with a vision guidance system and an automated end actuator, it realizes pillar erection work according to the specified position. Adopting two machine vision cameras as guides to complete the automatic gripping and accurate alignment of pillars, it realizes the automation of pillar erection.

(3) *Multi-degree-of-freedom control of high-precision heavy-duty robotic arms*
For different types of columns, we perform gripping, attitude adjustment, and alignment operations, realizing the precise alignment of column

flanges and concrete foundations. In the process, the maximum column mass is 2 t, and the alignment deviation is ±3 mm. Different slots are set on the clamping jaws for different types of H-type columns to ensure that they are tightened in the appropriate slot position to realize a safe and reliable column gripping process. Under heavy load conditions, the robot arm realizes any attitude adjustment of 6+3 degrees of freedom within the reachable range in three-dimensional space.

The HSR H-pillar intelligent erection technology equipment has the full working conditions of self-propelled autonomous operation, reduces the input of construction personnel, improves construction efficiency, ensures the safety of on-site construction work, realizes remote control of the equipment through the BIM platform, and improves installation accuracy.

3. Contact network wrist arm installation

Through contact network wrist arm installation intelligent equipment with contact network wrist arm pre-loading, alignment, and installation functions, the construction process is mainly mechanized, with manual assistance acting as a supplement, which enhances the level of mechanization of the wrist arm installation equipment and reduces manpower. The equipment takes the railway flatbed truck as the carrier and has the functions of adapting to multiple working conditions (railway, highway, or track board), automatic positioning and installation of mechanical arm based on visual guidance system, modular transfer and automated installation of wrist arm, mechanized installation of wrist arm based on the existing process, and protection of multiple safety measures, etc., as shown in Figure 3.84.

The contact network wrist arm installation intelligent equipment mainly consists of an engine system, an aerial work platform, a mechanical work arm, a wrist arm assembly area, a wrist arm material placement area, and other parts. Through multi-degree-of-freedom mechanical arm high-precision control, machine vision guidance, spatial multi-dimensional path planning, and other new technologies, it can be completed by automatically aligning with the installation of the wrist arm.

Key technologies include: high-precision wrist arm automatic alignment installation based on machine vision guidance, parking guidance technology based on BeiDou + Real Time Kinematics (RTK) + Inertial Measurement Unit (IMU), end-of-arm control technology based on articulation decomposition, and mechanized continuous operation of wrist arm installation under the existing process flow.

Figure 3.84. Contact network wrist arm intelligent installation operation diagram.

The HSR contact network wrist arm intelligent installation technology and equipment, with all working conditions of self-propelled autonomous operation, reduces construction personnel, improves installation efficiency, ensures the safety of on-site construction work, and improves installation accuracy.

4. String calibration

The contact network string calibration intelligent equipment is oriented toward the on-site construction and operation environment of the HSR contact network, realizing the automatic positioning and calibration of the installation position of the string, the automatic measurement of the geometrical parameters of the contact line, and realizing the high-altitude operation through the carrying of the dual-use platform of the public railway. It mainly consists of a traveling operation platform, a control system, a contact network parameter monitoring system, a Cartesian coordinate manipulator, a distance measuring and calibrating system, a code spraying system, and other parts. The intelligent equipment is divided into two working states: measurement and calibration. Through laser and image fusion detection, three-dimensional dynamic following, and other technologies, it realizes the accurate calibration of the installation position of the suspension string and the accurate measurement of the contact line height, pull-out value, and other parameters under the walking state, as shown in Figures 3.85 and 3.86.

Figure 3.85. Intelligent equipment for string calibration.

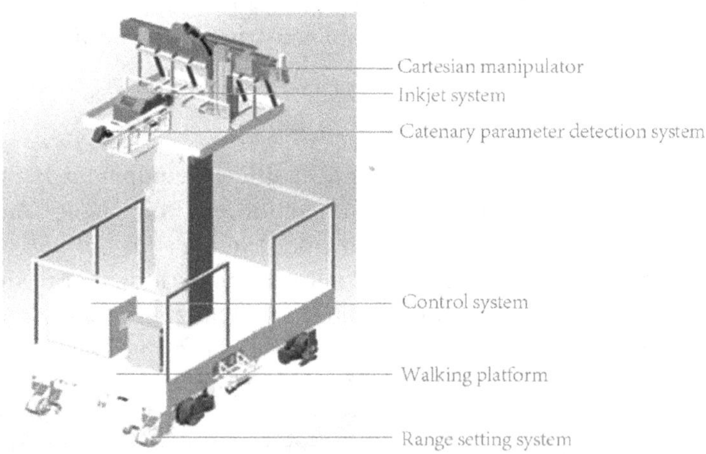

Figure 3.86. Schematic diagram of the overall structure of the intelligent equipment for string calibration.

The working principle is to use a combination of laser detection and image detection to improve the accuracy of contactless measurement of guide height and pull-out values in outdoor environments. Applying 3D dynamic following technology with a single-span wrist arm positioner as the starting zero position for calibration, the contact lines are calibrated one by one according to the calculated parameters of the lifting chord installation position.

Key technologies mainly include real-time following technology of contact lines by manipulators based on a traveling operating platform, zero-point positioning technology based on contact line measurement, high-precision distance measurement and calibration technology based on independent traveling, and continuous automatic measurement technology of geometric parameters of contact lines.

String calibration equipment has the functions of automatic positioning and calibration of string installation positions, automatic measurement of geometric parameters of contact lines, span value measurement, etc. It can effectively improve the efficiency of the contact network construction work, reduce the number of personnel, and reduce the labor intensity of manpower, and it can meet the needs of high-speed contact network construction line straight line sections, curved sections, and other on-site working conditions. It is possible to realize the calibration of the contact line at different heights of the conductor by adjusting the height of the lifting platform.

5. Automated Installation of Contact Grid Bollards in Tunnels

At present, the installation of the railway contact network superstructure adopts the manual method involving multiple people for lifting the suspension column (200 kg) to the top of the tunnel and then installing it on the scaffolding in the right position, which has the shortcomings of high labor intensity of the workers, low working efficiency, and the dangers of manual work. Intelligent installation equipment for the contact network columns in the tunnel completely replaces manual labor, completing the installation and inspection of the superstructure columns.

Intelligent installation equipment for contact network lifting column consists of a self-propelled vehicle, a zero-gravity robotic arm, a multi-joint main robotic arm, a horizontal adjustment robotic arm, a base plate module, an electric-hydraulic power module, an operation platform, etc. It realizes the functions of lifting column grasping and lifting, automatic walking, precise positioning, bolt tightening, remote monitoring, and remote control, as shown in Figures 3.87 and 3.88.

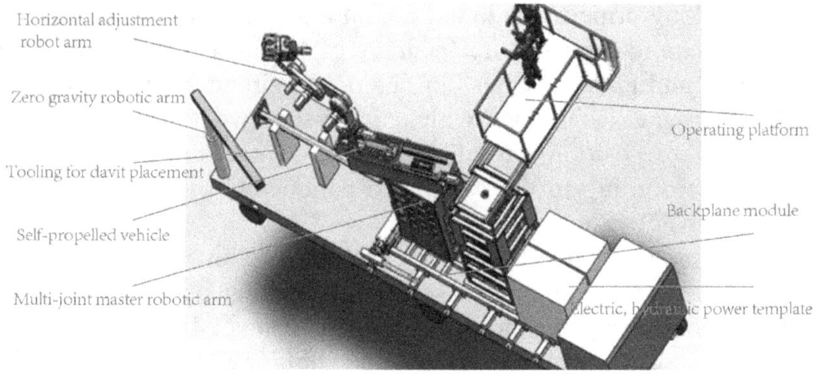

Figure 3.87. Schematic diagram of contact network bollard intelligent installation equipment.

Figure 3.88. Schematic diagram of intelligent installation of bollard.

The working principle is to pre-position the tunnel lifting column on the tooling. The robot is controlled by the control system to automatically grab the lifting rod and move the lifting column to the target position, and a vision system is installed on the robot to accurately position the screw and, at the same time, detect whether the relative position of the screw is correct or not. The front end of the horizontal adjustment robot arm is equipped with a dual-axis inclination sensor to monitor the status of the boom in real time, and the control system controls the robot arm to adjust the boom to the desired state (vertical and downward line angles). The horizontal adjustment robot arm is equipped with a laser displacement sensor to detect the distance to the root of the bolt and manually load the appropriate number of adjustment shims according to the detection result to ensure the vertical installation of the boom; after the installation of the boom is completed, the parameters of the installation limit of the boom and the slope of the plumb and downstream lines are detected and transmitted to the control system. The data after the installation can be stored, exported, and transmitted remotely. The automatic servo torque wrench is installed on the operation platform and can work with the manipulator to tighten the nut automatically.

Key technologies mainly include: remote monitoring and remote control of equipment, self-diagnosis of equipment status, rapid arrangement of measurement and control network based on track control network, and autonomous precise positioning of column installation.

The application of this equipment reduces the labor intensity of workers, reduces personnel investment, improves work efficiency and quality, and ensures on-site construction safety.

3.4.6.2 *Communication signal simulation optimization and intelligent construction*

3.4.6.2.1 BIM+GIS-based network optimization

The mobile communication system of HSR is the transmission platform of the train control information, which is an important guarantee for the safe operation of HSR. Therefore, avoiding weak coverage and communication interruptions is the primary task and basic requirement of HSR mobile communication network planning. In the face of an increasingly dense railway network, the complexity of the construction of a railway digital mobile communication system (global system for mobile communications — railway, or GSM-R) is getting higher and higher, which poses a great challenge to the planning of the network.

At present, GSM-R system wireless coverage design should reasonably select the propagation model according to the terrain, but the Okumura-Hata propagation model is usually used in practice. After simulation planning based on the empirical model, the simulation precision and accuracy cannot meet the requirements, which leads to the need for fine optimization work during the commissioning stage of HSR, and the existing means mainly rely on the low-speed network optimization by carrying rail cars and the high-speed network optimization by carrying high-speed inspection trains, which has shortcomings such as adjustment and verification workload, efficiency, and waste of resources. The existing means mainly rely on carrying railcars for low-speed network optimization and high-speed detection trains for high-speed network optimization, which have the shortcomings of a large workload of adjustment and verification, low efficiency, and waste of resources. Therefore, it is extremely necessary to study the intelligent coverage prediction technology of the HSR mobile communication network.

(1) **Function and composition**

The railway scene is complex and distinctive, containing a large number of special and irregular structures with high spatial heterogeneity and frequency dependence. Therefore, the construction of a high-precision railway scene 3D model library that affects the propagation characteristics of radio waves becomes the foundation and prerequisite for wireless network planning and optimization. First, the BIM+GIS convergence database of the railway scene is built, and the method of accurately and efficiently expressing the propagation characteristics of radio waves into the multi-dimensional characteristics of multipath signals realizes the "precision, accuracy, and speed" of wireless network planning and optimization and significantly improves the quality of the GSM-R wireless network in the complex sections of crossing and merging lines and comprehensive hubs, as shown in Figure 3.89.

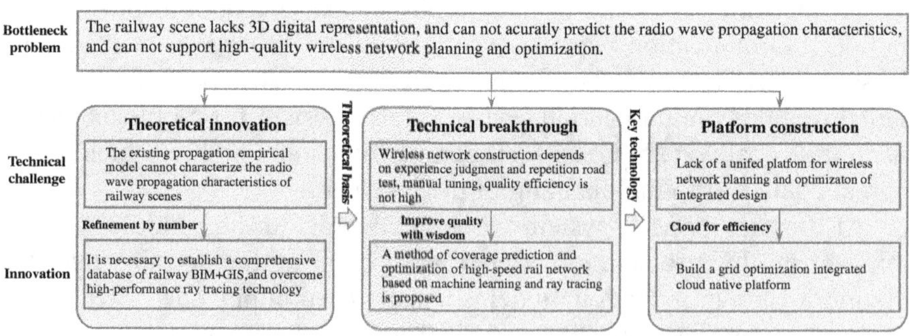

Figure 3.89. Wireless network planning and optimization.

(2) **Key technologies**

When adopting the high-precision intelligent network planning and optimization technology of railway mobile communication based on BIM+GIS, the core of direct concern is the reception level and the load-to-dry ratio (C/I) of the train onboard module. In the complex section of HSR, for a specific location point of a railway in the region, under a certain frequency point, ray tracing simulation prediction is carried out for the base station of a useful signal source and the base station of an interfering signal source, firstly, to ensure that the coverage level is required, and at the same time, the width of the overlapping area and switching position is considered, and the same-frequency interference and neighboring-frequency interference C/I value is calculated, so as to determine whether it meets the index requirements in the specification.

(a) *GSM-R ray tracing technology based on BIM+GIS*: Adopting multi-origin heterogeneous cross-domain scene generation technology, the raster, vector, tilt photography, satellite pictures, and other heterogeneous data from BIM, GIS, CAD, and other "multi-origin" systems are fused and optimized in a unified three-dimensional space. The "cross-domain scene" basic database covering typical railway structures such as electric traction frame, road rift, station, bridge and building from outdoor to indoor is fused and optimized in a unified three-dimensional space to build a "cross-domain scene" basic database covering from macroscopic to microscopic and from outdoor to indoor. On this basis, according to the specific requirements of wireless network planning and optimization, spatial and temporal segmentation and tailoring are carried out to dynamically generate a multi-granularity 3D model of the target scene, and a BIM+GIS converged database of the railway scene is built.

The ray tracing technique based on scattering center sparsification and high-performance computation is adopted to accurately characterize the embedded relationship between the structure and the propagation mechanism, and the average error of the measured power in many HSR such as Beijing–Hsiunggung and Beijing–Shenzhen is less than 3 dB, and the root-mean-square error is reduced by about 78% compared with the Hata model, COST231 model, and other commonly used models in existing wireless network planning. This provides a key common technology to improve wireless network planning accuracy in complex railway scenarios, as shown in Figure 3.90.

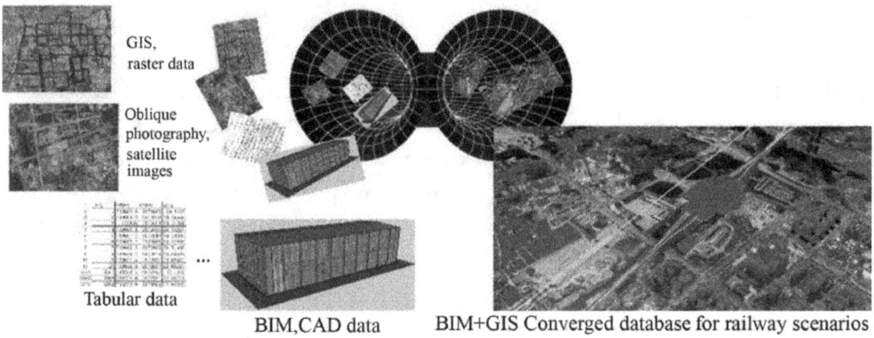

Figure 3.90. Constructing a BIM+GIS converged database for railway scenarios using multi-source heterogeneous cross-domain scenario generation technology.

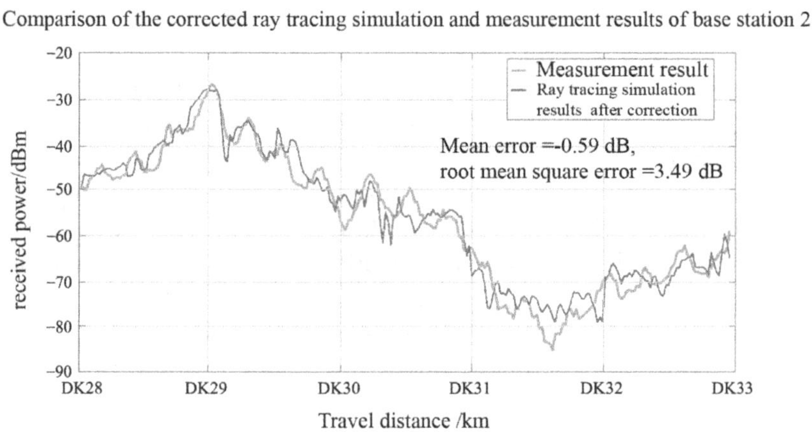

Figure 3.91. Comparison of high-precision railway wireless network coverage prediction results with measurement results.

(b) *A method for railway network coverage prediction and planning optimization based on machine learning and ray tracing*: Aiming at the possible errors between the initial prediction value of ray-tracing simulation and the actual measurement value, network coverage prediction and planning optimization are realized by adopting the fusion scheme of artificial intelligence and ray-tracing based on the field measurement data and the machine learning reinforcement training technology of neural network fusion.

Figure 3.91 shows the predicted results versus the measured power in the Beijing-Xiongan HSR scenario. The average error between the prediction results and the measured power is −0.59 dB. This is in agreement with the average error of −0.59 dB and a root-mean-square error of 3.49 dB, which is better than other existing models and methods.

Using the corrected ray-tracing simulator, the target scene is simulated and computed to obtain the multipath information of each measurement point, which includes the complex electric field, propagation path, propagation delay, and large and small structures related to the path for each path, so as to realize the accurate received field strength prediction. If the predicted value does not meet the requirements of field strength coverage and same-neighbor frequency interference index, it will gradually meet the index requirements by continuously adjusting the base station deployment point and antenna angle setting.

At present, the high-precision intelligent network planning and optimization technology of railway mobile communication based on BIM+GIS has been successfully applied, which improves the quality of network delivery, reduces the frequency of post-adjustment and ride-testing, saves labor and testing costs, and achieves significant economic and social benefits.

3.4.6.2.2 Accurate measurement of railway signal project based on BeiDou system

For the first time, BeiDou satellite navigation and positioning technology was applied to railway signal engineering measurement and positioning, and a positioning system architecture based on the BeiDou navigation system was established, which improved the measurement accuracy, measurement efficiency, and installation accuracy of railway signal equipment.

(1) **Function and composition**

It realizes the transmission of satellite differential signals of the BeiDou navigation system and the chance continuous operation reference station (CORS) system and transmits the corrected satellite differential signals to the handheld RTK mobile station through the connection between the handheld terminal and the server of the cloud platform, so as to make the RTK mobile station equipped with the ability of on-site measurement and fixed measurement and sampling, and has the functions of tilt measurement and corrective positioning, batch importing and exporting of measurement data, measurement of track segments, limit measurement and fusion measurement, etc., which improves the accuracy of positioning and measurement.

(2) **Key technologies**

(a) *Multi-parameter synchronized measurement technology of track sections based on BeiDou system*

The corrected Beidou differential signal can be used for railway signal equipment to coordinate acquisition and sampling. The mobile measuring device installs the RTK measuring instrument on both sides of the mobile device and the laser rangefinder on the central slide in the center of the mobile device so that various measuring data, such as the length of rail section, limit of signal trackside equipment, and coordinates of signal equipment, can be acquired simultaneously during the line measurement, as shown in Figures 3.92 and 3.93.

(b) *RTK measurement device tilt measurement corrective positioning technology*

By adding a gravity acceleration electronic bubble sensor and an attitude sensor to the iRTK2 receiver control panel, the center coordinates of the

Figure 3.92. BeiDou RTK measurement of track section lengths.

Figure 3.93. Signal equipment limit measurements.

signaling device points can be accurately acquired during alignment bar tilt measurements.

(c) *Techniques for the application of the seven parameters of the regional coordinate system*
Apply the seven parameters of the regional coordinate system to the project and convert the parameters of the measurement area, and modify the residual values of the measurement accuracy to get the minimum plane residuals and elevation residuals so as to achieve the minimum cumulative error data of the signaling equipment measurement.

3.4.6.2.3 Simulation of indoor and outdoor equipment simulation tests

Equipment simulation debugging is very important in railway signal construction. Railway signal debugging is faced with new station construction, old station reconstruction, new equipment, and other working conditions. The use of computers, communications, power electronics, and other technologies to build a simulation test system, replacing the traditional simulation equipment, to achieve intelligent engineering testing, test operation with less manpower, on-site handling and portability, and functional enhancement of the integration of comprehensive equipment debugging, has become an important direction of development.

(1) **Function and composition**
The main functions of the railway signaling indoor and outdoor equipment simulation test system include: multi-mode combination based signaling indoor equipment debugging, integrated indoor rutting machine circuit debugging, interlocking mode based outdoor signaling machine debugging, combined outdoor rutting equipment simulation debugging, and independent outdoor track circuit equipment simulation debugging. Among them, the operation test of the upper computer equipment and the configuration of signaling machine types are shown in Figures 3.94 and 3.95, respectively.

(a) *Multi-mode combination based signaling indoor equipment debugging technology*
The analog commissioning systems and equipment for commissioning indoor signaling equipment realize the electrification of analog boards and the modularization of electronic components.

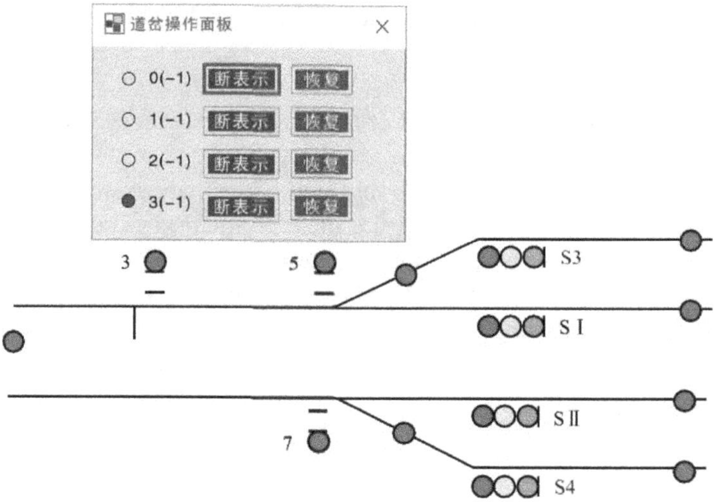

Figure 3.94. Upper computer equipment operation test.

(b) *Integrated indoor rutting machine circuit debugging technology*
Considering that the traditional debugging mode test is not thorough, the safety factor is low, the display is not intuitive, and so on, it is important to realize the indoor turnout start circuit and indicate the circuit synchronous debugging.

(c) *Outdoor signaling commissioning technology based on interlocking mode*
Aiming at the problem that the outdoor signaling machine can only use a 220 V power supply to carry out a single transmission test when the indoor circuit does not have the debugging conditions, which are incomplete and have a low safety coefficient, it realizes the debugging of the outdoor signaling machine based on the interlocking mode.

(d) *Simulated commissioning technology for combined outdoor rutting equipment*
Owing to the constraints of the indoor circuit conditions, only a single motor outdoor debugging is possible, the debugging efficiency is low, and the phenomenon of the simulation of indoor rutting machine devices is incomplete.

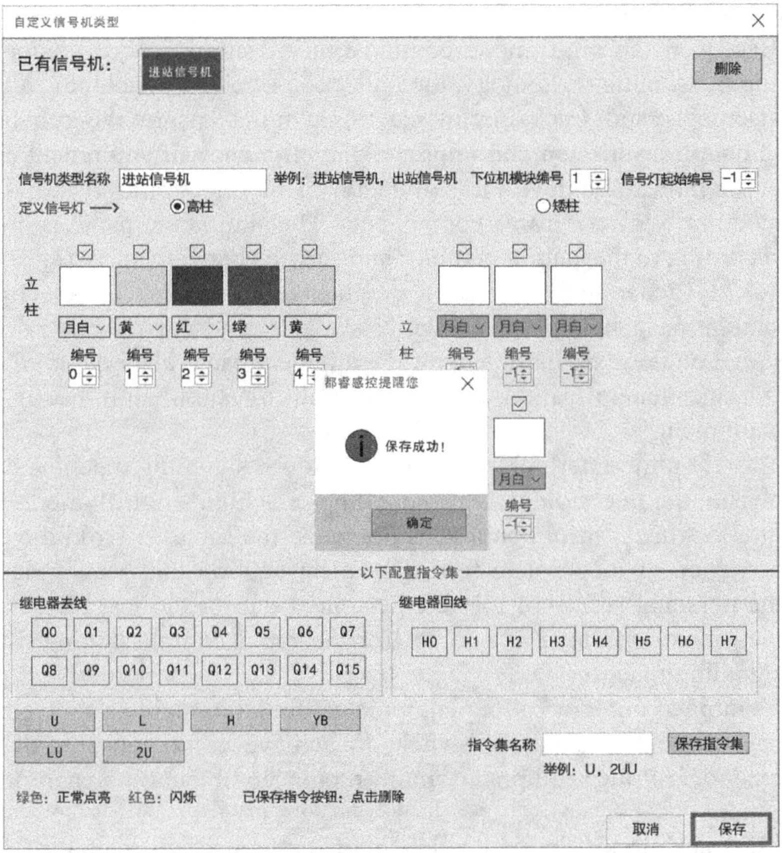

Figure 3.95. Signal type configuration.

(e) *Independent outdoor track circuit equipment simulation and commissioning technology*

The traditional outdoor track circuit equipment debugging is limited by indoor circuit conditions, especially for the transformation of the station, can only be at the skylight point or the opening point of the equipment debugging, occupies a lot of time, and is not conducive to the normal opening of the equipment operation. Therefore, the development of independent indoor track circuit simulation equipment is needed, which can individually debug the outdoor track circuit equipment without relying on indoor circuits and equipment.

(2) **Key technologies**

(i) Based on the multi-mode combination of signal indoor equipment commissioning technology, the collection of signal machines, rutting machines, and track circuit electronic units ensure the reliability of data transmission and improves the efficiency of equipment commissioning to achieve the simulation of the electronic disk and the modularity of electronic components. The simulation panel is drawn according to the station graphics and can be changed at any time, thus quickly generating the station graphics and realizing the electronic simulation panel. Controllable, cascadable, and expandable railway signal outdoor simulation control equipment can realize the miniaturization, generalization, and expansion of simulation debugging equipment.

(ii) Based on the interlocking mode of outdoor signaling machine commissioning technology, the signaling machine control mode and interlocking control commands are made the same to avoid the phenomenon of incomplete testing or electrocution that may endanger the personal safety of the test personnel during the test of a single lamp position alone and to realize the rapid commissioning of the signaling machine.

(iii) Combined outdoor rutting equipment simulation debugging technology, through the matrix switch, to achieve combined control and outdoor rutting equipment multi-motor combination synchronous action.

(iv) The track circuit simulation debugging equipment, through the 25 Hz phase-sensitive signal phase difference measurement method and the ZPW2000 frequency shifting signal detection algorithm realizes the correctness verification of the outdoor track circuit channel and equipment and solves the problem that the outdoor track circuit equipment cannot be debugged in advance. The application of indoor and outdoor equipment simulation test technology improves the debugging efficiency and quality, reduces the workload of debugging personnel, saves labor costs, and shortens the project duration. Especially for the new indoor and outdoor signaling equipment added in the reconstruction and expansion project, the equipment can be debugged without relying on the operating equipment, and the debugging of equipment can be carried out outside the sunroof point, which reduces the influence of debugging the new equipment on the operation, and guarantees the safety of the operation.

Figure 3.96. Railway track plate automatic punching device.

3.4.6.2.4 Automatic drilling for equipment installation

(1) **Automatic punching device for railway track plates**
A railway track plate automatic punching device (Figure 3.96) can solve the problems of low precision, high cost, and low efficiency of traditional punching and semi-automatic mechanical punching methods. Through the structural research, control research, and power supply research of the track plate punching device, combined with the dimensions of the track plate, the mechanical frame is made. Through the analysis of the punching process and punching standards, the control module is selected to realize the three-dimensional spatial movement of the impact drill, which improves the accuracy of capacitor, transponder, and other signaling equipment for the construction of punching holes in the track plate, reduces the labor force, and improves the operation and construction efficiency.

(2) **Automatic drilling equipment for crash barriers**
The automatic crash wall drilling machine (Figure 3.97) has a compact, modular design that can be quickly disassembled and brought online for easy assembly. The machine can travel on 150 mm and 200 mm thick crash barriers and can be used for drilling applications up to 100 mm in diameter. It reduces the labor intensity of field workers and improves operational efficiency.

3.4.6.2.5 Intelligent wire welding robot for signaling rooms

The intelligent wire welding robot for the HSR signal room can replace manual work to complete the automatic wire welding function of multiple

Figure 3.97. Automatic drilling equipment for crash barriers.

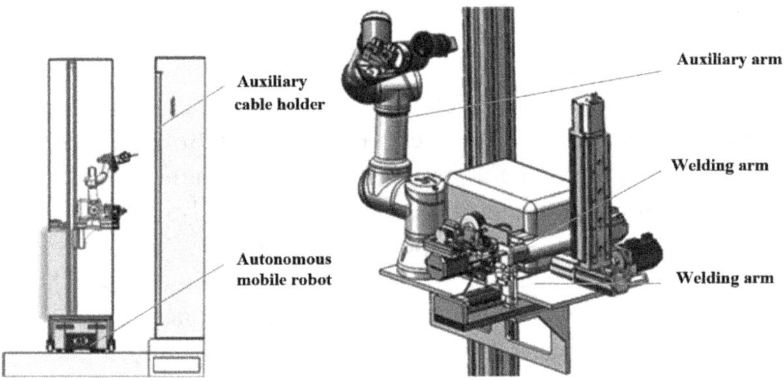

Figure 3.98. Schematic structure of intelligent wire welding robot in signal room.

combination cabinet cables in a railway signal room. It can automatically classify, grab, and weld the cable, improve construction quality and efficiency, reduce the labor intensity of the operators, improve the reliability of the construction quality, realize the high degree of automation and intelligence, and at the same time, ensure the safety of traffic and operation.

The intelligent wire welding robot for the signal room of HSR includes an auxiliary wire frame and wire welding robot system, in which the wire welding robot system includes the upper part of the intelligent wire welding robot and the bottom part of an autonomous mobile robot, and the two parts can work as a whole or can be separated and work independently. The structure of the intelligent wire welding robot in the signal room is shown in Figures 3.98 and 3.99.

Figure 3.99. Intelligent wire welding robot in signal room.

3.4.6.2.6 Key technologies

The key technologies include a modular tandem robot design method for the combined use of autonomous mobile robots and wire welding robots, kinematic analysis of a new six-degree-of-freedom wire welding robot for the gripping, positioning, stripping, sleeving, and welding of flexible wires and cables, the design of six-degree-of-freedom parallel flexible hinged micromanipulator grippers for precision positioning and gripping of flexible objects, and machine vision and deep learning technologies for weld quality inspection.

The application of the intelligent wire welding robot in the signal room combines the autonomous navigation robot with the wire welding robot, realizing autonomous movement; intelligent grasping of cabinet cables, intelligent disconnection, and intelligent welding; intelligent detection of the quality of the welded joints, which ensures the quality of the welded wires; and replacing manual work, which improves the standard of the wire welding process.

3.4.6.3 *BIM-based four-electricity equipment management*

Through electronic tags, two-dimensional codes, and other means of IoT on the four power equipment contracts, factory inspection, admission acceptance, installation, commissioning, and other whole processes of data collection, the quality management of the four power equipment factory, installation, commissioning, and other equipment is implemented. The attributes of the data dynamic means of expansion are used to meet the needs of the various phases of data collection to achieve the four power equipment information retrospective management for the formation of the four power professionals in line with the needs of operation and maintenance. Provide informationized management means for the formation of equipment history ledgers, such as “one pole, one file” and “one unit, one file,” which meet the requirements of operation and maintenance in the four-electricity profession.

1. System functions and components

The four power equipment management system includes order management, delivery management, on-site inspection management, equipment installation (commissioning) management, equipment attribute management, and other functions. The functional framework is shown in Figure 3.100.

(i) *Order management*: The construction unit is in accordance with the actual equipment needs of the site, in batches to the supplier under the order list, agreed on the name of the equipment, model, quantity and equipment arrival time, arrival location, delivery method, etc.

(ii) *Shipment Management*: After confirming the order online, the supplier of the four electric equipment will produce and manufacture the equipment according to the order agreement and provide the information on the equipment manufacturer, production batch, production number, etc. on the

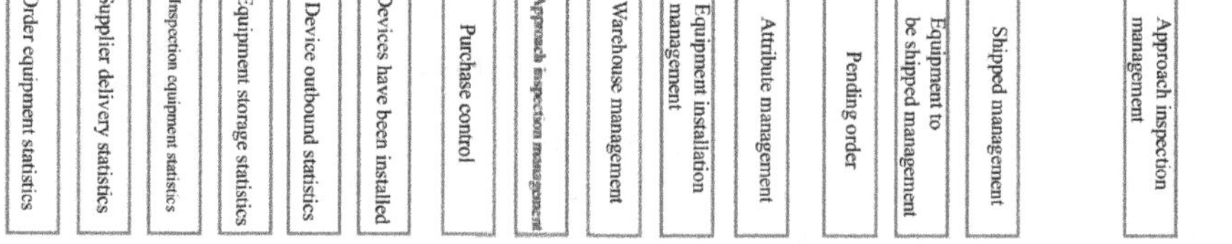

Figure 3.100. Functional framework diagram of four electricity equipment management system.

system, upload the certificate of conformity, inspection report, and other attachments, and generate a two-dimensional code, which will be pasted on the equipment to be shipped along with the equipment.

(iii) *Field inspection management*: After the equipment is sent to the site or to the warehouse, the construction unit (construction unit) and the supervision unit scans the two-dimensional code of the equipment to complete the equipment inspection, and in the system, the time of the inspection, the supervisor, whether qualified, and the disposition of the situation record is entered.

(iv) *Equipment installation (debugging) management*: After the equipment installation is completed, by scanning the QR code, the equipment installation site and location are selected, and supplementary installation (adjustment) information is added, such as the equipment installation time, the person in charge of the installation, and the supervisor.

(v) *Equipment attribute management*: After the equipment installation is completed, the construction unit supplements the attributes of the four power equipment according to the requirements of the operation and maintenance stage through the Excel table importing method to form the equipment ledger with the attributes of the four power equipment in each stage.

2. Key technologies

Through the IoT means to realize the four power equipment contract, factory inspection, admission acceptance, installation, commissioning and other whole process of data collection and information traceability management, the formation of the four power professional is set in line with the needs of operation and maintenance of the equipment resume ledger for the operation and maintenance of the maintenance of the basic data. It also realizes "one map" integrated management of BIM+GIS at each work point of the four power projects, which provides information means for the management of the whole life cycle of the four power equipment, as shown in Figures 3.101 and 3.102.

Through the application of BIM-based four-electricity equipment management technology, we can effectively improve the quality and efficiency of management, reduce project costs, enhance the level of informationization of management, and facilitate the query and traceability of information.

Figure 3.101. BIM and GIS scene modeling interface diagram.

Figure 3.102. Electronic sand table.

3.4.6.4 *Factory pre-wiring technology for computer room cables*

The computer room cable factory pre-wiring technology completes the pre-wiring in advance in the factory and optimizes the construction process through the plugging process, research and development of supporting pre-installation, and pre-wiring equipment and materials, deepening the design of the construction drawings or even modifying the engineering design drawings, so as to achieve the reduction of the number of workers in on-site construction and the construction period and, at the same time, to maintain the quality of workmanship and further reduce the risk of construction accidents.

1. Factory-based prefabricated workshops

The factory pre-wiring workshop is equipped with a cabinet wiring area, a pre-wiring operation platform area, and a material storage area. The cabinet wiring area should meet at least one standard four-strand station pre-wiring space, as shown in Figure 3.103. The pre-wiring operation platform area can complete cable prefabrication without cabinets. The material storage area is used for cable prefabrication without cabinets. The cabinet wiring operation platform can be used to complete the cable prefabrication without cabinets, as shown in Figure 3.104; the material storage area is used for storage of materials before pre-wiring and packing of finished products after pre-wiring.

Figure 3.103. Cabinet wiring area.

Figure 3.104. Prewiring operation platform area.

2. BIM technology-assisted factory material scheduling

BIM technology is applied to deepen the design of construction drawings, carry out virtual construction of indoor construction, accurately realize inter-frame wiring and cabinet wiring, establish a factory model library and realize platform management, form programmed modeling methods applicable to factory construction, guide the customization and scheduling of pre-wired equipment and materials according to the results of BIM output, and guide the factory construction and on-site construction, as shown in Figures 3.105 and 3.106.

3. Factory construction

Factory construction includes three parts: pre-assembly construction, packing and transportation of materials, and on-site construction.

(i) *Pre-wiring construction process*: Construction preparation → indoor cabinet installation → inter-row wiring channel connection → cable placement → cable tying and wiring → conduction and verification → cable extraction, organization and packing → clean up the scene.

(ii) *Packing and transportation of materials*: No need to dismantle after factory prefabrication, individually packaged after completion of prefabrication and transported to the site.

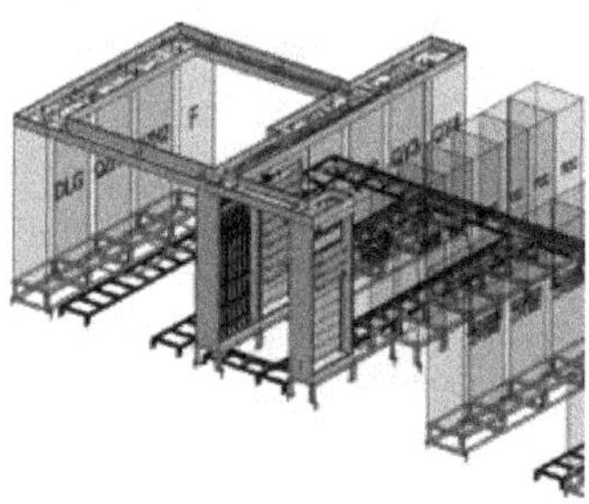

Figure 3.105. BIM modeling of equipment and wiring in a computer room.

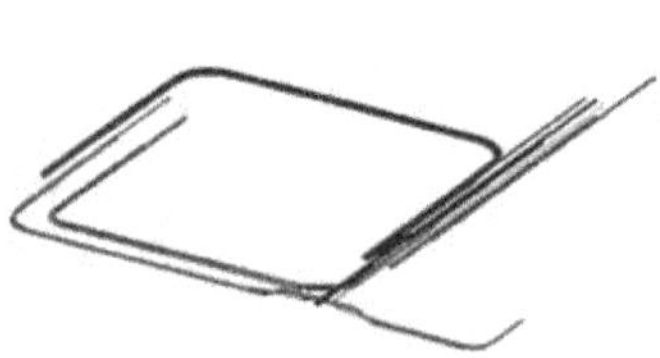

Figure 3.106. Interframe wiring separated from BIM model for cut-to-length production.

Figure 3.107. Simulation of cable laying on the low-voltage side of the main transformer.

(iii) *On-site construction*: pre-wired inter-shelf cables are installed in the on-site cabinets, and then the pre-wired cables are plugged in; cabinets that have been both pre-wired and not pre-wired are installed and wired on-site; equipment is installed in the cabinets (e.g., relays, transformers, transmitters, receivers, and analog network boxes); and equipment debugging is performed.

4. BIM technology-based cable simulation laying path optimization
Through BIM technology, we create a three-dimensional cable model, accurately connect the cable starting point equipment and end point equipment, and complete the initial model of cable laying, as shown in Figure 3.107. Combined with the requirements of the four power professions, the simulation of laying path optimization is carried out for each professional cable direction, placement location, space state, etc., before the cable laying construction planning for the four power professions cable laying details to meet the four power professions cable laying construction process standards, avoid cable crossover, and ensure that the cable bending radius and the specific professional requirements of the full reserve are met, so as to achieve the purpose of improving the level of the cable laying process.

3.4.6.5 *Intelligent wiring process robots*

Intelligent wiring process robots solve the problems of traditional manual production of signal wires, such as time-consuming and highly technical level requirements for personnel, greater uncertainty in the process, and lower quality and accuracy. It can realize signal wire handle production mechanization and automation in order to ensure uniformity in the process

Figure 3.108. Intelligent wiring process robot.

Figure 3.109. Main components of an automatic wiring machine.

of wire handling and quality standards, and at the same time, the efficiency of the work is greatly improved.

The intelligent wiring process robot consists of three workstations, namely, automatic wiring machine, automatic tying machine, and automatic winding machine, to form a production line, as shown in Figure 3.108.

After the operator sets up the wiring template program through the operation panel according to the type of wire handle to be made, the wiring process robot will start the assembly line work of automatic wiring, automatic tying, and automatic winding.

The automatic wiring machine mainly consists of a wire feeding, cutting, and stripping machine, a wiring robot arm, and a positioning robot arm, as shown in Figure 3.109.

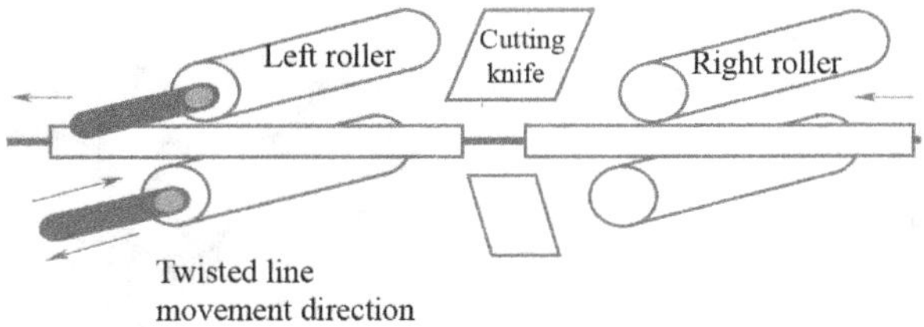

Figure 3.110. Wire feeder and cutter stripper structure diagram.

Figure 3.111. Wire feed process.

(1) **Wire feed:** The wire on the cable tray is first fed into the wire feeding, cutting, and stripping machine (Figure 3.110). According to the preset program, the cable is cut to length, ring longitudinal cutting, stripping of the two ends of the cable, and then sent out from the feeding port, and the cable skin waste can be automatically recycled, as shown in Figure 3.111.

(2) **Cabling:** In fixed-length cable delivery, a wiring robotic arm waits at the delivery point. The cable will be automatically set into the wiring robotic arm of the cable ferrule and then by the wiring robotic arm to the cable wiring start point. The robotic arm is positioned according to the preset procedures in the start position, and the cable is delivered to the location of the robotic arm and will be fixed by the positioning arm to clamp. Next, following the preset trajectory and the guidance of the positioning stakes (stainless-steel columns) on the cabling model, the cabling robot arm places the cables onto the cabling template and is held in place by the clamps (blue plastic posts) on the template, as shown in Figure 3.112.

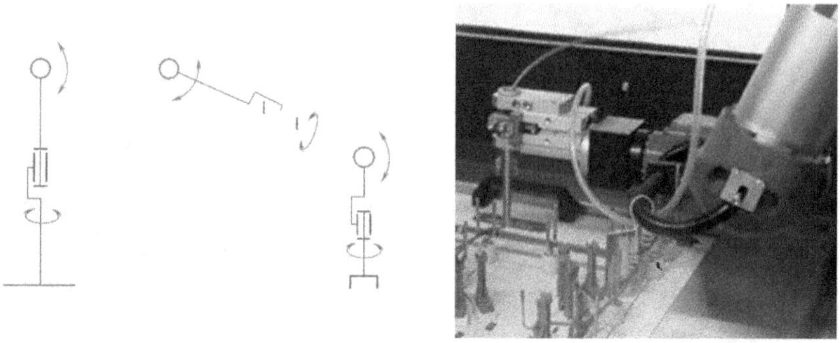

Figure 3.112. Wiring robot arm structure schematic and physical drawings.

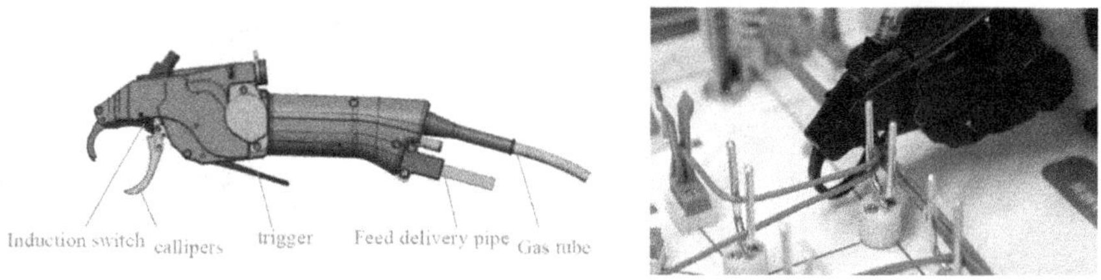

Figure 3.113. Automatic taping machine structure schematic and physical drawings.

(3) **Automatic taping:** The tying work is completed by the automatic tying machine, which can realize the tying tape blowing into the tying gun head automatically, carry out the tying, tying tape tail cutting, and waste recycling functions, as shown in Figure 3.113.

(4) **Automatic winding:** The winding work is completed by the automatic winding machine. The operator aligns the end of a single cable with the center hole of the winder and inserts it into a fixed position. Then, the automatic winding machine is started to automatically wind the ring wire around the outside of the core wire. Once the winding is completed, the operator manually cuts off the ring wire after bending it into a circular shape, and the end of the cable is set with the casing on the cover to complete the production of a single cable winding. Then, the operator continues with the next cable winding until all the wire handle end production is completed. The automatic winding machine's automatic centering accuracy of up to 0.05–0.15 mm, with a resonance suppression function, can ensure that the copper wire coil is tight, beautiful, and in line with the

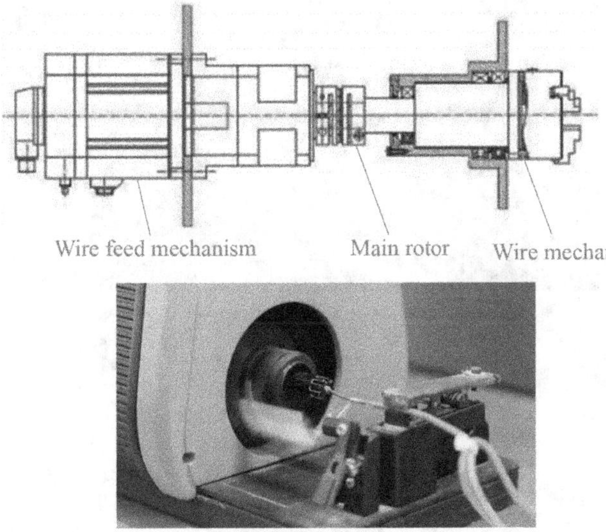

Figure 3.114. Schematic structure and physical drawing of automatic winding machine.

process quality requirements, while solving the traditional manual winding problems of low efficiency, being prone to error, and unstable quality. Automatic winding machine structure schematic diagram and physical figure are shown in Figure 3.114.

3.4.6.6 *Automated laying of optical cables*

The optical cable line is the carrier of electric energy transmission and signal transmission and is one of the most important projects of the four electric integrated systems. The stable and reliable operation of optical cable lines are closely related to the safety of train operations. As the optical cable line has obvious advantages over the overhead line, such as small influence from external factors, a small maintenance and management workload, hidden engineering, no influence on the environment, and personal safety, the use of the all-cable laying method for power lines has become the main technical standard of high-speed rail. However, the optical cable laying quantity is large and the laying environment is complicated, making the optical cable laying process and ensuring its quality one of the major difficulties in the construction of four power integration projects.

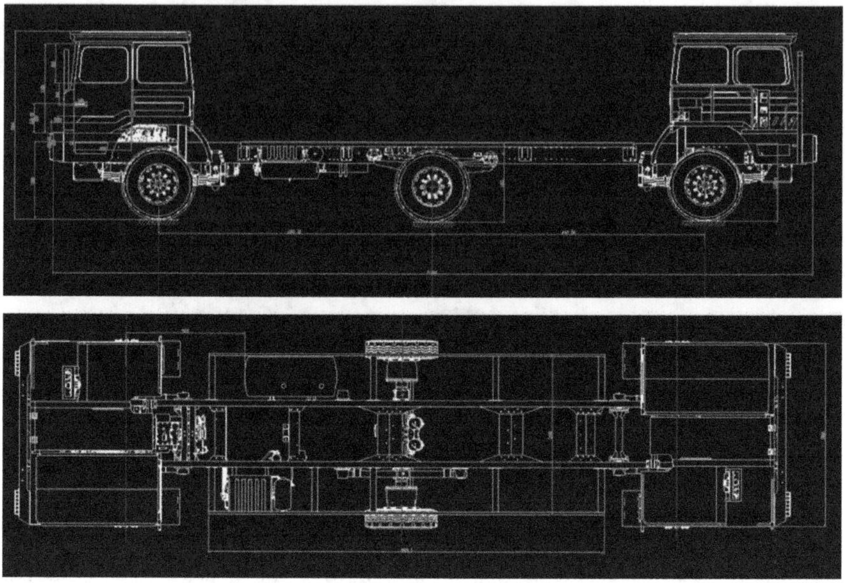

Figure 3.115. Side and top views of automatic laying operation of two-way cable for HSR.

A high-speed rail two-way cable automatic laying operation vehicle solves the problem of high requirements on the working surface conditions in the process of high-speed rail cable laying, reduces labor costs, and improves the cable laying rate and process quality. The two-way automatic cable laying operation vehicle for high-speed railroads consists of an operation vehicle body and power equipment, a horizontal cable payoff frame system, an automatic cable laying traction system, a cable laying guiding system, and other top-loading equipment. This offers two-way maneuvering, automatic laying traction, automatic control of the laying speed, and other functions, which are mainly applied to the automatic laying of power cables for high-speed railroad bridges and tunnels, as shown in Figure 3.115.

High-speed rail bi-directional cable automatic laying operation vehicle technology has been practically applied in the construction of the Xicheng Hitchhiker Specialized Four Electricity Integration Project, which has ensured the construction quality, shortened the construction

Figure 3.116. Automatic cable laying operation diagram for HSR bidirectional cables.

time, reduced the number of construction personnel, and saved the construction cost in the construction of the whole line, as shown in Figure 3.116.

3.5 BIM-Based Railway Project Management Platform

The BIM-based railway project management platform (hereinafter referred to as the platform) is built according to the mode of "one portal, six systems, and *N* applications" by taking standardized management as the handhold, BIM+GIS technology as the core, cloud computing as the platform architecture, perception technology as the foundation, mobile internet as the transmission structure, and construction projects as the carrier. It establishes a three-dimensional spatial data system for railways to support collaboration and application, establishes a business management model with mutual calibration of progress, quality, and safety, realizes progress management with visualization and automatic deduction, safety risk management with automatic perception and prediction, and quality traceability management with continuous tracking of men, machines, and materials. It also improves the application of comprehensive analysis of engineering and construction data from multiple angles, dimensions, strengthens the management level of all elements such as "man, machine, material, method, and environment" on site, forms digital assets for completion and delivery, and realizes the transfer

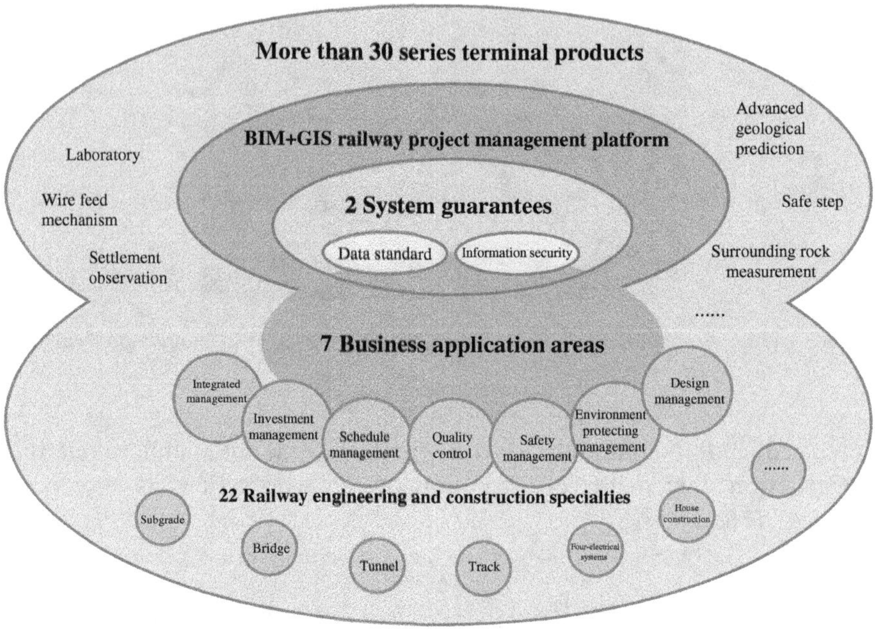

Figure 3.117. Functional system of the platform.

of digital assets from construction to operation. The functional system of the platform is shown in Figure 3.117.

3.5.1 *BIM-based design management*

3.5.1.1 *BIM-based data delivery and communication*

1. Functional composition

Railway design involves many specialties, is difficult to coordinate, and requires close cooperation among multiple (participates) participating units. The application of this module enables multiple units to work together, realizing seamless multi-stage connections and maximum information sharing. Through the component library to improve the design efficiency and through the model library to achieve the management of design results, BIM-based construction information data collection, archiving, data management, and achieving the standardization of the construction process data and standardized management to provide technical support for the smooth transfer of railway project data from the

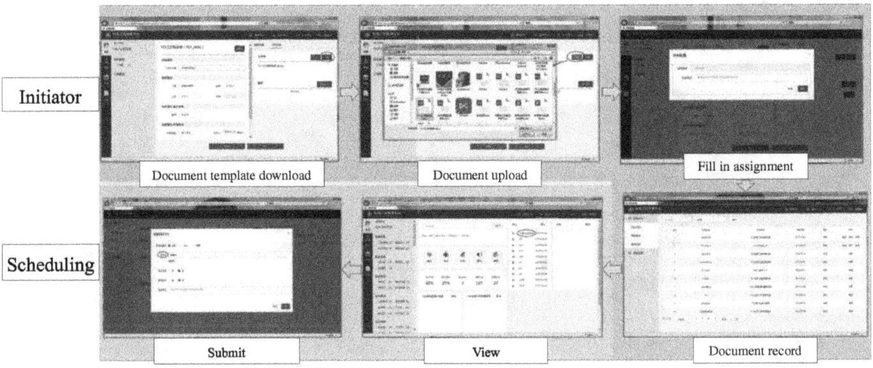

Figure 3.118. BIM-based data delivery and transmission flow.

construction stage to the operation and maintenance stage. BIM-based data delivery and transfer based on BIM contains five aspects: data description, data delivery, data storage, data organization, and data utilization. Among them, data description researches how to define the data through the attribute information of the documents. The attribute types include entity attributes, basic attributes, and business attributes. Information data delivery mainly studies the collection of data files, quality checking, and the dynamic generation of information catalogs according to document attributes. Information data storage mainly researches the storage methods of files and data, including database sub-tables, attribute dynamic expansion, and file calling interfaces. Data organization mainly studies the multi-dimensional classification of data, dynamic column classification, and statistical methods. Data utilization mainly studies document security solutions and document browsing queries, etc., of which document security solutions include physical security, flow security, and personnel authority control. The BIM-based data delivery and transfer process is shown in Figure 3.118.

2. Technical principles

Breakthrough technologies in cross-protocol data access and autonomous loading, Edge Computing gateway fieldbus integration, and Digital Twin construction information integration are implemented. Research is conducted on the multiple protocal fieldbus architecture for multiple protocols for the collection of typical equipment data in the digital construction process, realizing the effective collection and rapid processing of on-site

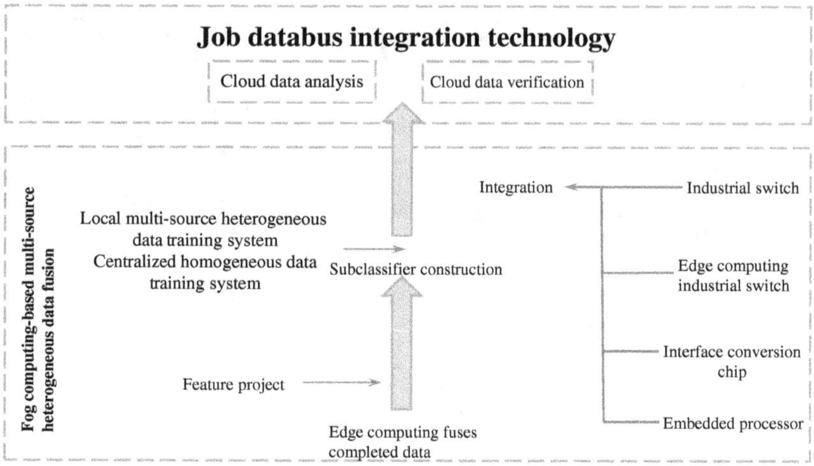

Figure 3.119. Multi-source data fusion delivery technology route.

perceptual data. Based on the effective information of the field edge computing gateway, study the layered information fusion strategy, design the open communication unified architecture, integrate the heterogeneous data from multiple sources and realize the flow transfer of the process data of multiple equipments, as shown in Figure 3.119.

3. Application effectiveness

It opens up the data transfer mode of design delivery, document bus, and basic data and realizes the offline process to online operation bus, electronic signature of engineering documents, terminal module automatic archiving, and pre-volume of engineering data. It makes full use of information technologies. Based on the standards of railway construction project data management, it addresses a series of problems faced by traditional project data management, such as lagging data collection, scattered storage, cumbersome organization, and difficult process supervision, enhances the authenticity, completeness, usability, and security of the project data, and improves the efficiency of collecting, integrating, and organizing the files of the completed project.

In addition, to address the problems of low automation collection rates caused by the non-unification of data sources and the low integration of data flow and data interaction in the railway construction process, research is required on layered information fusion strategies, the design of field bus architecture, open communication unified architecture for multi-protocol collection of typical equipment data in the digital construction process, and the integration of heterogeneous data from multiple sources. This can realize the effective collection and rapid processing of field sensing data and the flow transfer of process operation data in multiple equipment processes.

3.5.1.2 *BIM-based construction drawing review*

1. Functional composition

The management functions of the system cover five design stages: prefeasibility study, feasibility study, preliminary design, construction drawing design, and design changes. It realizes the management of consulting reports and weekly reports in each stage so that construction units, design units, and consulting units can work together through this system to improve work efficiency and promote the standardization of design work.

The BIM-based construction drawing review management system (Figure 3.120) realizes the formal review of BIM models and construction drawings of important and difficult projects and online construction drawing reviews, in addition to the management and maintenance of consulting reports, construction drawing management, and weekly reports of consulting work. It provides the functions of reviewing, tracking, and archiving the electronic documents of the design results in the feasibility study and preliminary design stage. In the stage of construction drawing

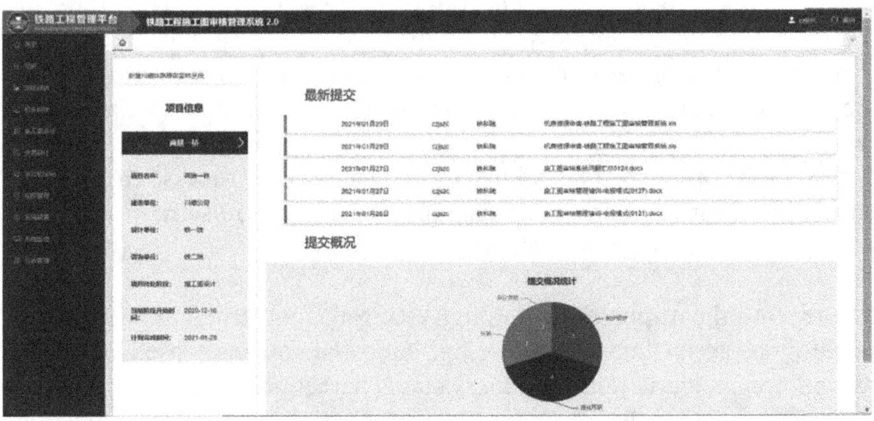

Figure 3.120. BIM-based construction drawing review management system interface.

design, it mainly provides functions such as planning the drawing supply, reviewing the process, tracking the progress of reviewing and important issues, inquiring about the implementation of reviewing opinions, and recording the reviewing process.

2. Technical principles
Based on BIM core graphics, it realizes BIM construction drawing review, BIM model preview, online annotation, review of attributes of the BIM model, and review of BIM model coding. By adopting the IFC standard to encode models and data, it can better match data with BIM models and effectively improve the delivery efficiency of construction data.

3. Application effectiveness
BIM-based construction drawing review improves the synergistic efficiency of construction, design, and consulting, strengthens the management of construction drawing review progress, records the review process and consulting opinions and implementation in the review process, realizes convenient query, traceability, and statistics, and has the ability of online review of construction drawings.

3.5.2 *BIM-based construction schedule management*

3.5.2.1 *BIM-based construction progress tracking*

Railway project construction is a long-duration, complex process influenced by many factors, and uncertainties are very high. Therefore, the traditional progress expression is difficult to accurately reflect the dynamic changes in the construction and the inherent relationship between the different construction tasks, and even more difficult to timely and effectively reflect the construction resources and project costs, so as to dynamically optimize and adjust the construction process. In order to ensure the normal course of events, it is necessary to use more effective construction management methods.

1. Functional composition: Through electronic construction logs, project scheduling, aerial image monitoring, and other means, based on image recognition, construction machinery status perception, and other means, it realizes comprehensive recognition and dynamic tracking of the construction progress of each specialty and each process, meets the management

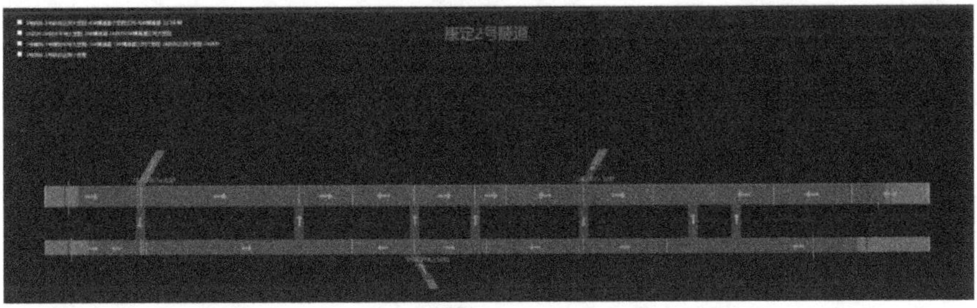

Figure 3.121. BIM-based 3D visualization of tunnel progress.

personnel's comprehensive mastery of the construction status of the workplace, and controls the construction progress by using the report output in conjunction with the graphic. The progress management and construction organization management, the collected data, compared with the construction progress plan, is combined with the BIM model to display the actual construction progress, accurately reflecting the current state of each part and achieving refined management. At the same time, the technical, safety, and quality logs are associated with the model to realize the expression of the progress status, which ensures that the construction management personnel can fully track the construction progress through the model. An example is shown in Figure 3.121.

2. Technical principles: The construction organization data update service receives design and construction organization information through a data interface and updates the railway engineering model in real time, driven by data, showing the latest construction plan, construction progress, design quantity, and other information. The above information is realized by using the JAVAEE system and the REST service architecture. The JAVAEE system simplifies and standardizes the development and deployment of multi-tier distributed application systems, which can significantly improve the system's portability, security, scalability, load balancing, and reusability. The REST service follows the principle of Create, Read, Update, Delete (CRUD), uses URLs to identify and locate resources, and completes the service through the standard action of the HTTP protocol. Through the standard action of the HTTP protocol to complete the relevant operations and processing, a unified interface to achieve communication between the client and the server. The dynamic update service of construction organization data is shown in Figure 3.122.

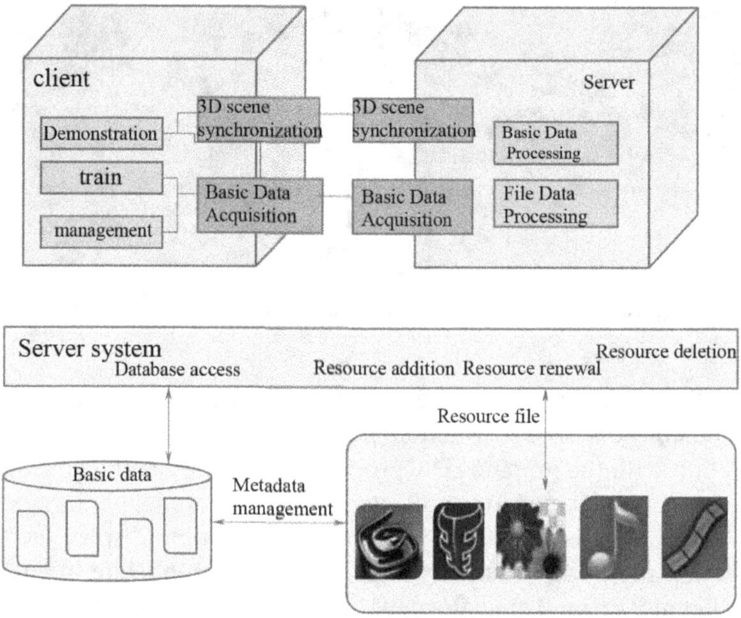

Figure 3.122. Construction organization data dynamic update service.

3. Application effectiveness: Using BIM technology, the whole construction process is simulated, and a reasonable construction plan is formulated. First, during the construction process, the IoT is used to collect the actual construction progress information on the site through the sensors and various types of terminal equipments arranged at the construction site. Second, it compares the construction plan with the actual progress and uses advanced algorithms to study the construction operation deduction and resource allocation plan adjustment. The interface of railway construction organization preparation and simulation is shown in Figure 3.123.

The progress projection function can realize the simulation of the construction period of key structures, such as bridges and tunnels, and predict whether the construction period will affect the next process according to the projection information and whether it can meet the requirements of the construction plan, so as to remind the construction management to carry out a reasonable allocation of resources to ensure that the corresponding structures can be completed in a timely.

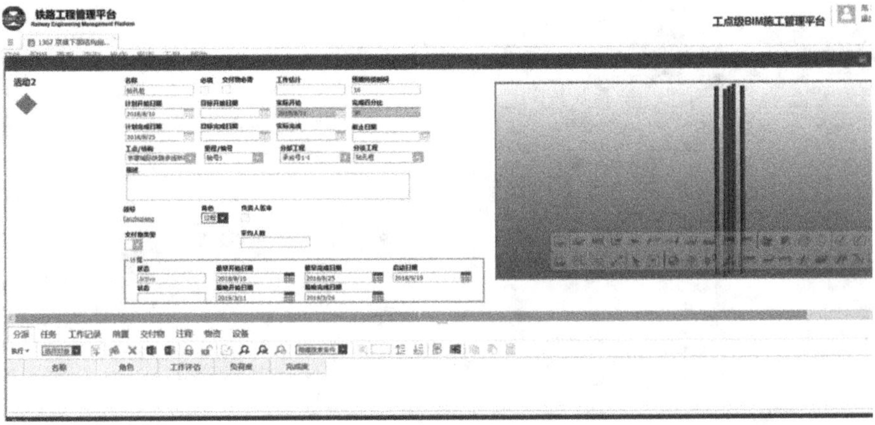

Figure 3.123. Railway construction organization preparation and simulation interface.

3.5.2.2 *BIM-based construction organization optimization*

1. Functional composition

Through the study of the coupling relationship between human, machine, material, method, environment, and other elements and on-site construction organization, construct the spatial and temporal projection model of construction progress driven by the information of key specialities; through the analysis of the deviation of the construction organization plan and the progress projection model, put forward the dynamic optimization technology of the construction organization plan with the constraints of the critical path to achieve the optimal allocation of resources and the effective control of the construction progress to guarantee the construction timetable.

The WBS of the project is carried out to meet the automatic association between the WBS and BIM model of the project, on the basis of which location-based planning and resource allocation are carried out, focusing on the analysis of the critical path, the study of progress deduction and red line alarm, the study of the closed-loop management of construction progress alarm, and finally the simulation of the research results. Thus, it realizes the simulation of construction operations, real-time tracking of construction operations, construction progress projection, and resource allocation plan adjustment. The construction organization simulation interface based on resource allocation is shown in Figure 3.124.

Figure 3.124. Resource allocation-based construction organization simulation interface.

2. Technical principles

According to the guiding construction organization of railway projects, the theoretical research on intelligent construction organization planning is carried out in combination with the actual situation on the site, construction organization plan management, progress tracking, progress projection, image progress, early warning, etc. By adopting particle swarm theory in combination with BIM lightweight technology, construction organization optimization algorithms are written for different working conditions of whether to set up transverse caverns, branch caverns, inclined shafts, and flat guides for tunnel projects. The optimal construction organization is loaded dynamically through data interaction.

Through BIM technology, the construction organization plan is simulated, unreasonable points in the construction organization plan are identified and optimized in time, and the optimized construction organization plan is used as the basis for the preparation of work orders. The work order is distributed to on-site construction personnel through a mobile application, and the construction progress information is collected in real time during the construction process through sensors and handheld devices. The collected construction progress information is used as the basis for construction simulation, thus improving the rationality of the construction organization plan. The algorithm principle of BIM-based construction organization optimization technology is shown in Figure 3.125.

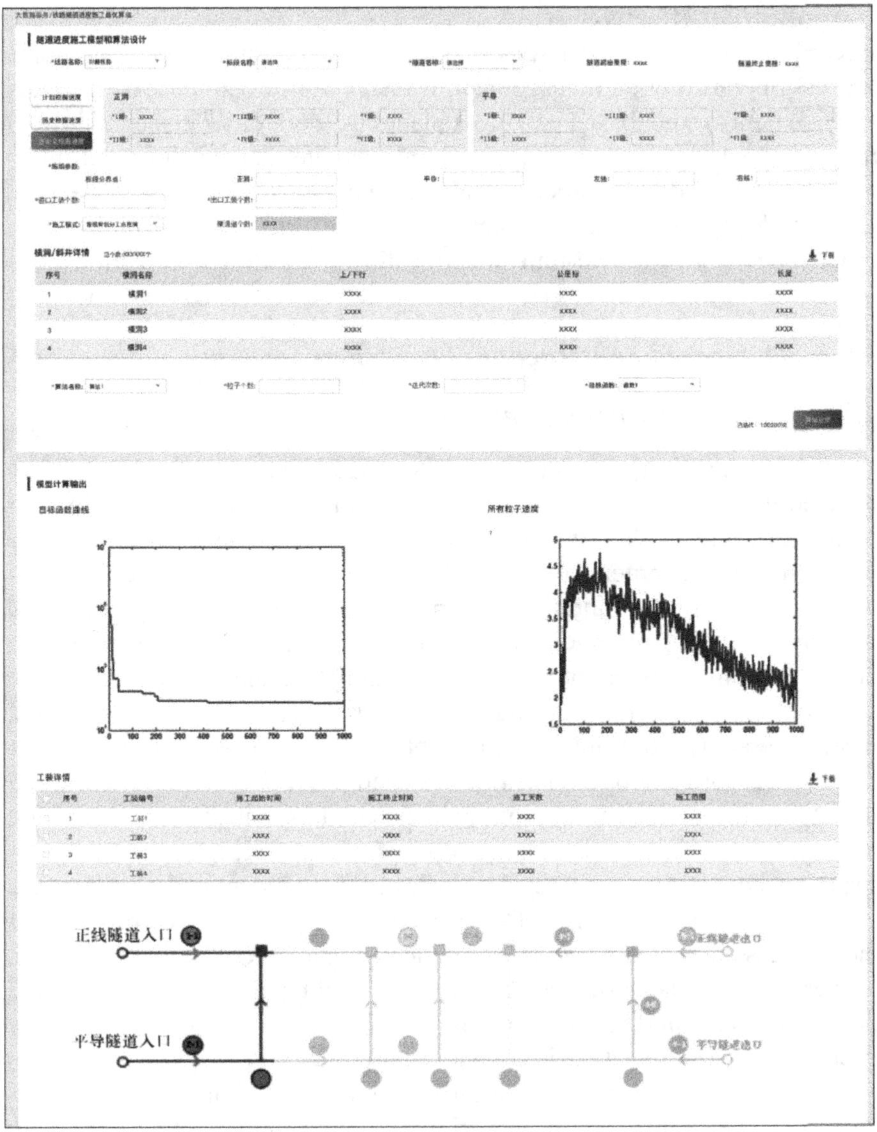

Figure 3.125. Principle of BIM-based construction organization optimization technology.

3. Application effectiveness

According to the characteristics of multi-party, continuous, and complex process of construction management, we have built a construction progress management system based on BIM's planning, work order dispatching, and progress tracking methods, which include on-site management and information entry, business process management and information collection, BIM modeling, and plan management, to meet the needs of multi-party, process development, real-time data updating and sharing, prediction, etc.

3.5.3 *BIM-based construction quality management*

3.5.3.1 *Cyclic process optimization*

For the construction of tunnels by drilling and blasting method in railway engineering, taking the smallest construction unit as the breakthrough point, the tunnel cycle process optimization system is developed to realize work order management for the first-line team in the construction operation — erection team, and the sequence of each construction process and the construction team are dispatched through the mobile applications. The person in charge of each process will arrange personnel to enter the site to carry out construction according to the work order and, at the same time, record the actual start and end times of each construction process; automatically record the operating time of each process, such as excavation, initial support, and spray mixing; provide an alarm mechanism according to the recorded time; and reasonably adjust the operating time of the construction equipments and the work-flow. This dynamically pushes the cycle time of each process to the terminal system to view the progress of the process in real time and automatically calculates the time savings, refines the material efficiency, safety, and quality indexes, optimizes the production plan, and realizes the fine management of the work point. The cycle process optimization is shown in Figure 3.126.

The tunnel cycle process optimization system calculates the reasonable time of each process through statistics so as to adjust the work plan. At the same time, it is able to check the completion of each process in real time, carry out key management on unreasonable situations such as overtime completion, and replace the teams with low working efficiency so as to improve the working efficiency. Based on the idea of "modulus-driven, axis-face synergy," the system realizes smooth logistics, multi-point synergy of human and machine, and information sharing, which ensures

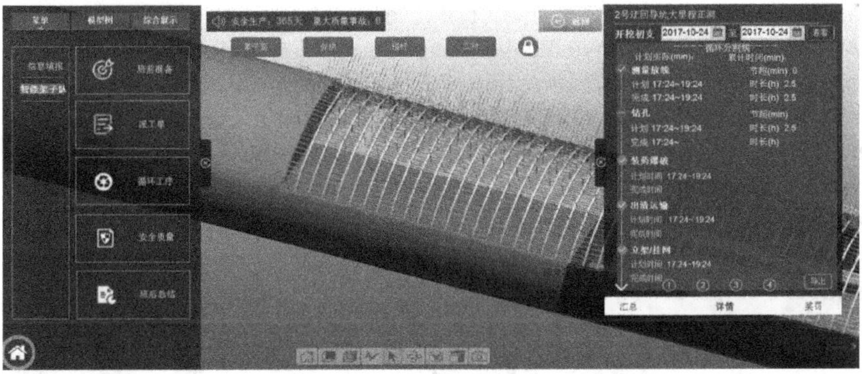

Figure 3.126. Cycle process optimization.

efficient advancement and docking of the construction of the underground stations and tunnels of the Beijing-Zhangjiakou HSR (13 working surfaces) and solves the problem of the multi-point construction organization of the single-channel.

3.5.3.2 *Section quality management*

In terms of tunnel section quality inspection, the monitoring is carried out with the help of informatization means to address the on-site quality control issues, such as over-under-excavation of the tunnel, thickness of the initial support, flatness of the initial support, thickness of the second lining, and headroom of the second lining. It uses 3D scanning to obtain the point cloud data of tunnel excavation, initial support, and second lining section, generating the actual contour scanning point cloud model of the tunnel, fusing the point cloud model with the BIM model, realizing the detection of the contour in the key process of tunnel construction, and combining the deepened model and the section monitoring data to realize the comparison between the design and the actual construction parameters so as to strengthen the quality control of the tunnel cross-section profile.

The section quality inspection system realizes the analysis of over and under excavation, thickness inspection, and flatness inspection of tunnel section quality. Through the fusion of point cloud data and BIM models, the over- and under-excavation position and volume of each tunnel section are calculated, and the analysis results are uploaded to the system in real time for early warning and control. Combined with the monitoring data, it can be used to regulate the amount of reserved deformation or optimize the blasting process so as to control the over-excavation volume and reduce the construction cost. At the same time, it can also establish three-dimensional models of initial support and lining thickness, accurately locate the area of insufficient thickness, carry out control before and during the incident, and also provide data support for the material preparation of the mixing station. After scanning the initial support, it can accurately assess the levelness of the initial support of the paragraph, quickly locate the unqualified area, and guide the construction remediation. It can also eliminate the hidden quality problems caused by the unevenness of the initial support, such as lining dehollowing or waterproofing slab rupture, and further improve the quality management of the tunnel section. The quality management of a tunnel section is shown in Figure 3.127.

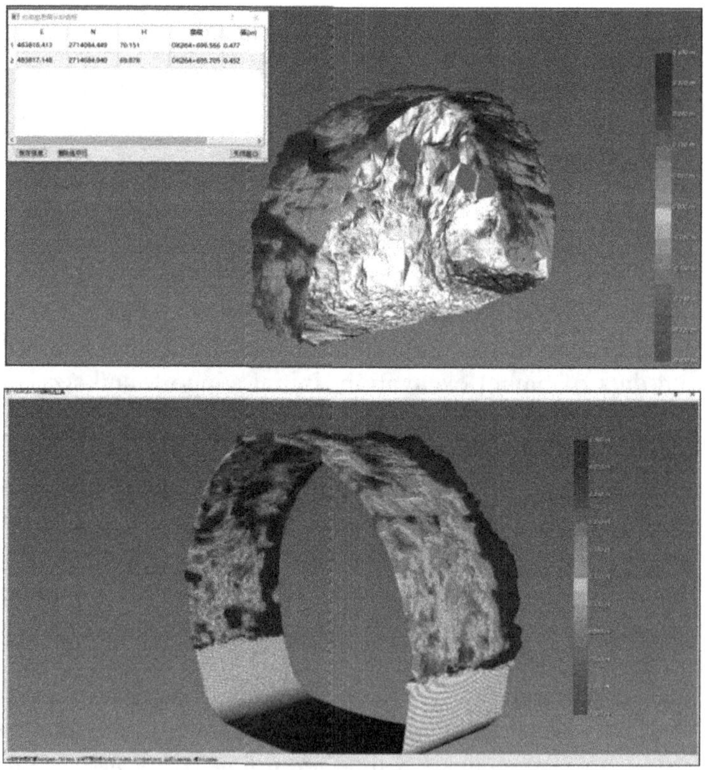

Figure 3.127. Section quality management.

3.5.3.3 *Three-dimensional technical presentation*

Relying on the visualization characteristics of BIM technology, combined with the construction process and method, the construction technical delivery library of key processes of various specialties is established by means of animation, simulation, and virtual reality, i.e., three-dimensional work instructions for the training of construction units. In order to intuitively guide the construction and avoid construction personnel's misunderstanding of the construction drawings, we rely on the established three-dimensional model, on-site simulation of key equipments, or special section installation and construction according to the construction demand, which helps construction forethought, discover program deficiencies, familiarize with the construction process, and improve the quality of construction. For example, the construction simulation of two different construction methods of tunneling is carried out in Tsinghua Park Tunnel and Badaling Tunnel.

1. **BIM-based construction simulation of Tsinghua Park Tunnel:** The shield of Tsinghua Park Tunnel crosses important structures, municipal roads, and pipelines along the tunnel route, and the areas it passes through are all primary risk source zones. Through the BIM virtual construction and simulation technology, it focuses on the simulation of the key processes of the starting shaft and the complex nodes of the starting section, which ensures the safety of structures, roads, and pipelines along the tunnel route during the tunneling process of the shield. With the characteristics and advantages of 3D visualization of BIM models, dynamic simulation of the construction process before and after the starting shaft of Tsinghua Park Tunnel 2-A is carried out to realize the application of simulation drills of the design scheme, construction technology briefing, and other processes to have complete control over the risk and optimization of the construction of complex nodes of the shield shaft and starting section and to realize the smooth boring of the shield tunnel. The simulation construction of shield tunnel starting shaft and starting section complex nodes of Tsinghua Park Tunnel is shown in Figure 3.128.

2. **BIM-based Badaling Tunnel construction simulation:** The largest span section of the Badaling Tunnel has an excavation width of 32.7 m. The construction plan was extremely complex, and after many discussions, it was decided to adopt the "talkative" construction method for the

Figure 3.128. Complex node simulation of shield starting shaft and starting section of Tsinghua Park Tunnel.

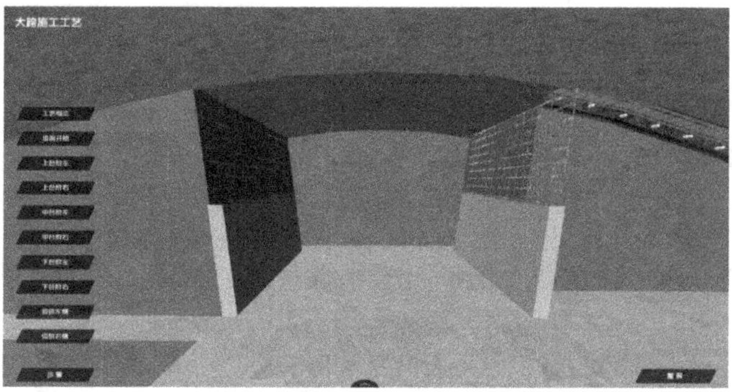

Figure 3.129. Badaling Tunnel Span construction simulation.

oversized section, to use anchor cable and anchor rods to support the perimeter rock of the arch roof (the key perimeter rock) in advance, and to simulate the construction process by using BIM technology. BIM technology is used to simulate the construction process, realizing a simple expression of the complex process and allowing users to fully understand the construction process. The simulated construction process of the large span of Badaling Tunnel is shown in Figure 3.129.

3.5.3.4 *Virtual assembly of bridges*

Using BIM section factory simulation technology, the industrialized construction of precast pipe piles and assembled bridge piers is realized, and the quality of concrete construction is improved. In order to carry out the integrated construction management of bridge assembly for the Guba Special Bridge, we propose a BIM+ precast assembled bridge pier construction plan applicable to HSR projects and carry out the simulation of the whole process of the construction process for the prefabricated assembled piers connected by sleeve grouting. The BIM+ prefabricated assembled bridge pier construction program for the HSR project simulates the whole construction process of the prefabricated assembled bridge pier with sleeve grouting connection and overcomes the key issues of large volume and large section assembling of railway bridge piers. This achieves the goal of simple assembling and fast construction and avoids the collision of prefabricated components in the process of lifting and installing so as to guide the construction. Through the BIM scene design, simulation of the assembled pier body, pier cap, and the bridge deck as a whole prefabricated cable groove prefabricated work process, the construction and installation efficiencies are greatly improved.

Because of the high production efficiency of integrated construction of assembled bridges, pre-stressed pile construction efficiency increased by 50%, and prefabricated assembled pier efficiency increased by two times. A large amount of on-site work “moved” into the prefabrication yard, reducing the risk of construction. In contrast to cast-in-place pier and girder construction, there were no slurry emissions, reducing the amount of on-site concrete poured on the site, saving energy, and protecting the environment.

BIM+RFID+IoT technology is incorporated into the bridge assembly production management, and RFID tags are implanted into prefabricated components to realize the quality tracking and tracing, from raw material purchasing, production, and transportation to installation, by collecting information throughout the whole life cycle of component design, production, inspection, transportation, assembly, and operation management, as shown in Figure 3.130.

3.5.3.5 *Customer service detail design and construction*

Using BIM to simulate passenger flow in the station, combined with optimizing the layout of customer service equipments, the travel experience of passengers in the station can be enhanced. Through accurate simulation and analysis of passenger flow in HSR stations for ticketing, entering, waiting, and exiting using BIM technology, deepened design by

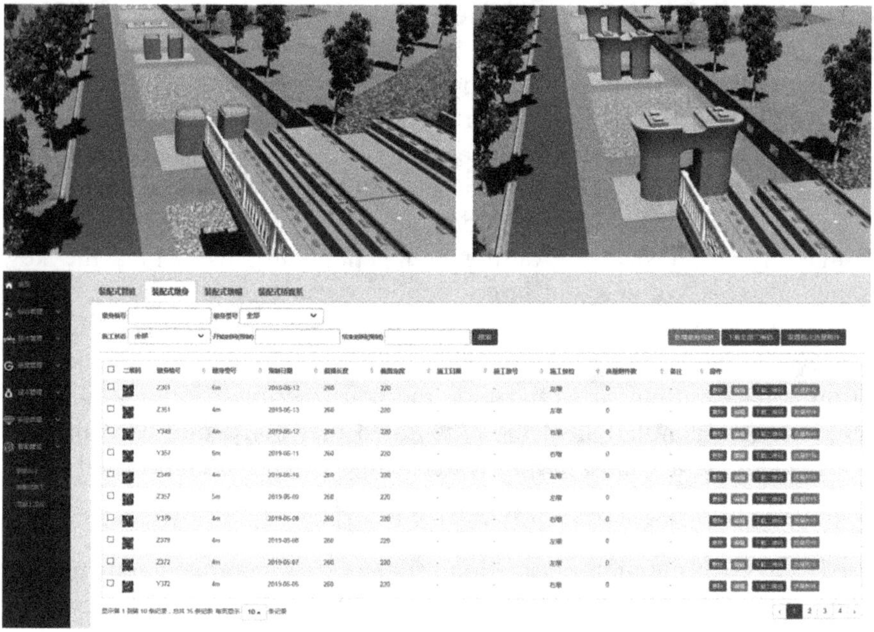

Figure 3.130. Integrated management of assembled bridge construction.

combining the distribution of functional areas in the station and the design layout of customer service terminal equipments in each functional area (Figure 3.131), we can achieve the optimally matching and integrating the allocation of functional space in the station, the number of customer service equipment installations, and their distribution with the path of passenger flow. The efficiency of customer service equipments in different functional areas of the station is maximized to better meet the refined demand for service information of passengers in the process of purchasing tickets, entering the station, waiting for the train, and exiting the station, and to enhance the travel experience of passengers at the station.

BIM+GIS simulation technology is used to simulate the sunlight in the station, analyze the influence of sunlight on the ticket gate equipments in the station, and provide a plan to avoid the influence of light on the ticket gate equipments, as shown in Figure 3.132. Sunlight shining through the station glass curtain wall directly on the ticket gate in the station will reduce the face recognition performance of the equipments,

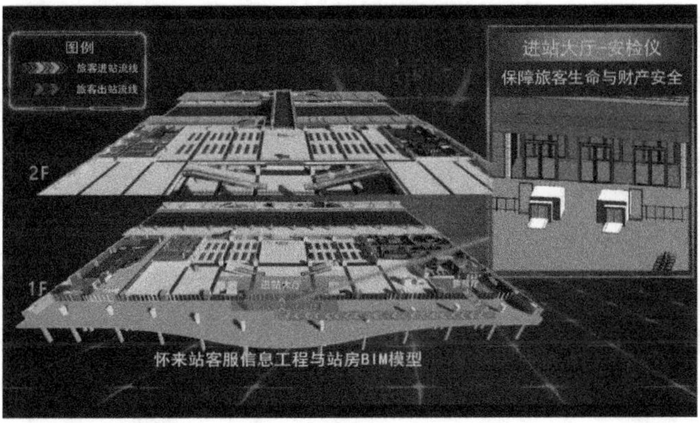

Figure 3.131. BIM simulation of passenger flow to optimize the layout of customer service equipments.

Figure 3.132. BIM-based simulation of daylighting at Beijing-Zhangjiakou Xiahuayuan North Station.

which can cause the problem of passengers not being able to check their tickets, which will in turn bring inconvenience to the passengers. The sunlight analysis based on BIM+GIS technology can simulate the light condition of the station at all times and provide a reference basis for the station operation unit to formulate a plan to avoid the influence of light on the ticket gate equipments.

Aiming at the complex structure of the information equipment room, multi-disciplinary intersection, huge number of equipments, complicated

Figure 3.133. Beijing-Zhangjiakou HSR BIM+VR information server room management system.

specifications, large number of pipelines, tight indoor space, difficult construction, etc., the research and development application of the Beijing-Zhangjiakou HSR BIM+VR information equipment room management system (Figure 3.133) based on the AR for the first-view angle control. Before the construction work of the information equipment room project, construction technicians can carry out a three-dimensional simulation of the overall layout, construction technology, work process, and work method based on this system so that they can have a deep understanding of the construction plan, and solve the difficulties of each stage of the work process at the site by comparing and analyzing the feasibility of different construction plans.

3.5.3.6 *One map for construction management based on BIM+GIS*

The comprehensive use of BIM, GIS, intelligent IoT (i.e., AI + IoT, or AIoT), and other technologies integrated with engineering construction technologies is employed to build a map system for railway engineering construction management (Figure 3.134). Through the integration of railway engineering line topography, railway structures model, various types of thematic information, construction business data, and other multi-dimensional information, we build a closed-loop system for the whole railway construction process, including engineering survey, design, construction, acceptance, safety, quality, and supervision of the

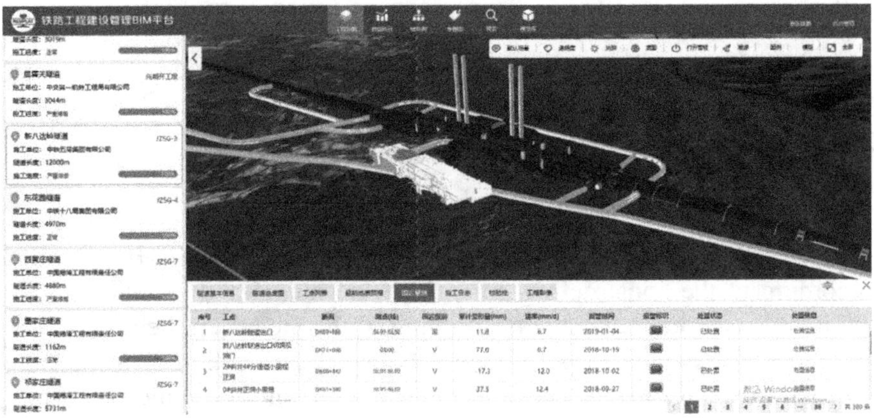

Figure 3.134. One map for construction management based on BIM+GIS.

construction process. We build a closed-loop system of life traceability; realize the fine and intelligent management of progress, quality, safety, and investment in the construction process; finally form a BIM model as a data carrier; integrate the design and construction process information; improve the quality and safety of the project; realize the fine management of the project in the construction stage; deliver the BIM-based integrated maintenance management system in the form of 3D digital assets; realize the development of the BIM-based integrated maintenance management system; and realize the transfer of digital assets from the construction process to the operation process of HSR.

The system establishes three BIM models of HSR by adopting a unified BIM standard design of railways from the design source and realizes non-destructive transmission of information from the design stage to the construction stage through the design delivery system of the engineering platform. By collecting the data of the construction stage and combining the BIM model and 2D design results, the system completes the virtual simulation of the construction plan for the key projects. By providing BIM+GIS services, the system realizes progress management with visualization and automatic deduction, safety risk management with automatic perception and prejudgment, traceability management of the quality of man, machine, and material, etc., which improves the quality and safety of the projects. It realizes the fine management of the project, makes the complex construction process transparent and visualized, and promotes railway construction from informationization and digitization to intelligence.

3.5.4 *BIM-based construction safety management*

3.5.4.1 *Security risk source management*

The relative spatial location of the risk sources are uniformly coded according to the construction mileage of the tunnel and associated with the BIM model to realize the positioning of the model. In the tunnel construction at Tsinghua Park, the shield tunneling direction is from large mileage to small mileage. For the same work point, the same risk types, according to the shield machine boring direction of the sorting number, through the entity structure based on the code, can be unified in the coding of risk sources. According to the coding instructions for the type of risk sources and the method of association between the model and coding, the risk sources of the shield tunnel can be effectively identified and located. The relevant professionals will assess and determine the risk sources and categorize them into relevant risk source levels. According to the distribution of environmental hazards in the Tsinghua Park Tunnel route and the surrounding area, they are unified and integrated in the appropriate location, as shown in Figure 3.135. During the construction process, the risk source identification can be continuously improved, and new hazard sources can be added according to the construction progress.

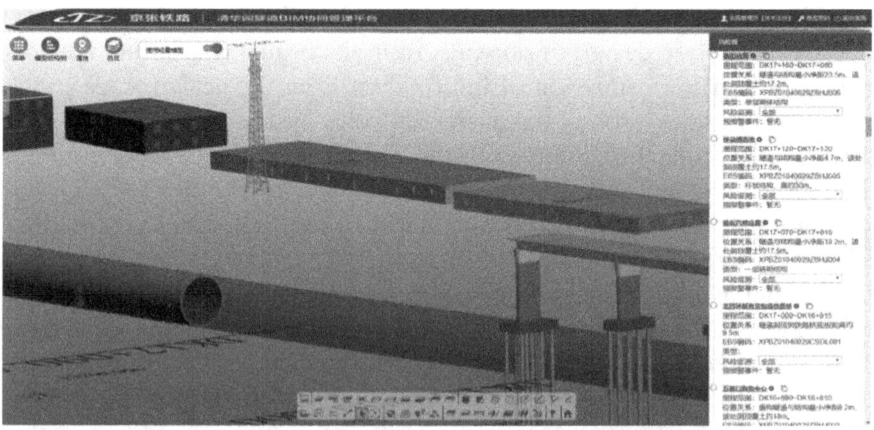

Figure 3.135. BIM-based security risk source management.

The risk source management function listed in detail under the current risk source status, technical indicators, construction disturbance range, early warning index, and overall risk assessment status can effectively realize the risk early warning, with the ring as the base of the warning. Combined with the monitoring and measurement data and dynamic changes in the curve, if the value is more than the threshold value in a certain shield machine construction mileage (ring number) outside the distance, the early warning notice will facilitate the warning program to be determined in advance. According to the emergency plan formulated in advance, proper preparation of personnel, equipments, and rescue materials are ensured to realize the timely warning and forecasting of risk sources and improve the site risk control capabilities.

3.5.4.2 *Engineering line transportation scheduling management*

In view of the fact that the railway engineering laying operation is in a state where monitoring and protection are performed mainly through human defense and supplemented by technical defense, the transportation production has been inefficient for a long time, and the safety risk is high. There are problems such as unplanned and black-planned construction. The research and development of the railway engineering line transportation scheduling management system (Figure 3.136), the establishment of the track-laying stage of the engineering line transportation scheduling command management information system, the application of construction and traffic plan management, wireless scheduling commands, traffic monitoring, plan monitoring, video monitoring, safety protection zone alarms, can assist in solving the problem of coordinating the organization of traffic and construction to reduce the people–vehicle and vehicle–vehicle collision. It can also promote the level of informatization of scheduling and command of engineering line management units.

Using the design drawings to generate the visual expression of railway stations, interval equipments, and other construction elements according to the day shift and dynamically generated safety protection areas and construction work areas, relying on Beidou positioning, locomotive proximity alarms, and other technologies, we can obtain a real-time visual display of the location of rolling stock, speed, and other information in the virtual line space. As a result, it is easy to grasp the real-time operating status of locomotives, track-laying, trains, construction

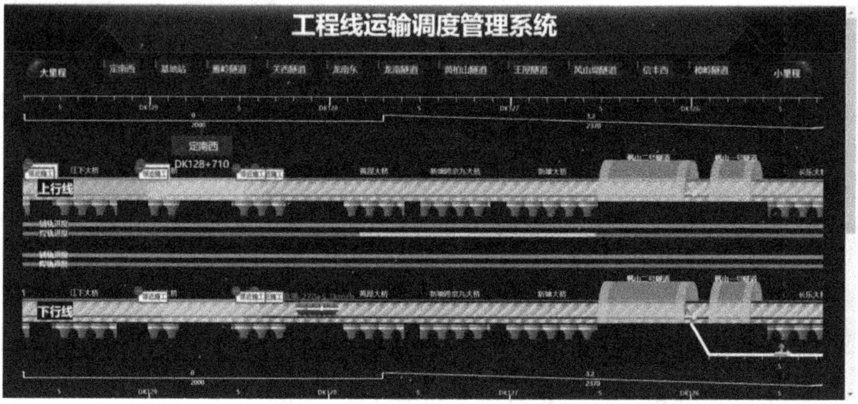

Figure 3.136. Project line transportation scheduling management system.

workers, etc. It establishes a three-level warning mechanism to further enhance on-site construction safety. Through on-site construction progress information, it drives the construction map elements to change color, dynamically display the construction progress of track-laying, turnouts, and crossings, etc., so as to realize the goals of informatization of construction scheduling and commanding, real-time monitoring of train operation, visualization of construction safety management, and intelligent analysis.

3.5.4.3 *Management of advanced geological forecasting*

By integrating the information of long-distance, medium-distance, and short-distance overrun prediction, the overrun geological forecasting management system are developed (Figure 3.137). With the BIM and geological models used to display the progress of the heading face, the design of the surrounding rock grade, the scope of the detection, the detection risk, auxiliary measures, construction methods, and other information, we can dynamically build the model of the overrun geological mechanism in front of the construction of the heading face so that the site managers and construction personnel can understand the geological risk of the heading face construction in time. Seismic wave reflection, heading face sketch, comprehensive conclusion of geological prediction, and other exploration results are comprehensively presented and analyzed so that site management and construction personnel can timely acquire the geological risks in advance.

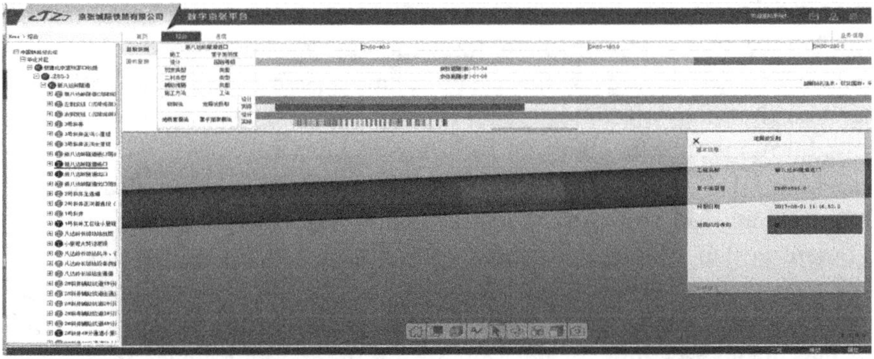

Figure 3.137. BIM-based-advanced geological prediction management system.

BIM-based over-the-horizon geological forecasting management is performed through the BIM model of the heading face with the color risk level labeling, the occurrence of warning points can be quickly located, and it can dynamically provide risk disposal measures, remind the construction unit of the existence of safety hazards, and as part of rapid resource deployment, develop targeted response plans to collaborate with construction management personnel, comprehensively master the safety situation at the construction site, in order to improve safety during the tunneling process.

3.5.5 *Development outlook*

The whole life cycle management of infrastructure involves different phases, different subjects, different levels, different fields, and other dimensions, as well as data and models through the whole life cycle management of all phases and dimensions. So, modeling and digital integration are at the core of design–build–operation, and the maintenance of infrastructure in the life cycle management will become an important direction.

The modal-digital integrated management technology of sensing, transmitting, knowing, and controlling railway infrastructure will undergo significant developments. Based on the full speciality expression of railway BIM, it integrates the application of IoT, Industrial Internet, and other technologies to realize a comprehensive perception of the state of bridges, tunnels, subgrades, tracks, and other structures. Based on the next-generation communication technology, it realizes the transmission of data from the edge to the data center, and based on the big data service platform, it realizes the aggregation of data, the association of digital models, the fusion analysis, and the optimization of decision-making, ultimately reaching the purpose of sensing, transmitting, knowing, and controlling the state of infrastructures.

With BIM as the main line, the railway intelligent construction of upstream and downstream industry chains has entered a new stage of efficient collaboration. It integrates the whole process of railway infrastructure design, production, construction, operation and maintenance, and dismantle, realizes new applications, such as forward collaborative design, visual assembly, project quantity estimation, cost accounting, and collision detection, integrates the whole industry chain of railway intelligent construction, and realizes the whole process and industry chain synergy.

The railway BIM model can be transferred to the factory without any loss, realizing the fine preparation of materials, construction, and preassembly and breaking the bottleneck of collaboration between the design information model and the factory automated production line. Integrating the construction progress plan into the BIM information model can intuitively and accurately reflect the construction process and guide the timely optimization of the comprehensive transportation systems. The intelligent construction of railways and related industries in the whole society will be more efficient. The synergy between intelligent railway construction and related industries in the whole society will be more efficient.

Review Questions

1. Briefly describe the meaning and main components of the intelligent construction of HSR.
2. What are the elements of an integrated air-space survey?
3. What are the advantages of BIM-based co-design over traditional design methods?
4. What are the core elements of a digital factory building?
5. Give examples of the application of intelligent construction in the construction of HSR.
6. How do you build a BIM-based project management platform?

Chapter 4

Intelligent Equipment for High-Speed Railways

The intelligent equipment for high-speed railways (HSR) integrates advanced technologies, such as cloud computing, the Internet of Things (IoT), mobile communication, big data, and artificial intelligence (AI), with HSR equipment. Through the integration of comprehensive perception, ubiquitous interconnection, active learning, and scientific decision-making, we have accomplished secure, efficient, and intelligent management of the entire life span of intelligent equipment for HSR. Among others, the key technologies of intelligent equipment primarily encompass HSR intelligent electric multiple units (EMUs), autonomous driving, intelligent traction substations, new-generation communication, and intelligent detection and monitoring.

Based on the architecture of the intelligent HSR system, intelligent equipment constitutes one of the three major sectors, which is further divided horizontally into four categories: mobile equipment, communication signals, traction power supply, and detection and monitoring. The field of mobile equipment comprises intelligent EMUs and intelligent integrated measurement EMUs. Intelligent EMUs consist mainly of innovative features such as intelligent monitoring, intelligent diagnosis, and intelligent services for EMUs. Intelligent integrated measurement EMUs, on the other hand, include intelligent detection equipment and intelligent detection data analysis as their innovative features. The field of communication signals encompasses various areas, including technology for signaling and communication. The signal area incorporates inventive elements, such as automatic control of station routes, train operation control, and automatic train operation (ATO). In contrast, the communication aspect includes forward-thinking concepts, such as intelligent communication carrier networks, intelligent mobile communication, and intelligent dispatching communication. The field of traction power supply encompasses intelligent traction power supply, etc., primarily including innovative features such as intelligent traction substations and intelligent power supply dispatching systems. The field of detection and monitoring involves intelligent detection and monitoring, specifically encompassing innovative technologies such as infrastructure detection and monitoring, natural disaster monitoring and early warning, perimeter intrusion intelligent monitoring, and environmental intelligent monitoring. The intelligent equipment sector can be categorized into four fields, six areas, and eighteen innovations. In the future, these innovative features can be developed to address emerging needs.

The intelligent equipment sector closely collaborates with intelligent construction and operation. Taking, for example, the application of detection and monitoring in the intelligent equipment sector, in the early stage, it is required to collect the design and construction data on lines, bridges, tunnels, tracks, and other infrastructures, including line slopes, curve elements, station turnouts, platform tracks, the allowable speeds of the lines, disaster monitoring, and video surveillance points. In the later stage, it is required to provide infrastructure and environmental detection data, monitoring data, alarm and prediction data on tracks, tunnels, subgrades, turnouts, and natural disasters for intelligent operation and maintenance applications in the intelligent operation sector. This includes the detection and monitoring data on the apparent state of tunnel lining, tunnel deformation, and longitudinal settlement, the vibration of lifting equipment inside the tunnel, the geometric dimensions of the lines, the status of track structure, contour clearance, turnout and rail flaw detection, wind speed, rainfall, snow depth, etc., and alarm data, such as lining faults, dynamic pressure alarms, structural safety parameter alarms, mortar layer separation, void track modes, typical safety hazards in the environment, and changes in surrounding environmental displacement.

The most typical applications for detection and monitoring in the intelligent equipment sector are modular driving and axial–plane collaborative systems. The models and data utilized in infrastructure detection and monitoring include the following:

(i) **Characterization model:** It involves a BIM model for completion and delivery, and corresponding changes are made to the BIM model after device modifications. It is further broken down into individual equipment components, including track fastenings for maintenance, bridge bearings, switch machines for signaling & telecommunication, track circuits, power supply overhead catenary system (OCS) struts, insulators, and other individual BIM models.

(ii) **Mechanism model:** It is based on the dynamic analysis of train–track–tunnel coupling, the theoretical analysis of crack damage propagation, image recognition, and the analysis of interactions between track (ballast bed) plate structure and various components. It also includes bridge vibration and lateral and longitudinal displacement analysis, tunnel stress structure analysis, the analysis of the tension of the OCS, the analysis of insulator contamination, and the analysis of tunnel aerodynamic effects and crosswind effects.

(iii) **Dynamic and static data:** Data on equipment for maintenance (lines, bridges and tunnels, subgrade), signaling & telecommunication (mainly signals), power supply (OCS, power transformation, and electric power), and discipline thematic maps (station wiring diagrams, line comprehensive diagrams, etc.); on-site inspection (visual inspection, small instrument detection, etc.), fixed monitoring (rail temperature and rail breakage, signal centralized monitoring, power supply 6C alarm, etc.), mobile detection (dynamic detection EMU, onboard line inspection instrument, portable monitor meter, signaling & telecommunication detection EMU, power supply 1C–6C detection defects, etc.), and other dynamic detection and monitoring data, as well as external meteorological information (such as regional historical rainfall, ground flash, extreme weather data, and real-time rainfall information). Using an integrated analysis of models and data, it provides early warning analysis for disasters, such as wind speed, rainfall, and snow depth. The system also analyzes typical environmental safety hazards and environmental displacement changes in the area, ensuring the operational safety of HSR.

4.1 Intelligent Electric Multiple Unit for HSR

The current global research on high-speed EMUs focuses on safety, economy, environmental protection, and intelligence. Ensuring the safe and reliable operation of high-speed EMUs is the primary goal of safety research. The aim of economic research is to decrease the total cost of ownership of high-speed EMUs and increase their operational efficiency. This includes developing modular and standardized designs for high-speed EMUs and their core components, creating low-cost and easily maintainable equipment and facilities, and implementing a highly mechanized and low-investment maintenance system. The objective of environmental protection research is to attain energy conservation, lowered noise levels, and completely emission-free operations. This is primarily achieved through technology upgrades that improve energy efficiency, train driving assistance, and hybrid and alternative energy sources. Additionally, noise reduction technology is also a key focus. The research on intelligence mainly includes intelligent monitoring, detection, and maintenance technologies for EMUs, such as self-monitoring, self-diagnosis, self-decision-making, onboard passenger information service platforms, seamless onboard door-to-door connection technology, and intelligent robots.

China Railway has successfully introduced, integrated, assimilated, and advanced high-speed EMU technology. While enhancing the speed capabilities of EMUs, China has achieved global dominance in high-speed EMU technology through major improvements in safety, reliability, economic efficiency, and convenience. In recent years, China Railway has spearheaded research on intelligent EMUs, incorporating comprehensive perception, ubiquitous interconnection, fusion processing, active learning, and scientific decision-making as their fundamental elements. The goal is to enable EMUs to have functions such as self-perception, self-operation, self-monitoring, self-diagnosis, self-decision-making, and self-protection while achieving the development objectives of creating safe, reliable, cost-effective, convenient, and comfortable high-speed EMUs.

Focusing on the complete life cycle management of EMUs, intelligent EMUs are divided into four key parts: an intelligent platform for basic technology support, intelligent R&D and manufacturing, intelligent maintenance and application, and intelligent operation, as shown in Table 4.1.

Table 4.1. Key technical composition of intelligent EMUs.

Intelligent technology support platform	
Intelligent R&D and manufacturing	Intelligent design
	Intelligent R&D
	Intelligent manufacturing
Intelligent application and maintenance	Virtual maintenance technology for intelligent EMUs
	Intelligent operation and maintenance management technology for power units
Intelligent operation	Automatic driving technology for EMUs
	Control technology of power unit
	Intelligent diagnosis and oriented security technology
	Intelligent service technology for EMUs
Communication information	On-board information IoT
	On-board Ethernet communication networks
	Ground cloud, on-board data storage computing centers
	Train-ground data communication networks based on 5G and quantum communication technology
	EMU Wi-Fi coverage based on advanced communication network
New technologies for EMUs	Permanent magnet motor technology
	Eddy current braking technology
	SiC (silicon carbide) technology
	On-board new energy system based on fuel cell technology
	3D printing technology

The intelligent technology support platform is a collection of technologies that offer common capabilities for intelligent R&D, manufacturing, application, maintenance, and operation. This includes new communication and information technologies, as well as new energy and materials for EMUs. Among the new communication and information technologies, the primary ones are the IoT for onboard information, onboard ethernet communication networks, ground cloud, onboard data storage computing centers, train–ground data communication networks based on 5G and quantum communication technology, and advanced communication networks providing EMU Wi-Fi coverage. New technologies of energy and materials for EMUs include onboard new energy systems based on fuel cell technology, permanent magnet motor technology, eddy current braking technology, silicon carbide (SiC) technology, and 3D printing technology.

Intelligent R&D and manufacturing mainly include technologies such as intelligent design, intelligent R&D, and intelligent manufacturing. Intelligent application and maintenance mainly include technologies such as virtual maintenance of intelligent EMUs and intelligent operation and maintenance. Intelligent operation mainly includes technologies such as intelligent control of EMUs, intelligent diagnosis and guidance safety, intelligent energy management and control, and intelligent services.

4.1.1 *Key technologies for intelligent design*

An intelligent EMU is a complex engineering system that encompasses multiple systems and disciplines. From a system perspective, the EMU includes subsystems such as the train body, end connection, bogie, traction high voltage, auxiliary power supply, air supply and braking, network control, passenger information, safety monitoring, interior environment control, water supply and drainage, hygiene, interior facilities, driving facilities, and train control. From a disciplinary perspective, the EMU includes disciplines such as transportation engineering, mechanical engineering, power and electrical engineering, electrical engineering, communications engineering, automatic control engineering, computer science engineering, chemical engineering, environmental science, safety science, and management.

As computer science, AI, and machine learning advance, computer-aided design technology continuously matures and becomes more sophisticated, playing an increasingly vital part in modern design. It progresses toward full-process autonomous intelligent design. Intelligent design utilizes computer-aided design technology and tools in conjunction with traditional R&D systems to replace human-led design tasks in both breadth and depth.

1. Key technologies of EMU design

China's HSR boast the world's largest operating scale, highest speeds, and most intricate operating conditions. Under complex operating conditions and dynamic application environments, high-speed EMUs encounter numerous obstacles in fundamental theory, application techniques, and engineering practices throughout their life span. The design of China's high-speed EMUs encompasses 10 critical technologies, namely pantograph–OCS coupling, traction and braking control, monitoring, safety, dynamics, aerodynamics, lightweight materials, service assurance design, and vibration and noise reduction, as shown in Figure 4.1.

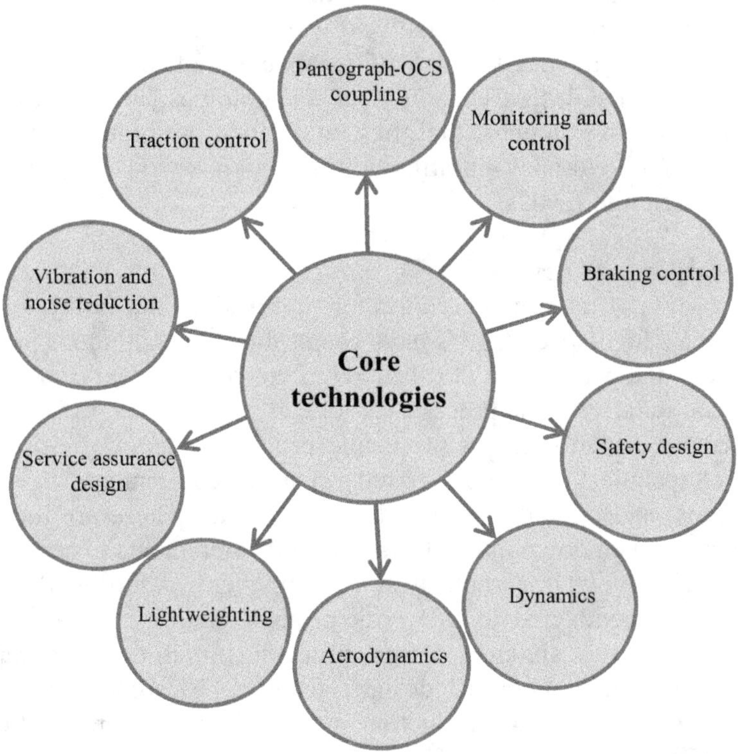

Figure 4.1. Key core technologies for high-speed EMU design.

Pantograph–OCS coupling refers to the relevant technology of coupling the EMU and power supply, which is closely related to the form and speed level of OCS. Traction control, braking control, and monitoring and control are related technologies for monitoring the train's own status and driving control. Safety design refers to related technologies such as structural safety, running safety, braking safety, electrical safety, fire safety design, and safety monitoring of EMUs. Dynamics refers to the integration of the running parts and the coupling of wheel–rail relationships in EMUs, which are closely related to the structure and parameter configuration of the running parts, wheel–rail coupling relationship, speed, etc. Aerodynamics is the technology that combines EMUs with infrastructure, including the environment and tunnels. It is intricately linked to the industrial design of the train's head form, body section and smoothness, and the connection method between train ends. Lightweighting technology involves weight management, structural integration design, and the use of lightweight materials for EMUs. Service assurance design encompasses technologies such as passenger compartment environment, facilities, and layout. Vibration and noise reduction involve modal vibration and matching, sound insulation, and noise reduction in EMUs.

2. Forward design process

In the process of introduction, digestion, absorption, and further innovation, China's high-speed EMUs have been subject to continuous practical application and evaluation. This has led to the development of a forward design and iterative cycle process.

Forward design involves proposing product system definitions and top-level technical indicators based on user requirements, laws and regulations, and operational boundary conditions. These are then progressively broken down into subsystems and components to create technical specifications for each system component. Additionally, R&D activities are conducted in line with product maturity. A new design product follows a six-step design process: initial decision-making, design planning, schematic design, technical design, construction design, and product trial production, verification, and confirmation, as shown in Table 4.2.

Table 4.2. Schematic diagram of the main flow during the design stage of high-speed EMUs.

Initial planning (decision-making)	Guidelines and intentions for scientific research project approval; Demonstration and analysis of scientific research project approval; Research project approval contract and implementation scheme; Technical conditions for bidding; Technical proposal for bidding; Technical clarification for contract negotiation; Final version of contract technical conditions.
Design planning	Preparation of a list of design inputs; Review of design inputs; Preparation of design plan; Review and approval of design plan; Top-level indicator decomposition.
Schematic design	Technical decomposition; Schematic design; Preparation of design plan; Review and approval of planning documents.
Technical design	Technical design; Technical design and review; User confirmation and approval.
Construction design	Construction design; Review of construction design; User confirmation and approval; Filing of construction drawing documents.
Product trial production, verification and confirmation	Trial production; Product testing; Design changes and reviews; Tracking test of service characteristics; Assessment of product application; Parsing check; Product finalization and qualification; Design changes and reviews; Finalization and mass production of products.

4.1.2 *Key technologies for intelligent manufacturing*

4.1.2.1 *Concept, characteristics, and architecture of intelligent manufacturing*

1. Concept of intelligent manufacturing
Intelligent manufacturing is a new approach to production that leverages technological processes and resources across both digital and physical spaces throughout all phases of design, manufacturing, logistics, sales, and services. It is equipped with capabilities for self-awareness, decision-making, execution, learning, and optimization, enabling the creation and delivery of novel types of manufacturing products and services.

2. Characteristics of intelligent manufacturing
Intelligent manufacturing is equipped with capabilities for self-awareness, decision-making, execution, learning, and optimization.

3. System architecture of intelligent manufacturing
The architecture of intelligent manufacturing systems encompasses three dimensions: life cycle, system hierarchy, and intelligent features, each containing five main components, which are illustrated in Figure 4.2.

The life cycle dimension consists of design, production, logistics, sales, and services.

The dimensions of the system level are composed of the equipment level, unit level, workshop level, enterprise level, and collaboration level.

The dimension of intelligent features consists of resource elements, interconnection, integration and sharing, system integration, and emerging business formats.

4.1.2.2 *Intelligent manufacturing of EMUs*

The core of intelligent manufacturing of EMUs lies in implementing digital twin technology, equipment interconnection and interconnectivity, structured process-driven production line operation technology, and other digital methods. With the aim of enhancing equipment intelligence, resource networking, and digitizing both business and production processes, lean improvements have been implemented to streamline operations, including planning and scheduling, process execution, logistics distribution, equipment management, and quality inspection. This has

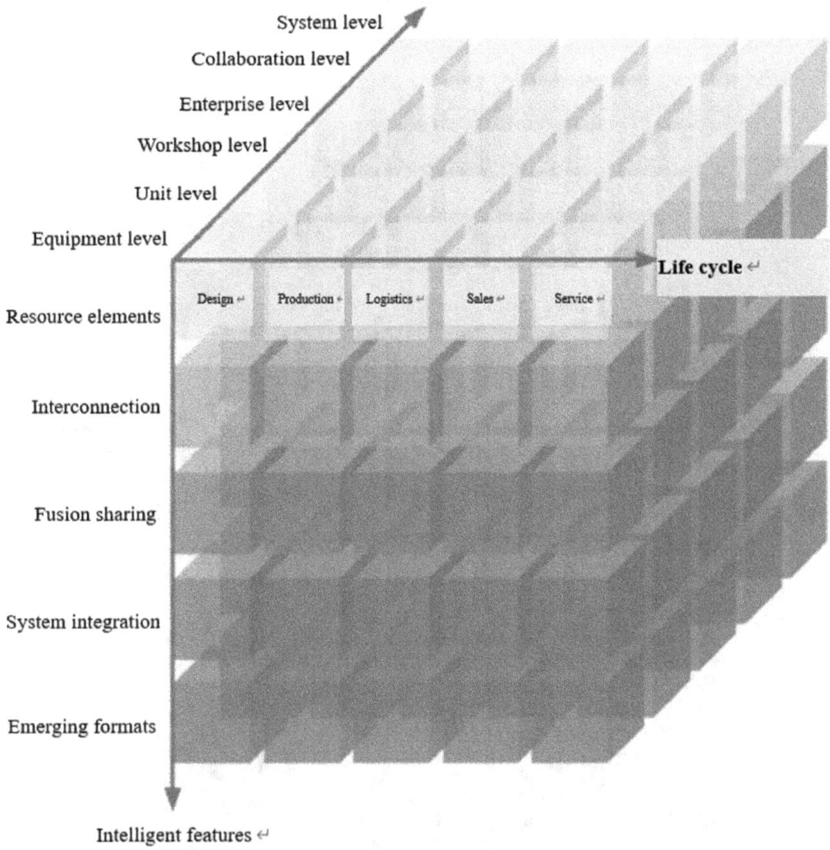

Figure 4.2. System architecture of intelligent manufacturing.

resulted in higher production efficiency, lower operating costs, better product quality, and an improved user experience during the manufacturing of intelligent EMUs.

The intelligent manufacturing architecture for EMUs comprises the edge, storage, capability, and application layers. Figure 4.3 depicts the architecture. The architecture comprises the manufacturing operation management system, intelligent factory digital twin technology, equipment interconnection and interoperation technology, structured process-driven production line technology, manufacturing process digital simulation technology, intelligent visual recognition and monitoring technology, manufacturing data value-added application, and more. Combined with advanced

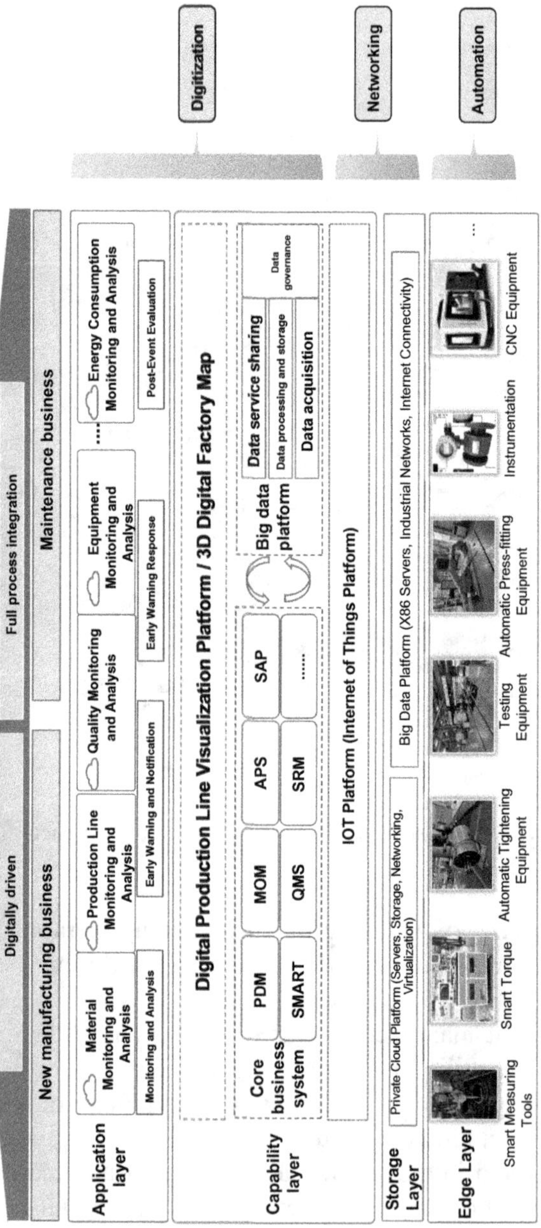

Figure 4.3. Overall architecture of intelligent manufacturing.

digital manufacturing resources, it establishes precise planning, efficient logistics, clear processes, stable quality, and intelligent equipment to create a transparent intelligent manufacturing system.

1. Manufacturing operations management system

The manufacturing operations management (MOM) system and advanced planning and scheduling (APS) system integrate R&D data from the PDM system, process planning and procurement data from the SRM system, quality data from the QMS system, and cost data from the SAP system to enable data connectivity and forward driving from an application perspective, coupled with the solidification site station-based rhythm management mode.

Plan management: Based on constraints related to limited production capacity, it comprehensively considers order priority, process parameters, material completeness, and availability and establishes a strong planning management model.

Material management: It drives material traceability and manages all logistics execution processes of materials in warehouses and workshops.

Capacity optimization: It displays the progress of multiple project orders and the actual load of equipment and workstations in real time online. It automatically adjusts the production line or resource load while taking into account the limitations of auxiliary production resources. This allows for the real-time reception of order completion status, feedback on plan completion status, and production anomalies. The ultimate goal is to optimize and improve workshop capacity.

Quality management: It creates comprehensive quality inspection standards and items, establishing the foundation for recording real-time quality inspection parameters while manufacturing and developing a complete analysis capability for quality.

Resource management: It efficiently aligns manufacturing resources, including personnel, equipment, tooling, and hardware, within the manufacturing management process for a variety of standard products to provide management and allocation, and it satisfies the solidification of information technology for day-to-day business processes, such as accounting, requisition, and distribution of physical resources.

Equipment management: It integrates the IoT technologies, gathers crucial equipment data in real time, tracks equipment status, and offers features such as daily and preventative equipment maintenance.

2. Digital twin technology for intelligent plants

A digital twin platform has been developed using new information technologies, such as spatial information, virtual reality, simulation, computer graphics, and big data. The platform boasts comprehensive functions, seamless integration, advanced technology, and dependable performance that meet the demands of intelligent manufacturing production control. This initiative eliminates the data silos between production and operation links and systems, achieving seamless integration of all factors within the factory. It enables control of the entire production and manufacturing process, leading to comprehensive enhancements in the enterprise's level of intelligent manufacturing.

Dynamic production line layout adjustment supported by panoramic 3D visualization: Through the visual display of the 3D model of the factory layout, it facilitates process technicians to identify problems and hidden dangers in the layout as early as possible during the initial stages of factory planning. This early detection allows for quick adjustments, reducing the risk of major adjustments needed after construction.

Virtual electronic map and material navigation planning: It utilizes the map analysis capability of 3D digital factories to accurately estimate the position, length, and other relevant information for each conveying path. Inertial navigation is achieved by monitoring the speed of conveying equipment, enabling the system to achieve precise material distribution within the factory area by determining the real-time material position in the monitoring system and the expected arrival time at the target location.

Monitoring of production and operation data and positioning of event alarms: It integrates the MOM system for production operation management and the IoT system for equipment interconnection, which enables obtaining real-time production data on site. These data can be synchronized with a 3D scene. In the 3D factory setting, personnel can view information pertinent to the current workstation, such as work tasks, personnel on duty, personnel load rate, material shortage, production progress, and completion rate. This enables remote monitoring of the production line and provides management and operation personnel with a visual-based system for real-time tracking of production execution. Simultaneously, sensors or other integrated platforms can collect information on equipment failures, material anomalies, production irregularities, quality inconsistencies, and related metrics. These data can be used to identify alarms within 3D scenes, visually represent the types and locations of anomalies, and generate practical solutions promptly.

3. Equipment interconnection and interoperation technology

Equipment interconnection: There is at least one physically compatible connection line between two pieces of equipment, and the communication protocols between equipment are compatible and can understand the information semantics of each other.

Equipment interoperation: Intelligent equipment exchanges information and applies it to its own functions to collaborate correctly with other intelligent equipment.

The entities involved in achieving interconnection within the intelligent manufacturing system of EMUs comprise intelligent equipment, integrated control systems that incorporate the intelligent equipment into production lines, software systems such as the IoT that gather and monitor intelligent equipment, distributed numerical control (DNC) systems that oversee CNC machine tools, and workshop-level manufacturing operation management systems, such as manufacturing execution systems (MES), MOM, and maintenance and repair operations (MRO).

Through digital upgrades and equipment transformation, equipment can be interconnected and interoperated. Integrating upstream information systems with intelligent devices enables automatic issuance of equipment operation programs, automatic collection of process data, accurate energy consumption data statistics, display of equipment utilization rates, fault trend analysis, predictive maintenance, and more. This effectively improves equipment OEE, lowers maintenance costs, boosts production line efficiency, and optimizes process flow.

4. Technologies for structured process-driven production lines

The structured process upgrades processes and quality planning data from offline to online mode. It also creates structured process guidance and quality inspection guidance based on action steps. These structured data drive the efficient operation of the production line, achieving an intelligent manufacturing model that uses standard data to drive production line operations.

Planning of structured processes: Structured process documents are planned through PDM and other systems. MBOM/BOP functions are utilized for new manufacturing processes, while MSBOM/SBOP functions are employed for maintenance processes. These structured process documents lead to the development of standard process knowledge bases, process resource databases, and other process information databases in a standardized way. The use of structured documents for data transfer results in the standardization of process document preparation.

Execution of structured processes: Structured process files are presented on site using MES/MOM/MRO systems. These files are combined with handheld, VR, and other terminal display devices to provide step-by-step guidance and control to operators in various job scenarios. Structured processes can guide operators to follow work steps, push work orders, and process information to intelligent equipment and measuring tools via the IoT system. This enables the automatic collection of equipment execution data. The quality information structured by work orders can also be automatically pushed to the quality management system for archiving, facilitating traceability and analysis of product manufacturing process quality data.

By using work orders to drive on-site material delivery and the industrial internet to control the on-site environment, equipment, and process elements, structured process data can be comprehensively driven and controlled for human–machine material environmental testing on the production line. This allows for the prevention of errors in the operation process, automatic backfilling of process data, consistent planning and execution, and paperless documentation. The result is the efficient automation of the production and manufacturing processes.

5. Digital simulation technology for manufacturing processes

Digital simulation of manufacturing processes typically involves simulating the processes themselves and the manufacturing systems. Process simulation is grouped into different types based on the discipline, including machining process simulation, assembly process simulation, and welding process simulation. Manufacturing system simulation can be categorized into layout simulation, logistics simulation, and dispatch simulation based on the intended simulation objective. The layout simulation analyzes collisions, interference, and safety distances between objects, equipment, and personnel workspaces, as well as material pathways. The logistics simulation simulates and analyzes logistics paths, efficiency, dispatching, strength, and line-side warehouse inventory. During simulation, it is necessary to ensure that the simulation model and results are accurate and reflect the actual situation of the simulated object. The process is driven by either process data or production data, and simulation efficiency aligns with actual production needs.

Digital simulation technology can evaluate product structure and process design in manufacturing, achieve static and dynamic interference checks, and ensure efficient assembly paths. Under a specific production plan, equipment selection can be optimized and parameters adjusted to achieve a balanced work environment while reducing production costs and cycles. Additionally, arranging equipment in flat or workshop positions can minimize material transportation costs, resulting in improved production efficiency and reduced logistics costs that meet production guidelines. The assembly process simulation diagram is shown in Figure 4.4.

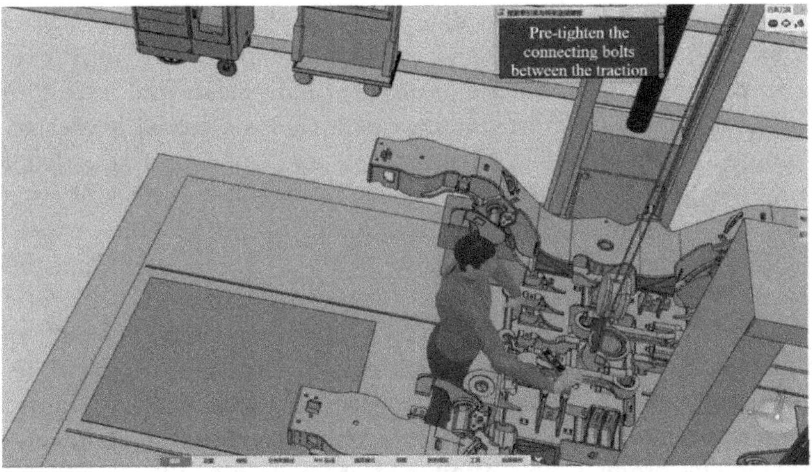

Figure 4.4. Simulation diagram of assembly process.

6. Monitoring technology for intelligent visual recognition

Visual recognition technology is used to monitor, identify, analyze, and alert different aspects of the production and manufacturing site. This includes personnel, quality, materials, and safety, achieving real-time positioning and display of personnel, materials, and equipment. It also intelligently captures and pushes alarms for abnormal information in the workshop and key workstations. Moreover, it includes features such as exchanging data with other systems, automated attendance tracking, automatic calculation of work hours, quality control of products, and alerting for irregularities.

Through the implementation of UWB positioning base stations and accompanying badges, the workshop can pinpoint the location of its workers and visitors. Real-time location tracking and display of personnel, alongside the storage of historical trajectories, can be achieved through the location calculation engine server. On-site video capture in the workshop can be achieved through industrial cameras, while the data storage of surveillance videos can be accomplished using a video storage server. Intelligent calculation and analysis of surveillance video data can be achieved through AI servers. High-precision location services and computer vision recognition technology can be utilized to accomplish real-time monitoring and alarms for product quality and on-site safety in the workshop.

7. Value-added application technology for manufacturing data

A manufacturing process big data analysis platform utilizes systems such as MES, MOM, MRO, and the IoT to gather data from diverse aspects of production, materials, quality, equipment, and technology. Utilizing this data, an individualized data analysis model is created to suit unique business needs. After analysis and calculation, the data are transmitted to a visualization terminal, as shown in Figure 4.5. This allows for real-time display of the status of production processes, including human, machine, material, method, environment, and measurement. It dynamically reflects various indicators throughout the production line. In addition, it creates a reliable and effective data mining and analysis system that consistently collects information from various aspects of the manufacturing process, improves and refines the data analysis model, and provides valuable support for management improvement, product innovation, and production quality, as shown in Figure 4.6.

Figure 4.5. Example of a visualization system for on-site manufacturing data.

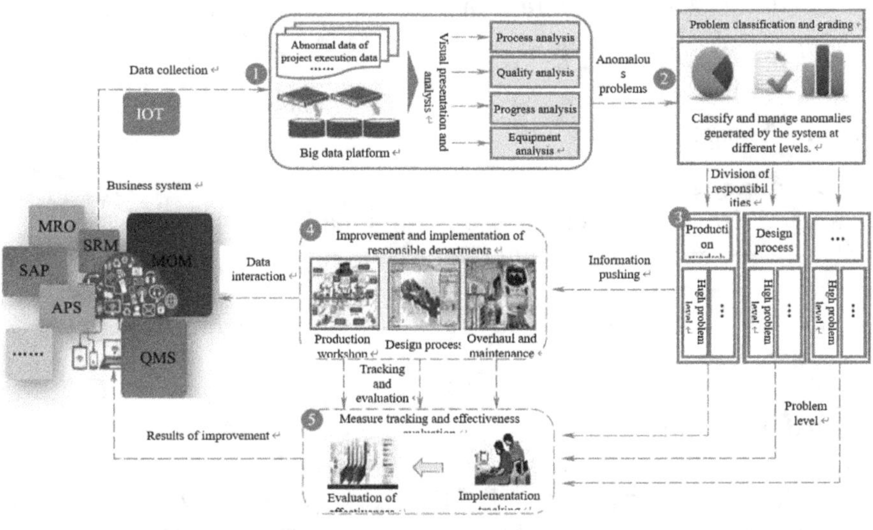

Figure 4.6. Architecture example based on data application improvement process.

4.1.3 *Key technologies for intelligent security*

4.1.3.1 *Real-time ethernet*

The train control network uses a real-time ethernet network that follows the IEC61375 standard. This network has a two-level topology structure, consisting of train level and vehicle level. Ethernet train backbone (ETB) is used for train-level communication, while ECN is used for rolling stock-level communication. Ethernet is used as the maintenance network for transmitting fault diagnosis, event recording, display information, software updates, and data downloads.

The ETB utilizes a dual-wire link aggregation architecture to handle train-level data communication. It operates over a transmission medium of Category 5 four-core shielded twist wire and achieves a communication rate of 100 Mbit/s.

Rolling stock ECN processes data communication at the rolling stock level and establishes an ECN network based on the power unit. To enable interaction between train-level communication and rolling stock-level communication, each power unit is equipped with two ETB switches, one operating and one on hot standby redundancy. The ECN switches create a linear or circular network for each power unit, while each rolling stock comes with two ECN switches. The rolling stock's ECN transmission uses a Category 5 four-core shielded twist wire with a communication rate of 100 Mbit/s.

The topological structure diagram of the train network.

4.1.3.2 *Security policy*

(1) **Boundary security**

Industrial firewalls are established individually between the train control network and the passenger information service network interaction interface, as well as the train–ground data transmission interface of the train control network, in order to prevent unauthorized network access.

(2) **Redundancy security**

(a) *Communication bus redundancy*

The train-level communication ETB's physical layer utilizes a redundant structure consisting of dual-channel link aggregation. In case channel A fails, the ETB switch's redundant control unit seamlessly transitions to channel B for communication without any negative impact on system performance. If both channels fail, the train's communication will be compromised by half, and it can operate in degraded mode or emergency traction mode. The rolling stock-level communication ECN utilizes a redundant ring network for communication. In the event of a single-point failure on the ring network, the network will automatically switch to redundant channel communication, without causing any disruption to network communication. In the case of dual faults, the train will run in degraded mode or emergency traction mode, depending on the fault location and driver's position.

(b) *Equipment redundancy*
Central control unit: Each network unit features two central control units in the head that are mutually redundant to ensure optimal performance and reliability. In the event that the master central control unit fails, the slave central control unit will take over automatically.

Redundancy of input and output devices: Key devices, including driver traction control handles and speed control handles, should incorporate device redundancy.

Redundancy of human–machine interface display screen: The cab control console of the driver is furnished with two mutually redundant display screens. If one screen malfunctions, the other screen can perform all the functions of the faulty screen.

(c) *Redundancy of subsystem communication interfaces*
All onboard subsystem ethernet communication interfaces are designed with dual port redundancy.

(3) **Ethernet security**
VLAN isolation technology, port speed limits, data prioritization, and other security measures have been implemented to ensure secure and high-priority transmission of control data, avoiding network storms.

(4) **Critical system fault-oriented safety**
When important systems such as traction, braking, and high voltage experience failures, the EMU has safety protocols in place, such as disconnection, isolation, and speed limits, to prioritize safety and availability.

4.1.3.3. *Onboard safety monitoring*

(1) **Safety monitoring of running gear**
Each coach of the EMU is equipped with an integrated onboard safety monitoring system. It consists of a unified system platform,

multifunctional modules, and various types of sensors. The system monitors the running gear's status, including temperature, vibration, and stability, as well as the train body's stability. It supports intelligent driving and maintenance operations. The system architecture is shown in Figure 4.7.

(a) *Axle temperature monitoring*
Temperature data are collected in real time by placing sensors on the journal box bearings, gearbox bearings, motor bearings, and stator of the bogie. The temperature signal is then converted into a digital signal. The digital signal is collected for channel fault identification based on the collection position, forming effective temperature data. The detection of abnormal conditions in rolling components is achieved through the use of predetermined warning and alarm thresholds and feature criteria. Subsequently, axle temperature warnings and alarms, along with speed limit control, are implemented.

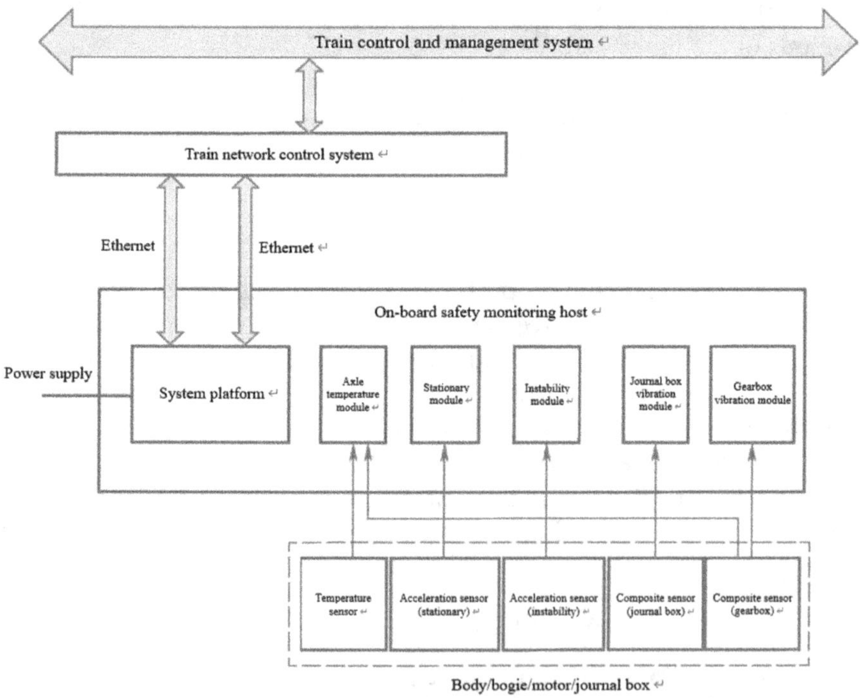

Figure 4.7. Architecture of onboard safety monitoring system.

(2) **Instability monitoring**

The 4–20 mA feedback signal from the bogie frame's lateral acceleration sensor is gathered. These analog signals are then converted into digital signals and filtered through data analysis algorithms to identify any acquisition channel faults and generate precise vibration data. Instances of rolling stock instability, such as snake-like movements, are identified through evaluation rules and alarm thresholds. These triggers, in turn, prompt alarms and speed limit control measures.

(3) **Stationary monitoring**

The stationary acceleration sensor collects the 4–20 mA signals for lateral, vertical, and longitudinal movements. These analog signals are then converted into digital signals and filtered through data analysis algorithms to identify any acquisition channel faults and generate precise vibration data. Then, the stationary evaluation index is calculated using the stationary logic model to detect anomalous conditions, such as body stability, shaking, and vibrations. The system then displays corresponding pop-up notifications on the intelligent monitoring screen in the control room.

(a) *Vibration monitoring*

The journal box vibration module collects both lateral and vertical acceleration signals, which are transmitted from the temperature vibration composite sensor. The signals are then converted into digital signals and analyzed using algorithms to determine acquisition channel faults and derive precise vibration and temperature data. Abnormal conditions, such as faulty bearings, polygonal wheels, and corrugated rails, are detected using feature criteria and highlighted with warning alerts.

The gearbox vibration module collects vertical acceleration signals, which are transmitted from the temperature vibration composite sensor. The signals are then converted into digital signals and analyzed using algorithms to determine acquisition channel faults and derive precise vibration and temperature data. Abnormal conditions of bearings are detected using feature criteria and highlighted with warning alerts.

(b) *Smoke and fire monitoring*

The EMU is equipped with an independent smoke and fire alarm system, consisting of a fire alarm controller and several fire detectors. The fire detector is installed in various areas, including the driver's cab, passenger compartment, PIS cabinet, electrical cabinet, kitchen, and toilet, to collect real-time data on smoke concentration and temperature. These data are then sent to the fire alarm controller. When the fire alarm controller senses smoke concentration and ambient temperature reaching the set alarm conditions, it will generate sound and fault notifications. Once the smoke and fire alarm deactivates, the alarm area must be manually confirmed and reset.

The network topology of the smoke and fire system is shown in Figure 4.8.

(c) *High-voltage insulation detection*

To accurately assess possible grounding or insulation issues between the high-voltage equipment in the roof pantograph area and the high-voltage equipment box section, the EMU is outfitted with a high-voltage insulation detection device. This device converts the 110 V DC power supplied by the rolling stock into 50 Hz AC power following inverter processing. When parking or pantograph dropping, any abnormalities in roof equipment insulation are detected by monitoring the voltage change on the low-voltage side of the high-voltage transformer.

(d) *Pantograph video monitoring*

The EMU is equipped with a pantograph video monitoring system consisting of an intelligent analysis host, video server, and camera for monitoring the pantograph video. The system uses the PIS backbone network

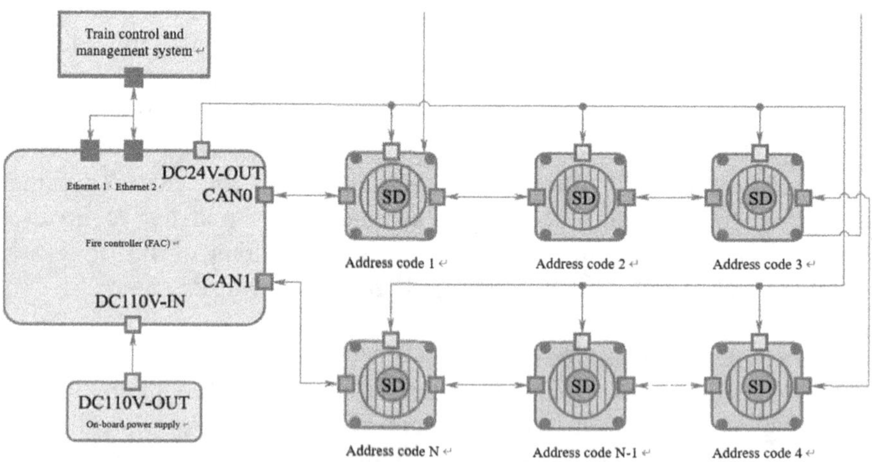

Figure 4.8. Network topology diagram of smoke and fire system.

to access real-time pantograph video, diagnose multiple types of failures (e.g., abnormal pantograph structure, foreign object invasion, pantograph arcing, and pantograph tilting), and display the pantograph video in real time on the intelligent monitoring screen in the monitoring room. When the intelligent analysis system detects a fault in the pantograph, the monitoring screen in the control room displays pictures of the equipment during the fault as well as videos taken before and after the occurrence.

(e) *Coach video monitoring*
The EMU has a coach video monitoring system that includes cameras and video servers. The system relies on the PIS backbone network to provide real-time monitoring of the interior, end corridors, and doors. Real-time video can be viewed on the intelligent monitoring screen in the monitoring room. If an overload alarm, fireworks alarm, passenger emergency alarm, or door abnormal alarm is triggered, the intelligent monitoring screen in the monitoring room allows viewing of the corresponding alarm area's video at the time of the alarm.

(f) *Medium voltage insulation monitoring*
The EMU utilizes a circuit breaker with leakage protection and an AC monitoring device to provide two-stage protection against electrical circuit leaks. The system can promptly alert and warn for ground leakage in each load or busbar of the auxiliary system, preventing further faults from occurring.

(g) *Earthquake warning and approach warning*
In addition to the safety monitoring mentioned above, the EMU also sets aside installation interfaces for earthquake and approach warning systems.

4.1.3.4 *Fault prediction and health management (PHM) of rolling stocks*

The EMU has developed an open platform for managing the health of train life cycles that provides control and PHM isolation. This platform is characterized by its openness, flexibility, and adaptability, enabling the integration of various models and boards from different manufacturers and companies. It incorporates key subsystem component models, including traction and door models, as well as train system models. It facilitates constant health monitoring, online fault diagnosis, and warning of trains and their critical components. Additionally, it provides guidance to onboard mechanics during emergency fault handling and depot maintenance, thus enhancing train operation safety and system reliability:

(i) **Predicting the temperature of the running gear:** Parameters such as journal box bearing temperature, gearbox bearing temperature, motor bearing temperature, and stator temperature are utilized to detect any abnormalities in the running gear.

(ii) **Diagnosis of abnormal vibrations, instability, and stability:** Vibration, instability, and stability parameters are used to determine whether there are abnormalities in the running gear, body, rail surface status, etc.

(iii) **Network voltage and network flow prediction:** The network voltage and flow parameters are analyzed to detect anomalies in the high-voltage traction system.

(iv) **Pre-judgment of abnormal door conditions:** The door opening and closing times, as well as the motor current, are utilized to identify any abnormalities within the sliding door system.

(v) **Prediction of abnormal voltage and current in the traction system:** The sensor parameters within the traction system are utilized to identify potential irregularities in both the traction and transmission systems.

(vi) **Diagnosis of the health status of critical loop relays and contactors:** The counting of action times is carried out on key loop relays, contactors, etc.

(vii) **Pre-judgment of abnormal air brake application and release:** The charging and discharging times during the application and release of air braking are used to identify any abnormalities in braking performance.

(viii) **Pre-judgment of abnormal parking braking:** The status parameters such as parking cylinder pressure are used to determine whether there is an abnormality in the parking brake.

(ix) **Pre-judgment of network communication anomalies:** The network bus communication quality is utilized to detect any irregularities in network communication.

4.1.4 *Key technologies for intelligent services*

The intelligent services of EMUs mainly include intelligent information prompts, service information fusion, intelligent environment regulation, service data transmission, and intelligent service facilities, as shown in Figure 4.9.

1. Intelligent message prompt

(1) *Multi-information split-screen display of high-definition TV on the train*

The passenger room's TV features an LCD split-screen display that merges guidance service and entertainment video content, effortlessly displaying real-time train operation status and local customs while entertaining passengers.

The antenna redundancy system, utilizing Beidou/GPS, collects train positioning data. The passenger information system controller stores pre-arrival, arrival, departure, and en-route information display content

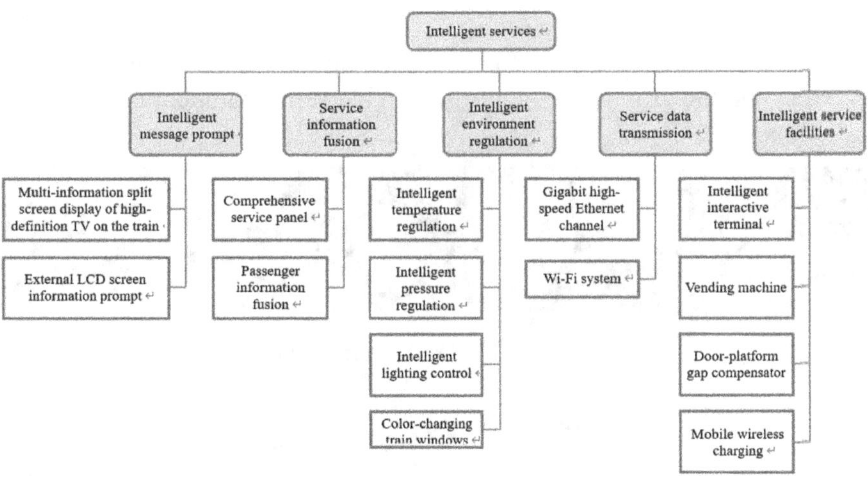

Figure 4.9. Schematic diagram of intelligent services for high-speed railway EMUs.

that is automatically triggered based on the positioning information. See Figure 4.10 for further details.

(2) *LCD screen information prompts inside and outside the train*
The train's LCD information display technology, combined with HTML5, achieves diverse and visually appealing content for both interior and exterior displays. The train's display provides information on train number, coach number, internal and external temperature, speed, time, and toilet occupancy. This allows passengers to easily understand their boarding information in a timely manner. The external information display now includes the seat number indication function. This feature indicates the position of seats near the door, guiding pedestrian flow and making it easier for passengers to find seats quickly. Additionally, this function prompts the optimal door position for boarding. The passenger information system controller controls the external LED information display, which can automatically turn on or off based on the train's speed. Refer to Figure 4.11 for a visual representation.

2. Service information fusion

(1) *Comprehensive service panel*
In the business area, a service panel is installed on the desk to centrally control lighting, air conditioning, volume, and service calls. This ensures privacy in the business area and independent control of the coach environment.

(2) *Passenger information fusion*
It offers a complete display of information about the interaction between the train and the ground, as well as local video entertainment information,

Figure 4.10. TV split-screen display.

Figure 4.11. Schematic diagram for LED seat number indication function outside the train.

including television, interactive terminals, and both internal and external displays.

The onboard 5G antenna and mobile data network enable access to live TV broadcast signals, webpage information, and other entertainment media. The intelligent interaction terminal of the business seat facilitates video on demand and mobile screen projection, along with live TV broadcasts and webpage browsing.

The toilet now features a voice player that provides intelligent prompts for smoking bans, service information, and more. Additionally, it is compatible with the silent coach functions.

3. Intelligent environment regulation

(1) *Intelligent temperature regulation*: Variable frequency control technology is utilized to enhance environmental state perception and optimize control strategies. When external environmental conditions or personnel load change, each sensor accurately provides temperature feedback to the controller. The controller then adjusts the working status in real time to achieve more precise temperature regulation in the passenger room, automatically adapting to changes in the rolling stock load. This reduces temperature fluctuations in the rolling stock, improving overall comfort. Additionally, the system includes features such as air filtration (see Figure 4.12), automatic abnormality alert, and redundant optimization control.

(2) *Intelligent pressure regulation*: The EMU is equipped with pressure wave sensors on both sides of the head of the train. When rolling stocks pass through tunnels or intersect at high speeds, sensors detect pressure changes outside and inside the rolling stock. The pressure signal is converted into a current signal and sent to the controller for analysis. The pressure wave controller receives feedback from the pressure sensor, taking into account the pressure difference and change rate, and activates the pressure wave protection system. Based on external tunnel information, the system can also close the fresh air and exhaust pressure protection valves proactively, achieving automated pressure protection inside the train, as shown in Figure 4.13.

(3) *Intelligent lighting control*: The EMU automatically adjusts the color temperature and brightness of the interior lighting based on operating status and environmental changes. The toilet features intelligent lighting that adjusts the brightness when someone enters, as shown in Figure 4.14.

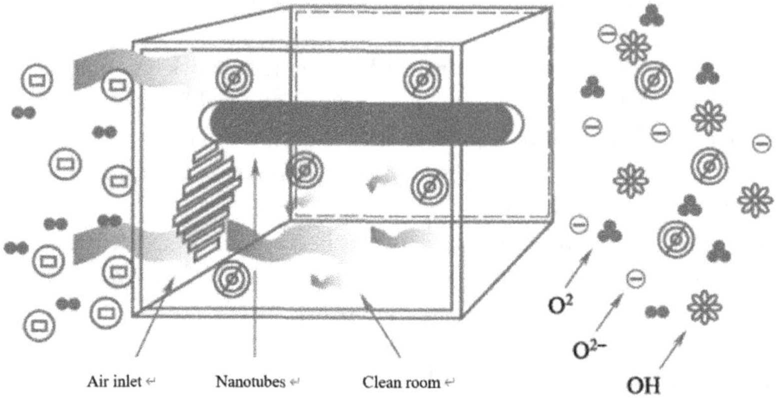

Figure 4.12. Schematic diagram of air purification device.

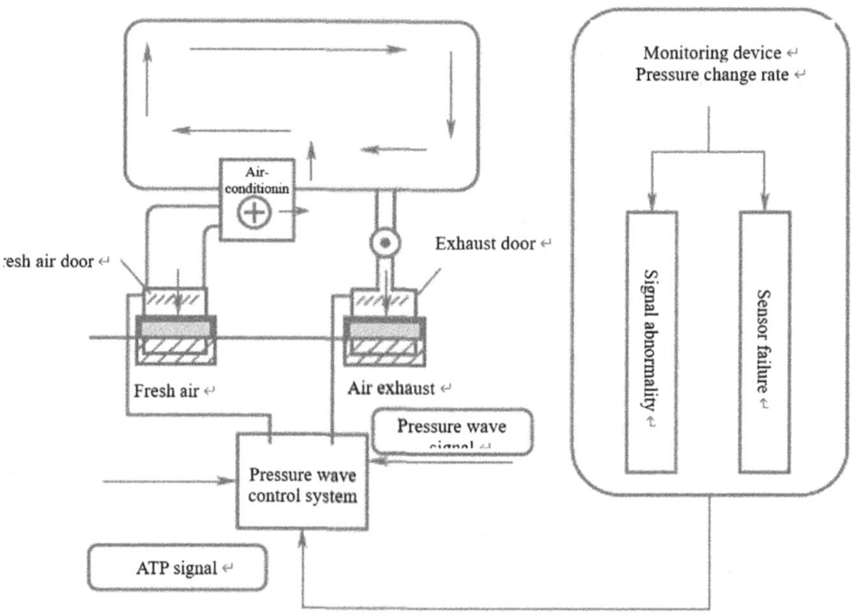

Figure 4.13. Schematic diagram of pressure wave control.

(4) *Color-changing train windows*: By setting up color-changing glass, integrating capacitive touch screens, and dye liquid crystal dimming film in the business area, color-changing can be carried out within the range of 0–30% transmittance. By adjusting the light transmittance of the window

Figure 4.14. Schematic diagram of intelligent adjustment function for toilet lighting.

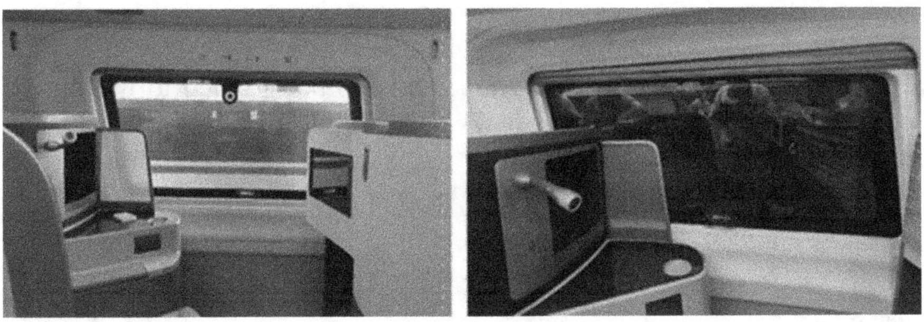

Figure 4.15. Comparison of color-changing windows before and after color change.

through buttons or hand-sliding touch, a light transmittance effect suitable for different passenger needs can be achieved, as shown in Figure 4.15.

4. Service data transmission

The onboard passenger information service adopts gigabit ethernet technology to achieve high-speed and high-capacity transmission of passenger information, audio and video signals, coach video monitoring information, pantograph-OCS monitoring information, and handheld terminal display information.

The EMU is equipped with 4G/5G antennas, access points, central servers, and bicycle servers to achieve internet access in the passenger coach. In conjunction with the wireless internet app, it provides passengers with multi-dimensional internet information, such as push information and travel service information. By utilizing onboard Wi-Fi coverage, passengers can access and achieve user real-name authentication through intelligent terminals with wireless LAN access functions. Separate VLANs are provided to avoid data interaction with other systems and ensure information security, as shown in Table 4.3.

5. Intelligent service facilities

(1) *Intelligent interactive terminal*: The EMU is equipped with intelligent interactive terminals that integrate functions such as information display, image viewing, and video playback. It can achieve functions such as passenger local media service on demand, television live streaming, local games, train information query, tourism information query, and calling crew members. It also has a mobile screen projection function to achieve interaction between onboard devices and passengers, as shown in Figure 4.6.

(2) *Vending machine*: A vending machine has been added to the dining coach. It can automatically sell snacks, fruits, beverages, etc., and has an automatic data statistics function. It can select meals through a mobile phone interface. Passengers can pay by scanning WeChat and Alipay, as shown in Figure 4.17.

Table 4.3. Wi-Fi system diagram.

Data-processing center
A station
Wired or wireless signal access
Leaky coaxial cable
Signal coverage of Mobile, Unicom and Telecom operators
3G/4G antenna
Central server
Ethernet bus
Server for a single train
EMU1; EMU2; EMU5; EMU3/4/6/7; EMU8
Gigabit Ethernet
Antenna feeder
Wi-Fi antenna

Figure 4.16. Schematic diagram of intelligent interactive terminal for business seats.

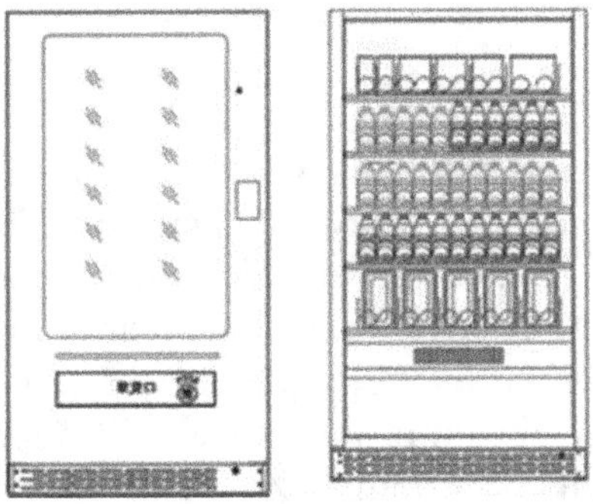

Figure 4.17. Schematic diagram of a vending machine.

(3) *Mobile wireless charging*: The business seat is equipped with a mobile wireless charging module to provide convenient charging services for passengers, as shown in Figure 4.18. The principle of wireless charging involves electromagnetic wave induction. There are two coils at the sending and receiving ends, and the sending-end coil is connected to a wired power source to generate electromagnetic signals. The receiving-end coil senses the electromagnetic signals from the sending end, thereby generating electric current to charge the battery. The wireless charging of mobile phones mainly adopts electromagnetic induction technology (Qi and PMA wireless charging standards) and electromagnetic resonance technology (A4WP wireless charging standard).

4.1.5 *Key technologies for energy conservation and consumption reduction*

The energy consumption for high-speed train operation mainly comes from the conversion of pantograph-OCS electrical energy into train

Figure 4.18. Schematic diagram of the installation location of wireless charging equipment.

operation kinetic energy, which is mainly composed of traction energy consumption and auxiliary energy consumption. The energy consumption of train traction (converted into the kinetic energy of train operation) accounts for about 60%, mainly depending on the efficiency of traction transmission system components, train running resistance, etc.; auxiliary energy consumption (air conditioning, lighting, control systems, etc.) accounts for about 30%, mainly depending on the power and efficiency of auxiliary equipment, such as air conditioning.

Train running resistance is the core factor affecting traction energy consumption, which is composed of mechanical resistance (including bearing resistance, rolling resistance, sliding resistance, impact, and vibration resistance. When the number of trailers, EMUs, and passenger capacity are fixed, the mechanical resistance remains basically unchanged), momentum loss resistance (when air-conditioners and other equipment take in and exhaust air, they exchange gases with the outside world, causing changes in momentum and generating resistance, which is proportional to the speed of the train), and aerodynamic resistance (mainly closely related to factors such as train shape, front slenderness ratio, rolling stock cross-sectional area, and train formation, which increases linearly with the square of speed). For high-speed trains, aerodynamic resistance dominates, and the energy consumption caused by aerodynamic resistance accounts for the largest proportion.

According to the analysis of factors related to train energy consumption, energy saving and consumption reduction of high-speed trains mainly focus on improving traction efficiency, reducing train running resistance, reducing rolling stock weight, and improving energy efficiency (improving the efficiency of auxiliary power supply equipment, adopting variable frequency air conditioning, etc.) to reduce auxiliary energy consumption:

1. Improving traction efficiency

The train traction system mainly includes traction transformers, traction converters, and traction motors, which improve efficiency and reduce weight by adopting new technologies and materials. Among them, the transformer can use running air cooling to improve the overall operating efficiency of the transformer; the converter uses silicon carbide, reverse conduction insulated gate bipolar transistor (RC-IGBT), etc., to improve equipment efficiency and reduce weight; and the motor uses a permanent magnet motor to improve equipment efficiency and reduce weight.

2. Reduce train running resistance

(1) *Train running resistance*: The frictional resistance is mainly composed of the mechanical friction of train wheel–rails, bearings, gears, etc. Generally, the heavier the train, the greater the resistance, so the resistance is related to the weight of the train. The momentum loss resistance of airflow in the train is mainly related to the adjustment of fresh air volume during train operation, train airtightness, vehicle speed, etc. The aerodynamic resistance (pressure difference and friction) is mainly related to the train shape, train smoothness, windshield design, skirt design, and bogie area design. Therefore, reducing weight and optimizing the aerodynamic shape of the body are also effective energy-saving and consumption-reducing measures.

(2) *Reducing air resistance*: The air resistance of high-speed trains is composed of pressure difference resistance and viscous resistance. Among them, pressure difference resistance can be controlled through the design of the vehicle's exterior, and viscous resistance can be controlled through the design of microstructure modifications and modifications on the train surface.

(3) *Improving auxiliary energy efficiency of trains*: In terms of lighting, constant temperature, and compartment ventilation, the EMU adopts intelligent equipment to reduce auxiliary energy consumption. The purpose of reducing auxiliary energy consumption is achieved through the

use of infrared detection of unmanned lighting, the use of intelligent adjustment equipment such as light sensing in the business area and other areas, and the use of variable frequency air conditioning.

4.2 Automatic Operation of HSR

The train operation control system is the main technical equipment of HSR and is responsible for directing train operation, ensuring traffic safety, improving transportation efficiency, and realizing train operation automation. It's the "brain and central nervous system" of a HSR system. The train operation control system performs control and decision-making for related equipment (including switches and trains) of the HSR system through the collection, processing, and handling of relevant information from trains, infrastructure, and environment. On the basis of traditional train control system, it is an important direction for the development of HSR train control system in the future to develop intelligent operation control by integrating new information and intelligent technologies, therefore improving train operation safety and efficiency, increasing system reliability, reducing the number of equipment, and reducing construction and operation costs.

In 2019, for the first time in the world, China's HSR realized automatic operation at a speed of 350 km/h based on CTCS 3+ATO, reaching the automation level of Grade of Automation 2 (GoA2). It has the functions of automatic departure of high-speed EMU stations, automatic operation in inter-area, automatic and precise stopping at stations, automatic and precise control of doors/safety gates, etc., which effectively ensures the safety of train operation, improves the efficiency of train operation, reduces the energy consumption of traction, reduces the labor intensity of drivers, and improves the experience of passengers. It has become another major landmark achievement of independent innovation by China's HSR.

4.2.1 *Structure and function of HSR ATO system*

Automatic train operation (ATO) replaces the driver's task of operating the train in an automated way and uses ground information to realize the operation control of tasks, such as traction, braking, idling, and automatic return, and carries out real-time calculations and optimization control by taking into account the operation plan, train performance parameters, line data conditions, etc., to make the train in the best operation state.

The ATO system of HSR is to add ATO on the basis of the automatic train protection (ATP) system of the train operation control system to realize automatic operation control, set up special precise positioning transponders on the ground to realize precise positioning, and realize platform door control, inter-station data sending, and operation plan processing through wireless communication. The main purpose is to realize automatic operation control of inter-area, automatic stopping and departure at stations, and door protection and platform door linkage control to improve transportation efficiency and safety.

As shown in Table 4.4, the international standard defines four levels of automation GoA, according to the degree of automation of rail transit

Table 4.4. Four-tier automation level of GoA.

GoA level	GoA name	Train operators	Brief description
GoA1	Non-automatic train operations	Train driver in the cab	Manual train operation with ATP in place; this GoA may also include providing auxiliary information to assist manual driving
GoA2	Semi-automatic train operations	Train driver in the cab	Automatic train operation with automatic control of stopping but requiring an onboard driver to initiate the automatic driving mode, where the driver can operate the doors (which can also be controlled automatically). The driver verifies the clear track ahead from the driver's cab and performs other functions. In emergency or degraded situations, the driver can take over
GoA3	Driverless train operations	Train without driver, with crew members	Automatic train operation, including automatic departure and automatic door control. In emergency or degraded situations, crew members can operate the doors
GoA4	Unattended train operations	No competent operator onboard	Unmanned train operation, where all train functions are automated, and in emergency or degraded situations, there are no onboard personnel responsible for control tasks

lines, and the commonly used automation levels are GoA1–GoA4, from low to high.

1. Architecture of ATO system for high-speed railway

The architecture of the high-speed railway ATO system is shown in Table 4.5. On the basis of the existing train control system, the onboard equipment increases the ATO unit and general packet radio service (GPRS) transmission radio, realizing automatic operation; the ground equipment increases the operation plan transmission, gate control, and other functions on the existing temporary speed restriction server (TSRS). At the same time, GPRS radio transmission equipment is added to construct an independent ATO and TSRS transmission channel between train and ground so as to realize the linkage control of the door/station door and the operation plan on board.

2. Composition of onboard ATO equipment on high-speed railway

The onboard equipment of the HSR train control system consists of ATP and ATO, and the equipment structure is shown in Figure 4.19.

(1) *ATP equipment*: ATP equipment consists of a vital computer (VC), GSM-R transmission unit (RTU), track circuit receiver unit (TCR), balise transmission module (BTM), judicial recording unit (JRU), driver–machine interface (DMI), train interface unit (TIU), and speed distance unit (SDU). Enhancements have been made to the existing ATP onboard equipment, including the addition of door protection functionality. This upgrade also includes precise parking error assessment and support for ATO control. The DMI has been expanded to display control-related information, status, alarms, and other details associated with ATO functionality, enabling support for ATO automatic driving and other features.

(2) *ATO equipment*: On the basis of CTCS-3, the onboard equipment enhances the onboard ATO system, GPRS radio, and related supporting equipment; the ATO onboard equipment receives the operation plan, inter-station data (including line basic data and temporary speed limit), and other information through the GPRS wireless communication unit with the permission of ATP to realize functions such as automatic control of train speed, automatic opening and closing of doors and platform door

Table 4.5. CTCS3 + ATO system architecture.

On-board Equipment	
GPRS Radio	
Driving Platform	ATO start button; DMI.
RTU	UEU; TIU.
Mobile Host	TCETU; Speed Distance Unit; Train Control Interface; New ATO Unit; ATP Unit; TCM Unit; TCR Unit; BTM Unit; SD Unit; MV8Interface.
GSM-R Radio	
Balise Receiver Antenna	
Track Circuit Receiver	
Vehicle Network Control System	
SDU	
GPRS	
BIS	
OTE	
OBE	
Stop sign	
PUC	
AH; JD1; JD2; JD3; JD4; J DS; AH	
Railside Equipment	
Station Equipment	
Active Balise	
LEU	
ZPW-2000 Track Circuit	
Central Signaling Monitoring System	
Platform Protect Door	
TCC	

(*Continued*)

Table 4.5. (*Continued*)

Station Interlock	
CTC	
Signal Centralized Monitoring Data Communication Ethernet	
Signal System Safety Data Network	
Central Traffic Data Communication Ethernet	
Central Equipment	
CSM Central Monitor System	
CTD Central	Dispatch Platform; Maintenance Platform; Server; Interface Server.
Temporary Speed Restriction Server	Temporary Speed Restriction Server; Interface Server; TSRM Maintenance Unit.
Radio Blocking Center	RBC Server; RBC MU; RBC Interface Unit; GSM-R Interface Unit.
GSM-Rcore Network	TRAU; MSC; GGSN; BSC; OTE.
Stopping Signage	
Active Balise	
Passive Balise	
Security/Screen Doors	

linkage control, energy-saving control of train operation, diagnosis, recording, and alarm of the equipment. It realizes automatic train speed control, automatic door opening, and door/platform door linkage control at precise station stops, energy-saving control of train operation, and equipment diagnosis, recording, and alarms.

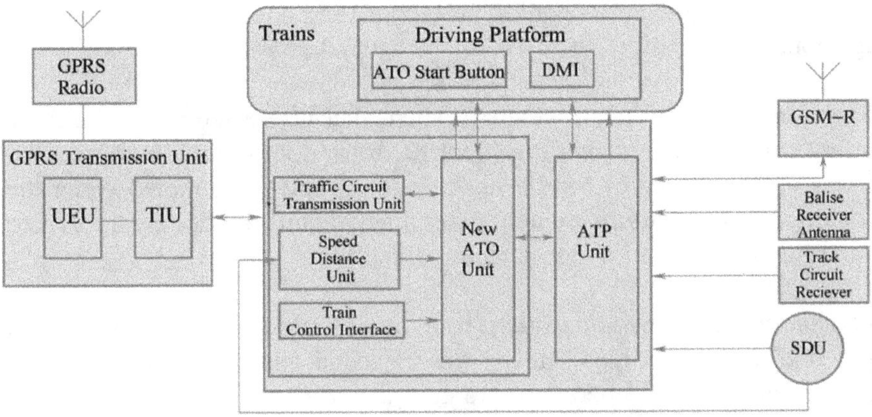

Figure 4.19. Structure of onboard ATP/ATO equipment for high-speed railway.

(3) *Track circuit receiver unit*: It receives information about the track circuit.

(4) *Balise transmission module and transponder antenna*: The transponder transmission module receives the information about the ground transponder by connecting with the transponder antenna.

(5) *Radio transmission module*: By connecting with the GSM-R onboard radio station, two-way information transmission between train and ground is realized.

(6) *Driver–machine interface*: Human–machine interface realizes information interaction between the driver and onboard equipment.

(7) *Train interface unit*: The train interface unit provides an interface between the safety computer and the train-related equipment.

(8) *Speed distance unit*: The speed distance unit receives signals from speed sensors and other equipment to measure the train's running speed and distance.

3. The main functions of ATO system for HSR

The functions of the ATO system for the HSR mainly include automatic departure at stations, automatic interval operation, automatic parking at stations, automatic door opening protection, and door/platform door linkage control:

(i) *Automatic departure at stations*: The driver is assisted in checking the safety conditions for departure (door/platform door closed, authorization status in front, control handle position, etc.) and is prompted when the conditions for departure are met. After manual confirmation by the driver, the train is controlled to depart automatically.

(ii) *Automatic interval operation*: It involves receiving the operation plan and line data in stations sent by the trackside equipment, selecting the appropriate interval operation strategy according to the operation plan, and controlling the trains to run on time and save energy.

(iii) *Automatic parking at stations*: Precise positioning and obtaining the position of the stopping point by means of the precise stopping transponder installed at the station to control the train to stop precisely at the right mark.

(iv) *Automatic door opening protection*: After the train stops at the platform, it will automatically determine the conditions for the train to stop stably and give the conditions for the door opening to ensure the safe opening of the door.

(v) *Door/platform door linkage control*: It automatically/manually opens or closes the doors according to the requirements of the operation plan and the driver's settings and sends platform door control instructions to the trackside equipment through train–ground wireless communication to realize the synchronized action of the platform door and train door.

4.2.2 *Key technologies of HSR ATO system*

The key technologies of the ATO system for HSR mainly include intelligent autonomous perception technology, train intelligent decision-making technology, train operation optimization control, and multimode train positioning technology.

1. Intelligent autonomous sensing technology
As the automatic driving level of HSR moves from GoA2 to GoA3 and GoA4, the intelligent autonomous sensing technology for various

complex internal and external environments, including line condition detection and obstacle detection, personnel detection, and onboard equipment status detection, needs to be researched in order to replace the drivers and crews.

2. Train intelligent decision-making technology
(1) *Train intelligent driving control evaluation system*: Its functions include: constructing an intelligent driving control evaluation system for high-speed trains with the goals of comfort, punctuality, energy saving, etc.; studying the correspondence between different scenarios and the optimization goals of train driving strategies (e.g., energy saving and punctuality under normal conditions and punctuality and delay reduction under unexpected conditions); and the coupling of the weights for different indicators in decision-making for multi-objective optimization, so as to provide a basis for online intelligent decision-making of high-speed trains.

(2) *Train driving strategy multi-objective optimization decision-making technique*: The process of high-speed train operation exists in special sections such as long grades and split-phase areas, which have a large impact on train operation, and it is necessary to study the optimization strategy of local driving control under safety constraints, and the optimization model of train driving strategy under a complex operation environment is constructed, with the corresponding optimization objectives selected according to the evaluation system of train intelligent driving control. The theories of AI and operation research are used to solve the problem of multiple objectives (punctuality, comfort, and energy saving) of train driving strategy within the global scope.

3. Optimized control of train operation
Optimized control of train operation is at the core of ensuring high-speed and smooth operation of trains, as well as one-time precise control (from the maximum operating speed of 350 km/h to 0 km/h when the train stops stably and accurately) under the conditions of multi-level switching and complex operation scenarios. Due to the complexity of the operating environment, traditional methods, such as PID, fuzzy logic control, neural network control, and iterative learning control, are difficult to apply, and it is necessary to improve their adaptive and self-learning characteristics. Curve optimization, on-time control, and tracking control are the key technologies to be solved:

(1) *Curve optimization*: The recommended speed profile is an ideal recommended speed that mimics the driver's experience when the train is autopiloted. Recommended speed profile optimization is often abstracted as an optimal control problem. The objective function is the train's energy consumption for a given running time, ride comfort, switching frequency, or a balance between these metrics. Comfort constraints, punctuality constraints, speed limit constraints, control variable limitations, etc., are considered in the optimization process.

(2) *On-time control*: The long distance between the stations of a HSR, the complex causes of operational disturbances, and the difficulty of controlling the running time between stations require the design of a closed-loop mechanism to control the deviation of train running time. Therefore, it is necessary to establish a high-precision train traction calculation model to ensure the accuracy of control. At the same time, a checkpoint-based on-time operation control scheme realizes closed-loop control by checking the on-time target at a predetermined point to ensure the on-time rate of trains.

(3) *Tracking control*: Tracking control is essentially an optimal control problem in multi-objective optimization under complex input conditions. The core of the problem is to seek the optimal interval operation method under the multiple objectives of train safety, punctuality, energy saving, and comfort. Intelligent interval operation strategy selection algorithms based on multi-objective optimization are commonly used.

4. Multimode train positioning technology

Safe, accurate, and highly reliable train positioning technology is the foundation and core of realizing HSR ATO. The single-speed measurement positioning technology cannot fully meet the needs of high-speed train operation. Combined positioning technology with multi-sensor fusion can realize redundancy and complementarity, providing more reliable and accurate train position and speed information. The technology can be utilized through ground positioning, active positioning, air and space positioning, relative positioning, and other multi-standard autonomous positioning of trains, realizing autonomous redundant positioning of trains, and canceling the ground track occupancy inspection equipment.

4.3 Next-Generation Train Operation Control

In order to keep up with the advanced level of the world's railways and maintain the leading position of Chinese railway technology, a next-generation train control system has been developed by Chinese Railway in line with the needs of transportation development. The system adopts the integrated design of air and space, realizes the train tracking operation in moving blocks, breaks through the key technologies such as integrated positioning of trains by integrating Beidou, autonomous train integrity safety checking on board, and realizes the simplification of trackside equipment, integration between stations and intervals, and intelligent equipment maintenance, as shown in Figure 4.20. The main technical features of the system meet the requirements of the CTCS-4 train control system, and the overall technical level is in the internationally leading ranks. It is a major breakthrough for the core equipment in our railway signaling field, from following to parallel running and leading.

4.3.1 *Structure and function of next-generation train control system*

The next-generation train control system is based on the CTCS-3 train control system and adopts technologies such as integrated positioning of

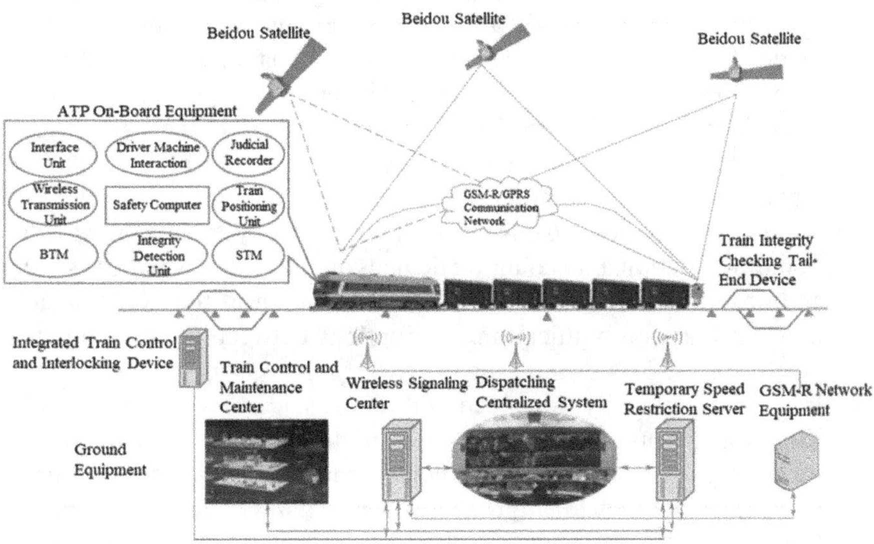

Figure 4.20. Schematic diagram of the next-generation train control system.

the train by using integrated BeiDou, autonomous integrity checking of the train, IP-based mobile communication, integrated remote control in station areas, and intelligent maintenance. It optimizes the signaling system structure, reduces trackside equipment, and realizes the functions of moving blocks, integrated remote control in station areas, etc. The main technical features are as follows:

(i) Less trackside equipment, high safety availability. The system eliminates track circuits for intervals and passing signals. The onboard equipment adopts BeiDou-based multi-source fusion technology to realize autonomous positioning and train integrity checking. The radio block center (RBC) combines the train position report (including integrity status) with station track circuit and axle counting equipment information in stations to realize safety control functions for train tracking operation in the moving block system. By setting up the virtual section, it realizes "partial blocking" of the section where the faulty train is located, thus improving the availability of the system under the premise of guaranteeing safety.

(ii) It complies with CTCS and is capable of interconnection and interoperability. The overall system design adheres to the "train-to-ground" control mode, as specified in the CTCS technical system requirements. The ground-based central equipment RBC calculates and generates the movement authority (MA) for operation in the mobile block control mode. This MA is then transmitted to the onboard equipment through the wireless communication system. The onboard equipment monitors the safe operation of the train based on the "target-distance" mode speed curve, according to the MA. Additionally, it is capable of seamless interconnection and communication with the existing CTCS system.

(iii) It has a backup mode to ensure operational reliability during degradation. Inter-station axle counting devices are installed to address safety and protection requirements for train operation in the event of failures, such as communication interruptions between the train and the ground or the loss of train integrity.

(iv) It is technologically advanced and aligns with development trends. The system's use of BeiDou-based multi-source fusion technology, train integrity inspection technology, and IP-based wireless communication technology all align with the current direction of train control technology development. Additionally, it provides a strong technological foundation for the advancement of intelligent railways.

4.3.1.1 *Architecture of the next-generation train control system*

The architecture of the next-generation train control system primarily consists of onboard equipment and ground equipment. The onboard equipment is composed of ATP and end of train (EOT), while the ground equipment comprises trackside and central equipment. Compared to the existing CTCS-3 train control system, the next-generation train control system has mainly made the following enhancements in terms of system structure and functionality:

(i) Onboard equipment is equipped with a satellite reception antenna to achieve integrated autonomous positioning with Beidou satellite system.
(ii) The newly developed EOT equipment for train integrity inspection is installed at the rear of the train, enabling integrated integrity safety checks for locomotive-hauled trains with the Beidou satellite system integration.
(iii) It further enriches the TSRS by adding features such as electronic map management and Beidou differential data transmission.
(iv) It utilizes an IP-based packet domain GPRS to meet the future requirements for train-to-ground communication upgrades.
(v) The RBC has been enhanced with the addition of mobile block signaling, train occupancy checking, and corresponding safety protection functions to achieve efficient and safe tracking of train operations.
(vi) The safe communication protocol adopts domestic encryption algorithms to enhance the system's information safety level.
(vii) The addition of the train control intelligent center, which comprehensively monitors and manages the operational status of the entire train control system equipment, enables the intelligent analysis of operational status and provides fault prediction capabilities.
(viii) The train control onboard and ground equipment have reserved ATO units and interfaces, laying the foundation for the subsequent implementation of ATO functionalities.

The structure of the next-generation train control system is illustrated in Table 4.6.

Table 4.6. Structure of next-generation train control system.

Onboard	
Rear-End Equipment	End-of-train Communication Antenna; Train Integrity Inspection End-of-train Equipment.
Onboard ATP Equipment	Communication Antenna; Satellite Antenna; Train Control Onboard Communication Radio Unit 1; Train Control Onboard Communication Radio Unit 2; Satellite Receiver Unit; Train Integrity Inspection Unit; Maintenance Data Unit; Train Interface Unit.
Other Interface Systems	ATP Main Control Unit; Wireless Communication Unit; Multifunction Control Unit; Electronic Map Unit; SDU; BTM; Track Circuit Information Reader; Driver Machine Interaction Unit.
Speed measuring equipment	
BTM Antenna	
TCR Antenna	
Ground	
Mobile Communication Base Station	
Onboard equipment	
Switch	
Passive Balise	
Trackside	Centralized Interlocking Station; CTC; Integrated Centralized Station Equipment for Train Control and Interlocking.
Remote Control Station	Integrated Remote Station Equipment for Train Control and Interlocking
Train Control Maintenance Data Network	

Table 4.6. (*Continued*)

Signal System Safety Data Network	
Dispatch Centralized Data Network	
Center	Train Control and Maintenance Center; RBC; CTC; Temporary Speed Restriction Server; Beidou Ground Augmentation System; GSM-R Network Equipment.
Note: The train control onboard equipment for locomotive traction trains may not be equipped with track circuit information reception units and antennas.	
Representation of Next-Generation Train Control System Equipment	
Representation of Optional or Reserved Configurations	
Representation of External Equipment for the Next-Generation Train Control System	
Representation of Optional External System Equipment	

4.3.1.2 *Next-generation train control system equipment composition*

(1) **Onboard equipment**

(i) ATP is primarily composed of the ATP main processor unit, multi-functional control unit, electronic map unit, DMI unit, BTM, maintenance data recording unit, train interface unit, speed and distance measurement unit, train integrity check unit, satellite receiving unit, wireless communication unit, communication radios, antennas, etc.

(ii) EOT should be installed at the rear of the locomotive-hauled train to cooperate with ATP for train integrity checking.

(iii) ATP Communication Radio 1 is used for communication with the RBC, while ATP Communication Radio 2 is used for communication with TSRS and EOT. When there is limited space for installing top antennas on trains, a combination antenna approach is advisable.

(2) **Ground equipment**

(i) Ground equipment consists of trackside equipment and central equipment. Trackside equipment mainly includes the train control and interlocking integration system (TIS) equipment and positioning balise, while central equipment mainly includes RBC, TSRS, train control maintenance center (TCMC), etc. To facilitate maintenance and localized management, the central equipment should be centrally located.

(ii) TIS is a part of the next-generation train control system and performs station interlocking and block signaling functions.

(iii) Stations are divided into centralized interlocking stations and remote control stations based on yard size, operational types, maintenance organization, etc., and are equipped with TIS equipment accordingly.

(iv) Stations should be equipped with track circuits or axle counters, signals at entry and exit points, and positioning balise groups. For example, for a three-track station, the configuration plan is shown in Figure 4.21. Depending on satellite signal coverage and project requirements, positioning beacon groups should be appropriately placed along the track sections.

4.3.1.3 *Key functions of the next-generation train control system*

The next-generation train control system is based on a wireless communication system to transmit traffic permit information and has a moving block function. In addition, ATP has multi-source fusion positioning and train integrity check functions, and it adopts target-distance mode curve monitoring for train safety operations. Its main functions include: IP train-to-ground bidirectional communication, temporary speed limitation, trip protection, automatic train protection, auto-passing neutral section, ATP reservation CTCS 2 and CTCS 3 level functions and

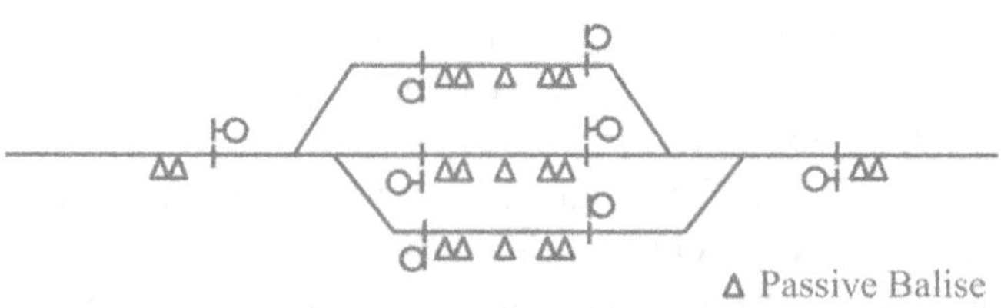

Figure 4.21. Station trackside equipment configuration diagram.

interfaces, shunting operation monitoring, visual check authorization and shunting protection, automatic inter-station blocking operation control, intelligent operation and maintenance, clock synchronization and unified interface standards to achieve interconnection, cyber-attack protection, and domestic cryptographic algorithms to meet information safety requirements.

4.3.2 *Key technologies of next-generation train control system*

The key technologies of the next-generation train control system mainly include a moving block, integrated autonomous train positioning with Beidou, train integrity safety checks, station area integrated full electronic control, IP train–ground wireless communication, etc.:

4.3.2.1 *Moving block*

Section block technology is a safety control technology that blocks the train running line in the section and does not allow other trains to enter, so as to prevent train collisions. This technology is the basic guarantee of railway train operation control and also an important means to improve transportation efficiency. According to different line conditions and transportation needs, section block technology can be divided into inter-station blocks, automatic blocks, and moving blocks. The three train operation modes under the block system are shown in Figure 4.22.

The next-generation train control system calculates the train permit according to the direction of the route, the direction of the section, the status of the line in front of the train, idle conditions, etc. The system can adopt a moving block or a virtual automatic block according to the operation needs so as to realize safe and efficient tracking and control of the train. As a new type of block technology, compared with inter-station block or automatic block, the most significant feature of moving block is that the fixed block section divided by trackside equipment is canceled, and the minimum running interval between trains is determined by the train position, speed, and other running states. When the train is tracking, the block area moves and adjusts continuously with the train running, and the driving permit of the rear train moves dynamically with changes in the position of the front train, so the operation is flexible and efficient.

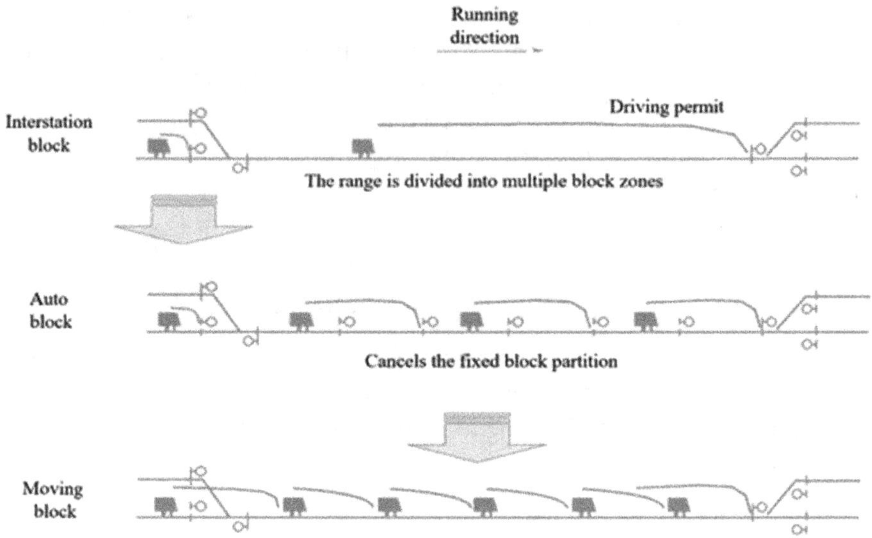

Figure 4.22. Train operation diagram under three kinds of block systems.

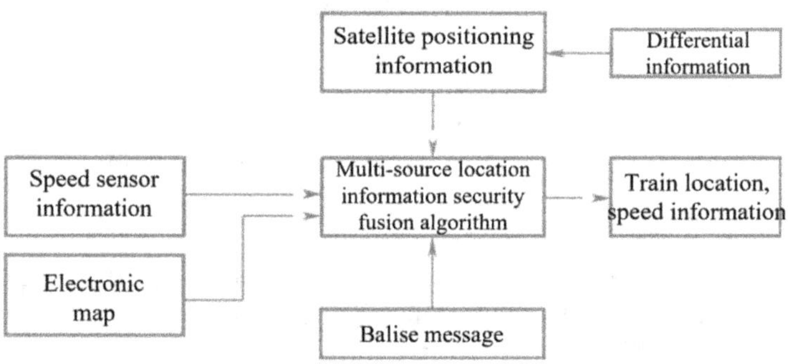

Figure 4.23. Train multi-source fusion location principle schematic diagram.

4.3.2.2 *Integrated autonomous train positioning with Beidou*

The next-generation train control system adopts the multi-source fusion positioning method based on Beidou satellite navigation, wheel speed sensors, and a small number of passive balises on the ground to realize high-precision autonomous train positioning. The principle is illustrated in Figure 4.23.

In areas with good satellite signal reception, trains rely on multi-source fusion methods, such as Beidou, for positioning and speed measurement. In areas where satellite signals are blocked (such as tunnels, mountains, and covered stations), accurate positioning can be achieved by relying on wheel speed sensors and passive balise calibration.

4.3.2.3 *Train integrity safety checking*

The next-generation train control system jointly completes the integrity checking function of the locomotive traction train through the onboard equipment of the train control system and the equipment of EOT. The functional structure is shown in Figure 4.24.

(1) For the locomotive traction train, ATP comprehensively implements the integrity inspection of the train in the following ways:
   (i) It compares and inspects the wind pressure at the head and tail of the train.
   (ii) When the EOT has a positioning and speed measurement function, the position of the head and tail of the train, the speed, and the train length input by the driver are compared and checked.

(2) After the normal operation of EOT, it should be able to continuously monitor the relevant information of the train integrity and periodically communicate with the ATP using a safe communication protocol to send wind pressure, position, and speed information at the tail of the train.

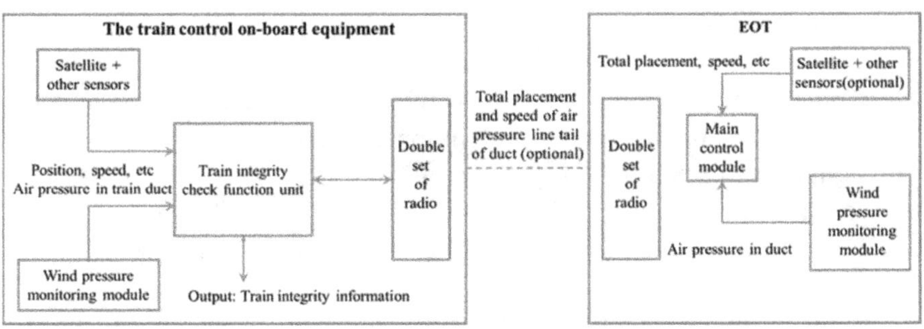

Figure 4.24. Train integrity inspection structure diagram.

(3) ATP periodically sends RBC the following train integrity status:

   (i) **Train integrity:** ATP check to confirm the integrity of the train;
   (ii) **Loss of train integrity:** ATP or manual checking confirms the integrity of the train;
   (iii) **Train integrity unknown:** ATP cannot confirm train integrity.

(4) In the case of loss of train integrity, the next-generation train control system should sound the alarm and implement safety measures; when the train integrity is unknown, the system should give an alarm and take safety measures when the duration of the unknown train integrity is exceeded.
(5) The train integrity checking function should comply with the fail-safe design principle, and the safety integrity level should meet the requirements of SIL4.

4.3.2.4 *Station area integrated full electronic control*

The next-generation train control system integrates all functions of the train control center and computer interlock through TIS, which realizes integrated control at the station and intervals, reduces the number of equipment and the number of interfaces between equipment, improves system reliability, and reduces the cost and workload of repair and maintenance. The use of full electronic modules aims to replace the vast majority of relays, occupy a small space, simplify the design, and ensure convenient site construction and maintenance. The overall structure is shown in Figure 4.25.

(i) The number of equipment is reduced, and the system structure of ground signal control system is simplified. The train control center, computer interlocking, and section block/section comprehensive monitoring (universal speed railway) are integrated into one system, greatly reducing the number of equipment, and the structure of station signal control equipment is relatively simplified and easy to maintain and manage.
(ii) The system architecture is optimized, and the number of interfaces is reduced. The 14 system interfaces are reduced to 8, which reduces the interface complexity of the whole system and facilitates better coordination and cooperation between system devices.

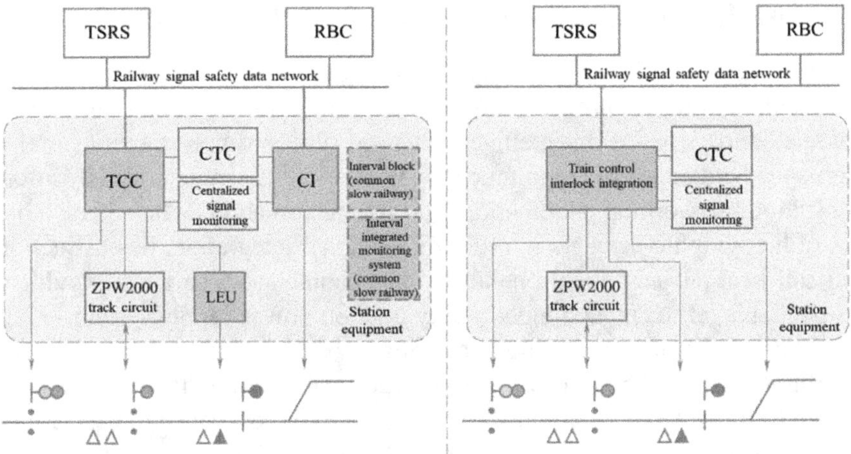

Station equipment when train control center and computer interlock are set up

Station equipment after integration of train control and interlocking

Figure 4.25. Train control interlock integrated structure diagram.

(iii) The indirect interface of the system is changed to a direct interface, which avoids circuitous data transmission and improves the real-time and reliability of communication. First, there is no direct interface between the train control center and RBC, and the occupied information of the block zone, foreign body intrusion limit information, and interval direction information need to be transmitted by the computer interlocking system. Second, there is no direct interface between the computer interlocking and TSRS. When the USU signal is opened by the interlock, the speed limit of the large number of switches needs to be obtained. The information needs to be processed by the train control center and sent to the interlock.

(iv) Data interaction links are reduced, and communication efficiency and safety are improved. The system operation data can be shared internally, which will effectively improve the operating efficiency and accuracy of the system. At the same time, the transmission of data is no longer limited by the transmission medium, reducing the risk of data lag and data error caused by communication delays, communication distortions, and other reasons in an independent setting and avoiding the safety risk during the tolerance period of communication interruption.

4.4 Intelligent Traction Power Supply for HSR

The intelligent traction power supply system utilizes modern advanced measurement, sensing, control, communication, information, AI, and other technologies and is based on intelligent traction power supply facilities and a high-speed bidirectional communication network. It is characterized by informatization, networking, automation, and interaction and has the characteristics of holographic perception, multi-dimensional integration, reconstruction and self-healing, intelligent operation, and maintenance so as to provide a safe, reliable, efficient, and high-quality traction power supply system.

The intelligent traction power supply system is mainly composed of intelligent traction power supply facilities, an intelligent power supply dispatching system, an intelligent power supply operation and maintenance system, etc., as shown in Figure 4.26.

4.4.1 *Intelligent traction substation*

4.4.1.1 *Intelligent traction substation general architecture*

An intelligent traction substation consists of intelligent high-voltage apparatus, wide-area protection, a measurement and control system, an auxiliary monitoring system, and an auxiliary facilities or systems. The overall architecture of an intelligent traction substation (Figure 4.27) is divided into three layers: process layer, interval layer, and substation control layer.

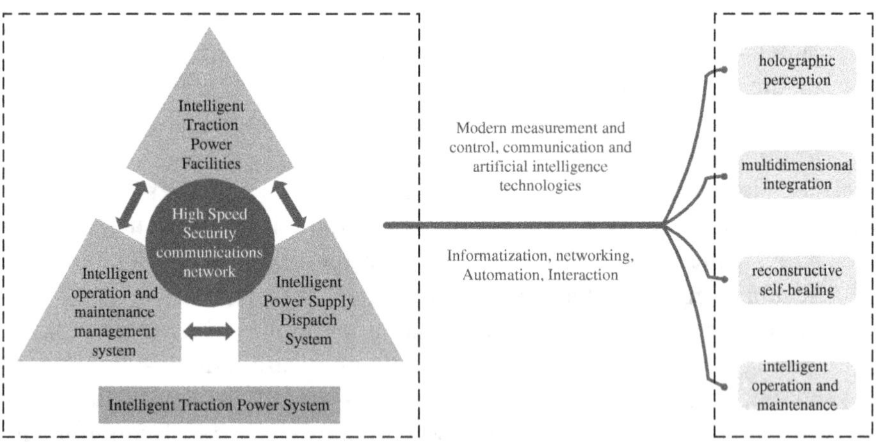

Figure 4.26. Overall composition of intelligent traction power supply system.

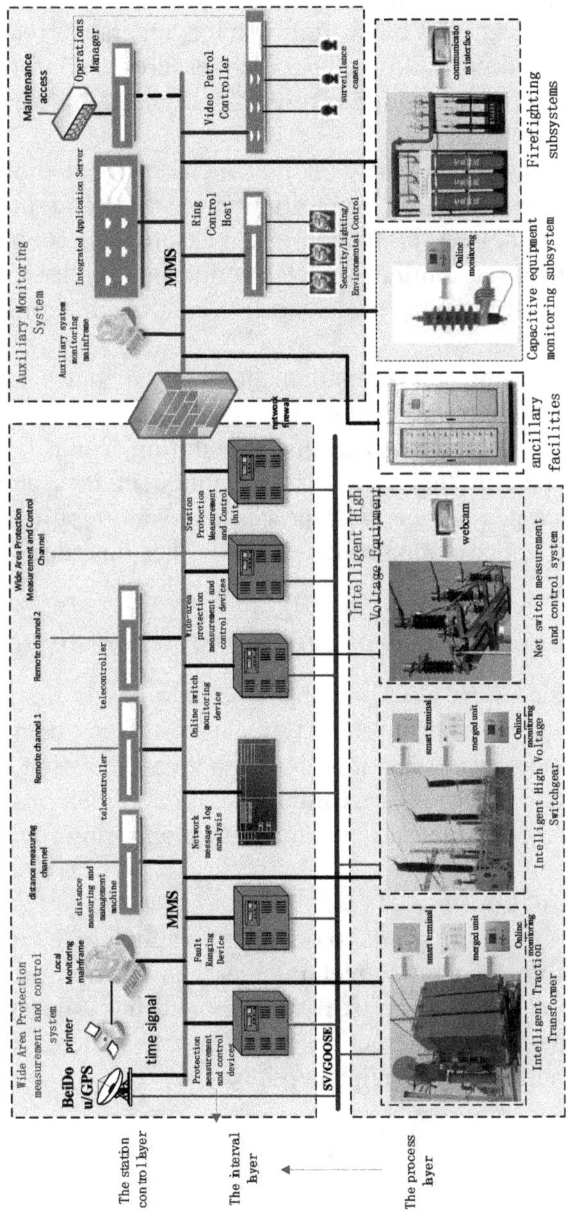

Figure 4.27. General architecture of intelligent traction substation.

The process layer equipment mainly includes intelligent traction transformers, intelligent high-voltage switchgears, voltage and current transformers, arresters, and other high-voltage apparatus, realizing real-time collection and transmission of electric measurement information and equipment state information, and receiving and executing various operation and control instructions.

The interval layer equipment mainly includes local protection and measurement control devices, substation protection and measurement control devices, wide-area protection and measurement control devices, fault location devices, online switch monitoring devices, and other devices, realizing protection, measurement, control, metering, monitoring, and other functions.

The substation control layer equipment mainly includes a local monitoring host computer, a remote control management machine, a time synchronization system, etc., realizing the monitoring, control, alarm, and information interaction functions for all equipment in a substation. It accomplishes data acquisition, data processing, state monitoring, equipment control and operation management, and other functions.

4.4.1.2 *Innovative functions of intelligent traction substation*

The basic function of the traction substation is to power the EMU train, achieve measurement, control, protection, signaling, and communication with remote systems, etc. As an intelligent traction substation, in addition to the above, it also performs a wealth of intelligent functions, which are divided into advanced functions and auxiliary monitoring functions.

(1) **Advanced functions**

Visualization of equipment state: To organically integrate the information, sensing, and automatic control technologies and traction power supply infrastructure, and to perform real-time monitoring, supervision, and operation information of the traction power supply system, realizing the informatization and holographic perception.

Alarm and analysis: To establish an integrated monitoring and control platform, integrating all data information about the substations, realizing standard access and transmission of all information based on IEC61850. In addition to normal monitoring of the operation state, it carries out online real-time analysis and reasoning of the operation state of the traction substation and automatically reports abnormalities of the traction substation with guidance for troubleshooting, realizing intelligent alarms, comprehensive data analysis, video linkage, and other functions.

System reconstruction and self-healing: The system adopts self-healing mode for transient and recoverable faults, while network reconstruction is required for partially recoverable and unrecoverable faults. The self-healing mode is categorized into fault prevention and fault self-healing. In terms of fault prevention, the traction power supply system upgrades its traction transformers and high-voltage apparatus by intelligentization and adopts prognostics health management (PHM), catenary 6C detection, and some other intelligent technologies. As results, fault warning for sub-station equipment and the OCS, quick fault diagnosis, equipment and system's health assessment, system reliability analysis, and risk assessment are realized. The substation equipment mainly relies on the switching of the main and backup switchgears to address the recoverable faults. The network reconfiguration mode can be subdivided into supply section reconstruction and supply segment reconstruction, which determines the appropriate operation mode of the power supply system according to the fault location and the state of the power supply equipment. Additionally, rapid reconstruction function for equipment is available.

(2) **Auxiliary monitoring function**

It realizes all-in-one monitoring and control for the whole substation, integrating the functions of video surveillance, environmental information monitoring, safety precaution, fire alarm, power lighting control, equipment inspection, etc. Data acquisition, access and storage, alarm processing, transmission and communication, linkage and monitoring, and some other functions are supported. The auxiliary system can interact with other systems, vertically exchanging information with the power supply dispatching system and superior video management system and horizontally exchanging information with the protection measurement and control system in the substation.

4.4.2 *Intelligent equipment*

4.4.2.1 *Intelligent equipment composition*

Intelligent equipment for electrified railways (Figure 4.28) consists of the equipment body, intelligent components, substation monitoring unit, and PHM system.

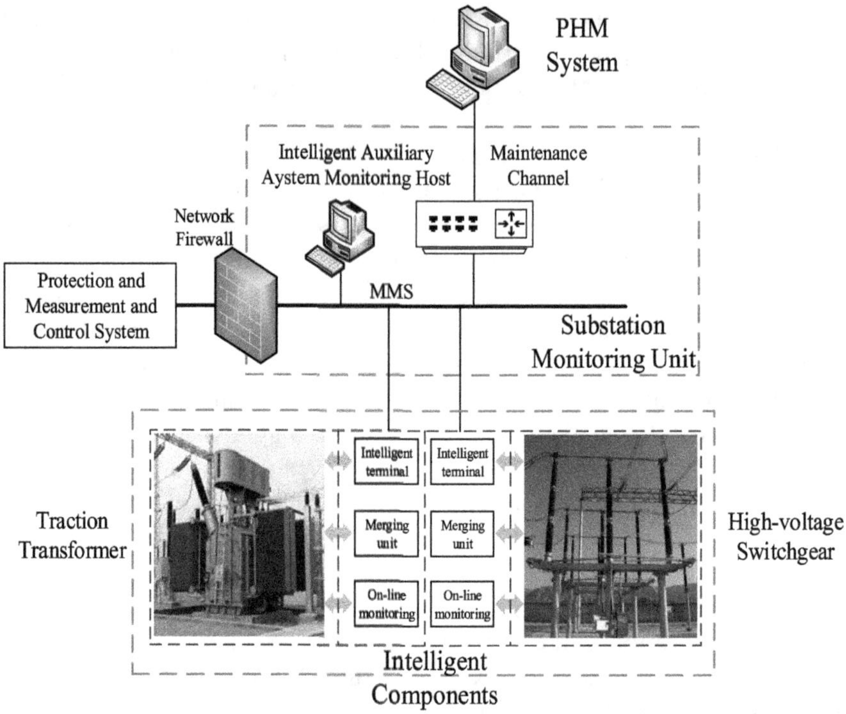

Figure 4.28. Composition of intelligent devices.

(1) **Equipment body:** The intelligent equipment bodies mainly include traction transformers, autotransformers, circuit breakers, disconnectors, voltage and current transformers, arresters, and 27.5 kV gas insulated switchgear (GIS).

(2) **Intelligent component:** The intelligent components are mainly composed of all or part of the intelligent electronic devices (IEDs) of the equipment for measuring, controlling, monitoring, protecting (non-electricity), metering, etc. The IEDs connect the sensors and control actuators of the high-voltage apparatus through cables or fiber-optic cables, forming an organic unit which realizes all or part of the functions that include signal acquisition, operation control, fault monitoring, and non-electricity protection of the high-voltage apparatus. Besides, the intelligent component provides various data to the substation monitoring unit.

(3) **Substation monitoring unit:** The substation monitoring unit is part of the intelligent auxiliary system of a substation. It is responsible for the access, data processing, and forwarding of all monitoring devices in a substation. It is responsible for the functions of data analysis and diagnosis, monitoring and warning, and comprehensive display of various information, etc., achieving the monitoring and management of all monitoring devices.

(4) **PHM system:** It implements equipment fault diagnosis, performance detection, fault prediction, health management, and life tracking.

4.4.2.2 *Technical characteristics of intelligent component*

(1) **Measurement digitization:** The intelligent component digitally measures relevant parameters of high-voltage apparatus or its components on the spot. The measurement results can be sent to the substation control layer network or (and) process layer network as needed to operate and control high-voltage equipment or its components.

(2) **Control networking:** It implements network-based control for high-voltage apparatus or its components with control requirements, including three methods: (i) local control of high-voltage apparatus or its components implemented by their self-contained controllers; (ii) intelligent component implements control through local controllers or actuators; (iii) the substation control layer devices implement control through intelligent components. Under regular operation, the priority order of networked control is substation control layer devices, intelligent components, and local controllers.

(3) **State visualization:** Based on self-detection information and other state information of high-voltage apparatus obtained through information interaction, the intelligent component represents its self-diagnosis results in a recognizable manner by other relevant systems, making the state of high-voltage apparatus observable in the power supply system.

(4) **Functional integration:** The functional integration of intelligent components is mainly reflected in three aspects at the design stage: integrating sensors or (and) actuators with high-voltage apparatus or its components to achieve specific monitoring or (and) control purposes; integrating

voltage and current transformers with transformers, circuit breakers, and other high-voltage apparatus to reduce land occupation in the substation; and integrating relevant measurement, control, metering, monitoring, and protection when designing the intelligent components.

(5) **Information interactivity:** Interacting with the substation monitoring unit, the intelligent devices submit the self-diagnosis results (including active and on-demand) to the substation monitoring unit, making it one of the basic information for dispatching decisions and developing high-voltage apparatus accident plans. Interacting with the wide-area measurement and control system, the switching state information and voltage and current transformer data of intelligent devices are submitted to the wide-area measurement and control system, achieving functions such as equipment protection, system reconstruction, and hierarchical locking.

4.4.2.3 *Composition and function of substation monitoring unit*

The substation monitoring unit consists of a monitoring unit and a communication management unit. The monitoring unit is a functional unit of the comprehensive application server of the intelligent auxiliary system, which is responsible for obtaining, processing, and forwarding all state monitoring data in the substation. The monitoring unit has the functions of data analysis and diagnosing, monitoring, and warning. Moreover, it shows various information linking to the comprehensive application server. The communication management unit is responsible for communicating with the monitoring devices of intelligent devices, obtaining monitoring data, and conducting standardized modeling. It transmits standard data to the monitoring unit through the communication protocol based on DL/T 860.

The interface of the substation monitoring unit is shown in Figure 4.34, which includes two parts: the internal system interfaces I0 and I1, and the external system interface I2. The functions of the substation monitoring unit are accomplished by the comprehensive application server sharing a human–machine interface with the intelligent auxiliary system. The I0 interface is the interface between the monitoring device and the communication management unit. If the communication management unit is not configured, this interface may not be set. The I1 interface is the interface between the communication management unit or monitoring devices that support DL/T860 and the comprehensive application server.

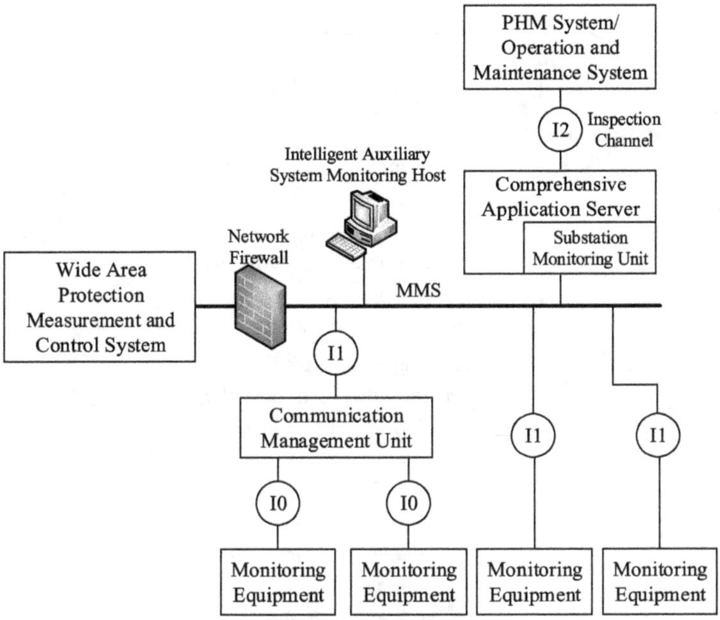

Figure 4.29. Schematic diagram of station end monitoring unit interface.

The I2 interface is the interface between the comprehensive application server in the substation and the remote PHM system, operation, and maintenance system master station.

4.4.2.4 *Prognostics health management (PHM) system functions*

The PHM system of intelligent devices mainly implements seven functions: fault diagnosis, fault warning, health assessment, reliability and risk assessment, residual life prediction, maintenance decision-making, and data management, as shown in Table 4.7.

4.4.3 ***Intelligent power supply dispatching system***

The intelligent power supply dispatching system is based on cloud computing, big data, and AI technologies, with the basic requirements of data panoramic visualization, dispatching collaboration, job automation, and decision-making intelligence. It realizes remote monitoring and control, dispatching operation management, auxiliary monitoring, and other

Table 4.7. Summary of fault prediction and health management (PHM) system functions.

No.	Function name	Time scale	Function description	Result
1	Fault diagnosis	Short term	By analyzing and processing protection action information, circuit breaker opening and closing information, online monitoring data, etc., intelligent algorithms or expert diagnostic systems are used to determine fault components, locations, types, etc.	Fault component, location, and type
2	Fault warning	Short term	Integrating equipment monitoring and detection information such as transformer oil chromatogram, temperature rise data, and insulation resistance values of high-voltage apparatus, analyze changes in equipment state, timely identify potential faults during the latent period or early stages of equipment faults, and determine possible types and severity of faults. Comprehensively judge the trend of fault development and change and timely provide fault warning before the consequences of the fault manifest.	Probability of failure, time of failure, number of faulty equipment within a certain period
3	Health assessment	Medium and long term	Analyze equipment's health state and development trends through offline detection, fault or maintenance record samples, and other historical data collected by substation monitoring units.	Health state, health value
4	Reliability and risk assessment	Medium and long term	Comprehensively considering equipment performance degradation, service environment, human maintenance activities, etc., the reliability and risk of equipment are analyzed and evaluated by processing and analyzing the online and offline data of equipment. The reliability indicators obtained through analysis and evaluation represent the reliable operating performance of the equipment, while the risk assessment indicators represent the possibility of equipment failure and the severity of its consequences and losses.	Reliability indicators, such as availability and failure rate; risk assessment indicators, such as average downtime and severity

Table 4.7. (*Continued*)

No.	Function name	Time scale	Function description	Result
5	Residual life prediction	Full life cycle	Based on the operation and service state and various parameters of the equipment, conduct comprehensive evaluation, analysis, and prediction to determine the residual life of the equipment.	Residual life
6	Maintenance decision-making	Full life cycle	Based on PHM results, maintenance regulations, and equipment health state, reasonable maintenance strategies are considered for yielding optimized maintenance cycles and suggestions for maintenance methods. Through subsequent summary and analysis, the maintenance plan is continuously improved.	Maintenance cycle and maintenance method
7	Data management	Full life cycle	Implement preprocessing functions such as denoising, filtering, and compression for the data of HSR traction power supply system and its equipment and be able to store and manage customized data after data analysis and signal preprocessing. Provide data of various indicators for online and offline monitoring of crucial equipment, and collect data from multiple sources, such as the equipment itself, adjacent equipment, and related monitoring devices, including offline detection data, historical fault record data, and real-time online monitoring data. Provide information exchange and visualization through data interfaces, meanwhile storing information related to equipment operation and maintenance procedures for maintenance decision-making function module.	Historical state information, visualization of traction power supply system and equipment-related data, and decision-making reports

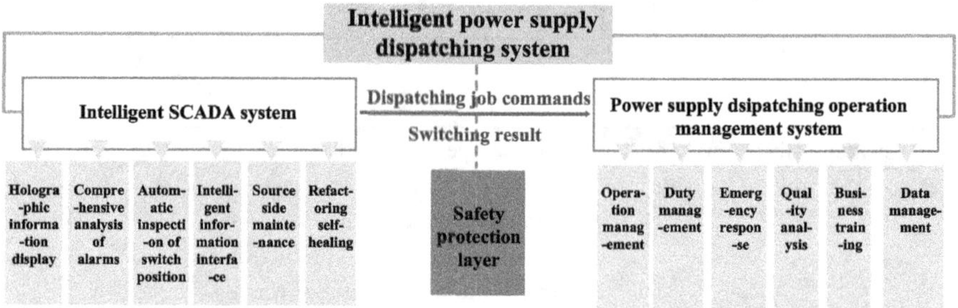

Figure 4.30. Schematic diagram of the composition of the intelligent power supply dispatching system.

functions for the traction power supply system, supporting coordination and linkage with other related systems, and improving the safe, reliable, and cost-effective operation ability of the traction power supply system.

4.4.3.1 *Composition of intelligent power supply dispatching system*

The intelligent power supply dispatching system (Figure 4.30) consists of an intelligent remote control system (SCADA) and a power supply dispatching operation management system. In addition to performing conventional SCADA functions, the intelligent SCADA system also has functions such as source maintenance, comprehensive analysis of alarms, and panoramic information display. The power supply dispatching operation management system has specialized management functions, such as automatic acquisition and statistics of railway power supply dispatching duty information, intelligent review and network circulation of work plans and orders, intelligent analysis and decision-making of emergency response, intelligent management of business training, and intelligent statistical analysis of operation and work quality.

4.4.3.2 *Innovative functions of intelligent power supply dispatching system*

(1) **Remote viewing functions for intelligent traction substation:**
- Real-time monitoring of multi-channel auxiliary monitoring image information for the same substation or multiple substations;

- Remote retrieval of image information from the auxiliary monitoring system of the substation; remote control of the video front-end equipment of the substation auxiliary monitoring system, with the control priority higher than that of the power supply section; linkage with remote control operation, switch tripping, fire alarm, and other major fault information, and automatically release corresponding auxiliary monitoring real-time images; store and query historical data of video information before and after faults; based on the electronic map and BIM model of the substation, the three-dimensional display function dynamically displays the state of power supply equipment and various auxiliary monitoring alarm information in graphical or animated form.

(2) **Source-end maintenance function with intelligent traction substation:** Import and automatically update the remote control/remote adjustment/remote signaling/remote measurement (four remote) point table, fault report, fault recording, and setting value static data model of the intelligent traction substation; automatically raise an alarm and synchronize the changes of the four remote dynamic data model during operation; automatically track and display the changes in the dynamic main connection diagram of intelligent traction substations during operation.

(3) **The main substation function of reconstruction and self-healing:** Remotely tune the reconstruction and self-healing execution mode (automatic execution mode of the controlled substation or manual confirmation mode of the dispatching end); receive reconstruction and self-healing requests from intelligent traction substation, and preview the reconstructed power supply mode in a graphical manner; confirm or revoke the reconstruction and self-healing request of the intelligent traction substation.

(4) **The function of achieving collaborative driving dispatching and optimizing driving command based on sharing information with the CTC interface:** Optimize the power supply operation mode during faults; optimize the train operation organization during faults based on power supply capacity; collaborative construction operation dispatching, auxiliary driving command, etc.

(5) **Automatic inspection function of switchgear:** Automatically inspect the remote signaling state of preset switches and provide alarm prompts

for abnormal states; automatic inspection of the preset switch remote control ready state and automatic alarm prompt for abnormal states; automatic inspection of communication and link state of the main and backup remote control management machines of the controlled station, and automatic alarm prompts for abnormal states.

(6) **Intelligent alarm and locking function:** Remote control operation automatically associates the power supply circuit current with the terminal network voltage detection, intelligently determines the remote control results, and automatically prompts for abnormal alarms; establish a unified alarm logic for various dispersed, isolated, and differentiated fault information, filter redundant fault information, and display fault information using the power supply arm as a unit; real-time online screening and inference for hierarchical classification of forecast information, and warning prompts for potential equipment hazards.

(7) **The function of panoramic display of information related to the traction power supply system:** The panoramic display of real-time train position information; power supply system operation mode, substation operation mode, and contact network operation state; real-time image information about the substation; natural disaster alarm information, such as wind, rain, snow, earthquake, and foreign object intrusion alarm information.

4.4.4 *Auxiliary monitoring system*

The auxiliary monitoring and control system integrates parts or all of the auxiliary equipment information in the traction substation, such as video monitoring and inspection, environmental monitoring, safety prevention, fire alarm, and power lighting control, to achieve information sharing, alarm linkage, and other functions, providing support for centralized monitoring and operation and maintenance of the traction substation.

4.4.4.1 *Composition of auxiliary monitoring system*

This system sets up an auxiliary monitoring system platform in the traction substation, section post, and AT post, collects auxiliary equipment information within the substation through standard interfaces, achieves information sharing and alarm linkage functions, and uploads monitoring and alarm information, as shown in Figure 4.31.

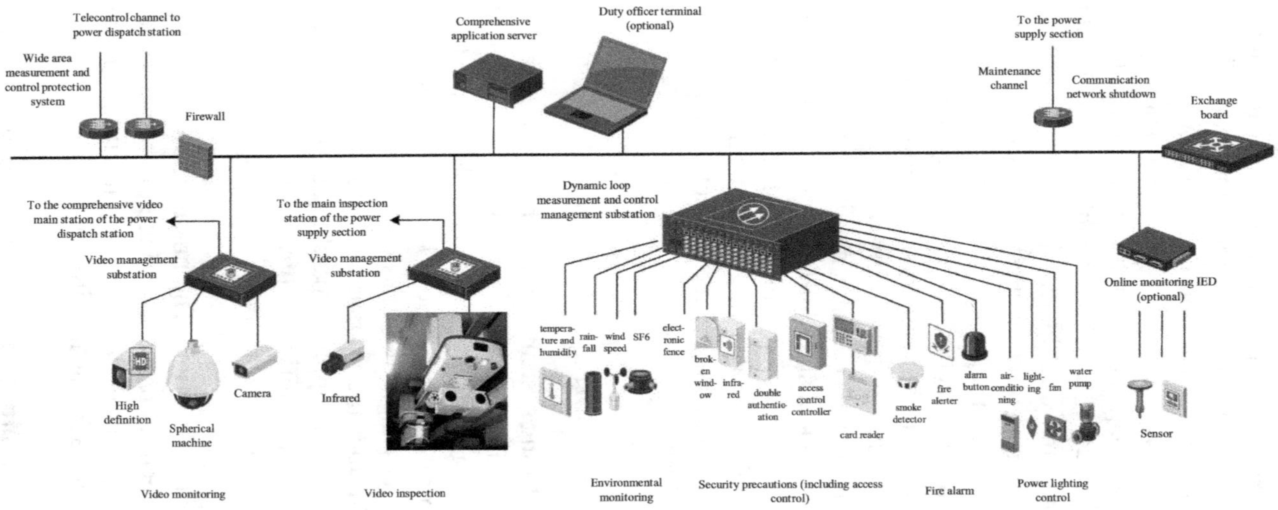

Figure 4.31. Composition of intelligent traction substation auxiliary monitoring system.

The auxiliary monitoring system consists of a substation-level auxiliary monitoring system and the main station of the auxiliary monitoring system for the dispatch center and power supply workshop. The substation-level auxiliary monitoring system performs configuration and integration of various subsystems, namely security prevention (including access control), environmental monitoring, power lighting control, automatic fire alarm, intelligent inspection, access to 27.5 kV cable monitoring, SF6 gas monitoring, and intelligent primary equipment monitoring. It has the conditions to access or expand other monitoring subsystems to achieve unified data collection for each subsystem and unified display and submission, based on which intelligent linkage functions of each subsystem can be achieved.

An auxiliary monitoring system master station is added at the dispatch end, where the safety monitoring information directly related to the dispatch is uploaded to the auxiliary monitoring system master station through the remote channel and replicated to the power supply workshop. At the same time, information can be uploaded to the power supply workshop-level auxiliary monitoring system through the operation and maintenance channel, achieving video inspection requirements and online monitoring data collection.

4.4.4.2 *Auxiliary monitoring system substation function*

The auxiliary monitoring system provides highly integrated and all-in-one monitoring of the entire traction substation, including monitoring, environmental monitoring, security, fire alarm, equipment inspection, and power lighting control. It has functions such as collection and access, data storage, alarm processing, transmission communication, linkage, and monitoring. In addition, the system can interact with relevant systems, vertically with the power supply dispatching system, superior video management system, etc., and horizontally with the protection, measurement, and control system in the substation.

(i) **Intelligent linkage function:** The linkage function is achieved among the subsystems of security prevention (including access control), environmental monitoring, power lighting control, automatic fire alarm, 27.5 kV cable monitoring, and SF6 gas monitoring.

(ii) **Intelligent control function:** Intelligently control the air conditioning and lighting, achieving green and energy savings.

(iii) **Visualization display function of monitoring information:** Visualization display and comprehensive alarm of monitoring information for each subsystem.

(iv) **Intelligent inspection:** Through video inspection, the main equipment operation state information is automatically monitored in each station, and combined with online monitoring state information of the equipment, abnormal equipment operation is timely detected to improve equipment operation reliability.

4.4.4.3 *Video inspection subsystem*

The intelligent traction substation achieves auxiliary monitoring by installing a video inspection subsystem, usually equipped with guide rail type inspection cameras in the control room and 27.5 kV high voltage room, inspecting secondary devices, and 27.5 kV high voltage equipment. Fixed inspection cameras are installed for indoor and outdoor traction transformers, ATs, 220 kV switchgears, 220 kV GIS, and 27.5 kV outdoor equipment. In addition, intelligent inspection robot technology has begun to be applied in railway traction substations. Table 4.8 shows the main equipment composition of the video inspection subsystem for a HSR line traction substation, AT station, and section post.

4.4.5 *Intelligent monitoring system for OCS*

The catenary intelligent monitoring system (named the 6C system) comprehensively applies sensing, communication, data analysis, and other technologies to detect and monitor the traction power supply system,

Table 4.8. Main equipment composition of video patrol subsystem.

No.	Substation	Fixed camera (outdoor)	Fixed camera (indoor)	Rail mounted camera	Intelligent inspection robot	Infrared online thermometer
1	Traction substation	7	0	2	1	2
2	AT post	2	0	2	0	0
3	Section post	3	0	2	0	0

providing a technical basis for fault analysis and maintenance of power supply equipment. The 6C system (Figure 4.32) is mainly composed of a high-speed pantograph and a catenary comprehensive detection device, a catenary safety inspection device, a vehicle-mounted catenary operation state detection device, a catenary suspension state detection and monitoring device, a pantograph sliding plate monitoring device, and a ground monitoring device for the catenary and power supply equipment.

1. 1C (Comprehensive pantograph and catenary monitor device, CPCM): The onboard catenary detection equipment installed on the

Figure 4.32. The 6C system.

comprehensive testing train conducts thorough testing of the parameters and state of the HSR catenary, as well as the relationship between the high-speed pantograph and the catenary. It mainly tests the basic parameters of the OCS, such as the height difference of the contact wire, hard points, pressure, wire height, and locator slope.

2. 2C (Catenary checking video monitor device, CCVM): The testing equipment is temporarily installed in the operating EMU and locomotive cabs to detect the OCS state, statistically analyze the technical state of contact and suspension components, and guide the maintenance of the contact network state. 2C, as a device monitoring and detection method, uses high-speed cameras to capture device photos and discover device defects by analyzing and retrieving device photos. It is a supplement to traditional inspection methods. It mainly has an apparent monitoring effect on abnormal traction power supply equipment and external environmental conditions (such as bird nests or trees posing a safety risk).

3. 3C (Catalog checking online monitor device, CCLM): Real-time displaying the images or videos of the operation state of pantograph and catenary and automatically recording the catenary pull-out values and the location where the hard point exceeds the limit are of great significance for the timely clearing up catenary faults, prevention of pantograph and catenary faults, and ensuring train operation safety. 3C data include fault images (recordings) and defect locations, explicitly monitoring the pull-out value and hard point defects of the OCS. It is also a further verification of 1C or non-contact detection data.

4. 4C (High precision catenary checking monitor device, CCHM): It is a device that uses high-definition imaging cameras to capture and record critical parts of the key components of the catenary suspension. It is more progressive than traditional detection and monitoring methods. It greatly improves labor efficiency and detection and monitoring efficiency, with the advantages of high accuracy and a short detection and monitoring cycle. More importantly, it has achieved local localization and detection of key equipment of the OCS, directly locating the type and location of defects, timely grasping the abnormal state and hidden defects of the equipment, and ensuring that the power supply equipment is under control.

5. 5C (Pantograph video checking device, CPVM): Through the installation of pantograph slide plate monitoring devices at HSR stations, where EMUs enter and exit the roundhouse, and station throat areas, it monitors the technical state of the EMU pantograph and slide plate, timely detecting abnormal conditions of the operating EMU pantograph and slide plate, guiding fault elimination, and ensuring the excellent operation state of the contact network and pantograph.

6. 6C (Ground monitoring device for catenary and power supply equipment (CCGM): Monitoring equipment are set up at special cross sections of the catenary (such as positioning points and tunnel entrances and exits) and traction substations to monitor parameters such as contact network tension, vibration, lifting amount, lead temperature, compensation displacement, the insulation state of power supply equipment, and cable head temperature and guide the maintenance of the contact network and power supply equipment.

Table 4.9 provides a summary of the functional classification of the 6C system.

Table 4.9. Functional classification of the 6C system.

6C	Carrier for building	Major function	Inspection cycle	Speed level
1C	Vehicle mounted (high-speed comprehensive inspection train)	The comprehensive inspection train conducts line real-speed detection of the catenary parameters and the operating state of the catenary. The main detection parameters include catenary contact force, catenary voltage, catenary high voltage, catenary dynamic pull-out value, catenary hard points, and catenary offline sparks	Ten days per cycle	200–350 km/h
2C	On board (operating EMU driver's desk)	The OCS safety inspection device refers to the use of portable video capture equipment to collect video of the state of the OCS and analyze the technical state of the suspension components of the OCS	Regular	200–350 km/h

6C	Carrier for building	Major function	Inspection cycle	Speed level
3C	On board (operating EMU)	Install OCS detection equipment on the operating high-speed trains to achieve dynamic detection of the contact network state of HSR	Online	200–350 km/h
4C	On board (OCS operation vehicle or dedicated vehicle)	Implement high-precision imaging detection of the components of the OCS suspension to guide the elimination of hidden faults in the overhead contact system	Regular (inside the sunroof)	100 km/h or specified speed
5C	Ground (station and entrance and exit lines of roundhouse)	Using video images to monitor the technical state of pantograph skateboards at stations and entrance and exit lines of the roundhouse	Real-time	
6C	Ground (OCS special cross-sections and power supply equipment)	Ground monitoring devices are installed at the special cross-section of the OCS and power supply equipment to monitor the tension, special section, vibration, lifting amount, thread temperature, compensation displacement, insulation state, and temperature of the contact network and power supply equipment and guide the maintenance of the contact network and power supply equipment	Real-time	

Various types of 6C devices and equipment have been widely used on site, leveraging their functions such as dynamic detection, constant velocity detection, and fixed-point monitoring to achieve full coverage, comprehensive and periodic detection, and tracking of power supply equipment. The intelligent 6C system integrated data processing center integrates and shares a large amount of detection and monitoring data, and automatically and intelligently analyzes the trends in data to better guide the operation and maintenance of power supply equipment.

At present, multiple data analysis, big data, cloud computing, service-oriented distributed architecture, and intelligent image recognition technologies have been widely applied in constructing the 6C system. The 6C system comprehensive data processing center (Figure 4.33) adopts a

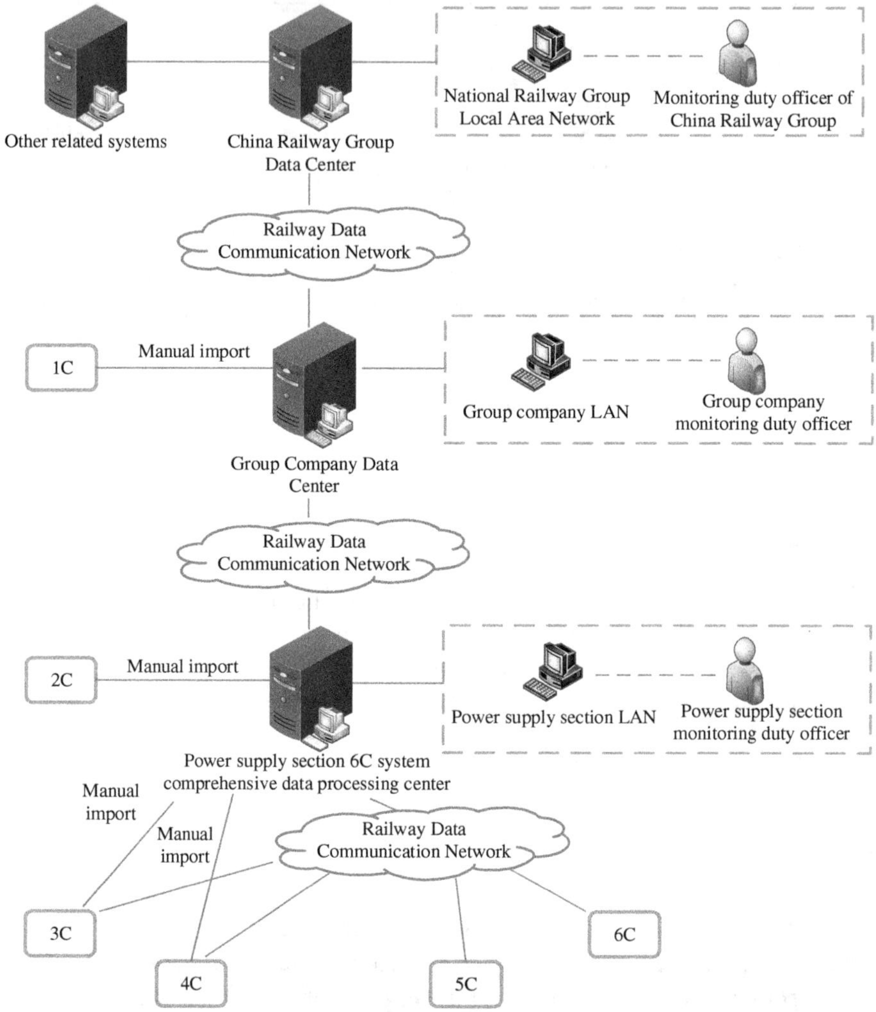

Figure 4.33. Overall architecture of the 6C system comprehensive data processing center.

service-oriented distributed architecture and a unified software platform. It forms a three-level system of headquarters, railway bureau, and power supply section. The software and hardware platforms of the data center adopt a centralized storage infrastructure, deepen the application of extensive data analysis, and use big data technology for hierarchical and distributed storage management of massive heterogeneous data, integrating intelligent image recognition and multivariate data analysis technology to enhance intellectual analysis capabilities. At the same time, it is supplemented by a unified and comprehensive system management and security guarantee system to meet the interoperability and interconnectivity of the 6C system's comprehensive data processing center, application scalability, and portability, as well as the requirements of data analysis integration and intelligence.

4.5 Intelligent Communications for HSR

HSR communication is a dedicated communication system that was designed to serve HSR transportation. It refers to the collective term for communication equipment and services that facilitate the transmission, exchange, and processing of operational and managerial information in HSR operations. Currently, the HSR communication system in China consists of the bearer network, service network, and support network. The communication equipment mainly includes communication lines, transmission systems, data communication systems, telephone exchanges and access systems, dispatch communication systems, dedicated mobile communication systems, conference television systems, integrated video surveillance systems, emergency communication systems, clock and time synchronization systems, and related devices.

To ensure operational safety, the development of a dedicated communication network (private network) for railways is a consensus in the global railway industry. The railway private network serves railway transportation, utilizing dedicated frequencies and other resources to establish an independent network that ensures high levels of safety and reliability. It is physically isolated from external networks, such as the internet. Measures such as deploying security equipment, implementing access controls, and conducting security monitoring are taken to isolate it from various application systems. This network provides communication means for transportation production, train dispatching, safety monitoring, and operational management. Adhering to internationally standardized technical systems and considering railway application characteristics, China has constructed a railway private network system architecture by incorporating network elements that accommodate specific railway functions.

HSR intelligent communications in the form of a private network serve the high-security and high-reliability data transmission requirements of key core operations in intelligent high-speed trains, including train-to-ground and train-to-train communications. It primarily carries out critical core operations, such as train dispatching and control, as well as operational maintenance. This includes functions such as dispatch communication, train control, synchronized locomotive operation, and autonomous driving.

Due to the rapid development of technologies such as 5G, big data, AI, cloud computing, the IoT, and IPv6, communication networks are moving toward the goals of intelligent network operations, intelligent resource management, intelligent system maintenance, and intelligent application of services. Digitization, high bandwidth, low latency, and wide connectivity have become the fundamental characteristics of intelligent railway communications. Communication networks are increasingly flexible in terms of service provisioning and scheduling, and they require higher levels of intelligent operation and maintenance. This section focuses on introducing key aspects such as 5G for Railway (5G-R), the transport network, and multimedia dispatch communication systems in intelligent railway communications.

4.5.1 *Private 5G network for railway communication*

Based on the 3GPP specifications for 5G systems, the user requirements are mapped onto the equipment to form the 5G-R system. 5G-R includes the core network, radio access network, user equipment, application access management equipment, and operation and support systems. The architecture of the 5G-R system is shown in Figure 4.34.

(i) **Core network:** The core network consists of both the shared core network equipment and the local core network equipment. It encompasses functions such as registration management, mobility management, session management, traffic control and processing, user plane management,

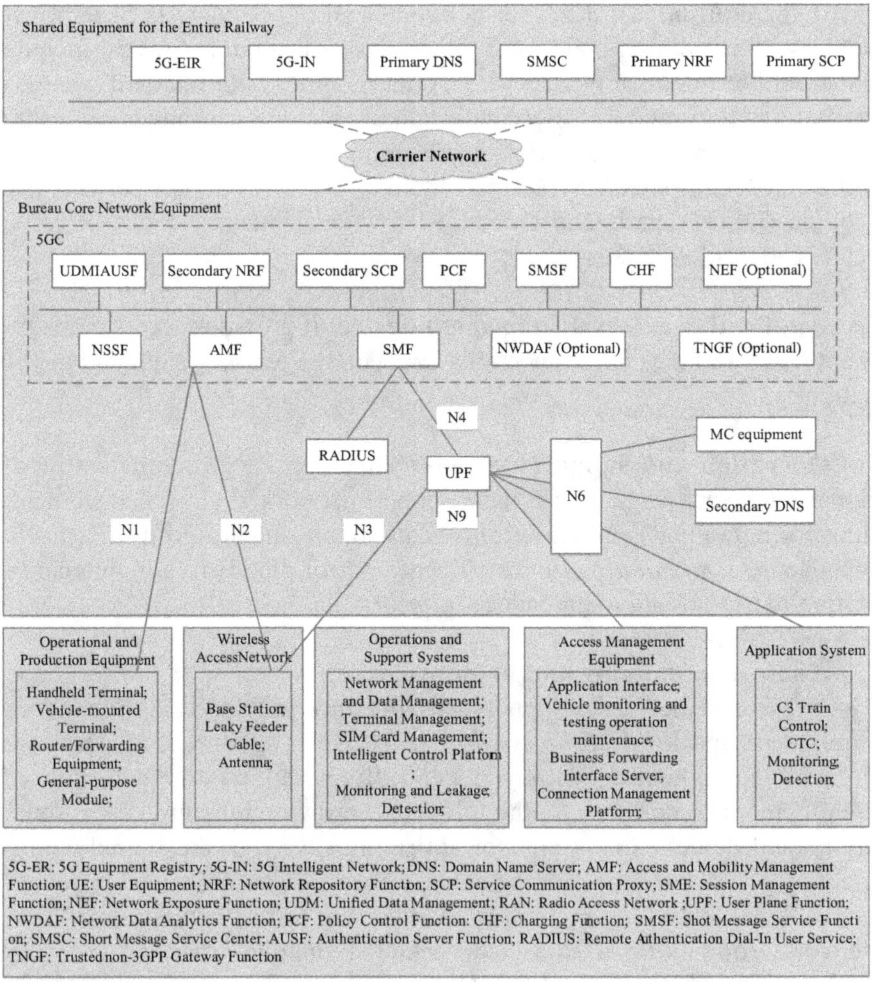

Figure 4.34. The architecture of the 5G-R system.

security, edge computing, network slicing, and quality of service (QoS) management.

(ii) **Radio access network:** The base station equipment is divided into macro and micro base stations, which consist of remote radio units (RRUs) and building baseband units (BBUs).

(iii) **User equipment:** The user equipment of 5G-R consists of terminal devices, service terminals, and SIM cards. Terminal devices include hand-held terminals, router devices, data modules, and onboard wireless communication modules. Service terminals include cab-integrated radio communication equipment (CIR) and onboard integrated wireless communication equipment. SIM cards include universal subscriber identity module (USIM) cards and embedded universal integrated circuit cards (eUICCs).

(iv) **Application access management device:** It provides access management and interfaces for connections between the 5G-R system and application systems.

(v) **Operation and support systems:** They consist of equipment/functional units such as network management, network slicing management, network data management, terminal data management, terminal security management, monitoring, an intelligent control platform, a maintenance testing platform, and maintenance tools.

Core network planning and deployment

The 5G-R core network adopts the architecture of "Shared core network Equipment and local core networks in 18 Railway Bureaus." The core network equipment is deployed centrally based on the principle of "deployment according to the different railway bureaus, with some equipment deployed centrally for entire railway." For the shared equipment of the core network for the entire railway, the remote disaster recovery backup networking method is adopted. For railway bureau-level core network equipment, a dual-data center redundant disaster recovery backup networking method is adopted with the data centers located in different locations of the same city. Edge computing node devices can be deployed at marshaling stations, major passenger stations, and other locations according to service needs to meet the requirements of high-bandwidth or low-latency applications. The links between core network nodes are protected using dual-channel and dual-route protection methods.

The planning and deployment of the 5G-R core network are depicted in Figure 4.35.

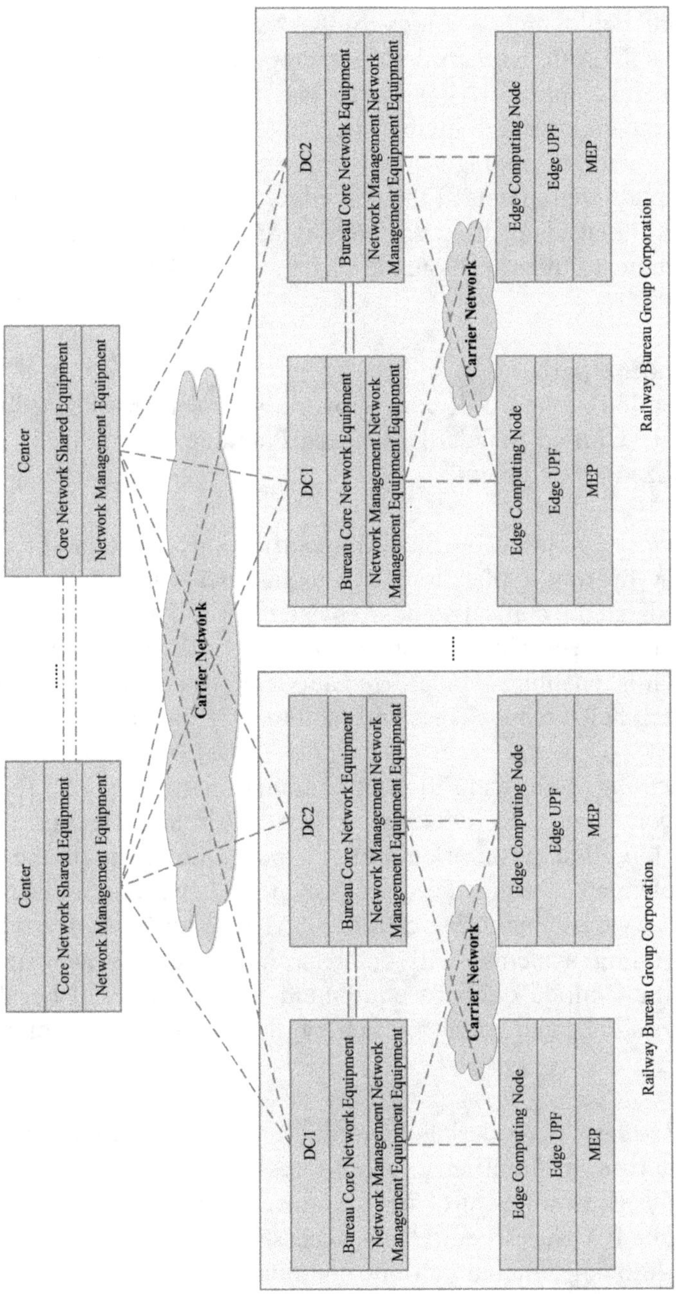

Figure 4.35. The architecture of the 5G-R system.

Radio access network (RAN) networking

The network deployment strategy of RANs should fully consider the requirements for system reliability, so it adopts a line-dependent and hierarchical coverage approach. For the railway lines with train control systems, a redundant coverage method is adopted. The RAN should ensure continuous coverage with high-quality radio signals. It is required to meet the throughput requirements at the cell edge. The predicted cell edge rate should meet the needs of train dispatching and control applications. It is also important to plan reasonable distances between different base stations.

Implementation path

The construction of the 5G-R core network is implemented based on the principles of "Unified Standards, Unified Planning, Centralized Access, and Centralized Procurement".

Unified standards: Based on the 5G specifications of 3GPP, combined with specific railway applications and requirements for safety and reliability, a dedicated technical standard system for railway 5G private networks is established. This system incorporates the smooth upgrade of 3GPP standards, enabling standardized applications and interconnectivity. This approach helps reduce costs and facilitates maintenance.

Unified planning: Following the implementation strategy of "Terminal First, Network Later; Core Network First, Wireless Network Later; Demonstration First, Promotion Later," the deployment of the 5G-R system is planned. During the transition period, various mobile communication systems, including 5G-R, GSM-R, and 450 MHz wireless train dispatching systems, will coexist along the railway lines. By deploying multi-mode onboard equipment first, trains will be able to operate across lines and networks, achieving a smooth transition for the system.

Centralized access: Considering the multiple and diverse requirements of railway applications, a unified ground access platform is constructed to enable various non-train control services to access the core network of the 5G-R system. It ensures centralized access, authentication, and secure control. Additionally, onboard mobile communication nodes are deployed to facilitate dispatching communication and non-train control applications for unified access to the 5G-R network. This solves the issue of limited space for communication modules of multiple different professional equipment. It can reduce the number of rooftop antennas and allow for a more efficient layout.

Centralized procurement: Drawing on the procurement mode and experience of telecommunication operators, joint centralized procurement solutions under large-scale deployment conditions are studied to reduce the comprehensive costs associated with equipment throughout its entire life cycle.

Evolution strategy

The 5G-R system will be implemented in an orderly manner through three stages: research and development, demonstration assessment, and deployment application.

During the research and development stage, several tasks, including applying for dedicated frequencies, addressing key technologies of the 5G-R system, developing major specialized equipment, formulating essential standards, conducting experimental verification, and carrying out trial assessments, will be completed. Lines with appropriate conditions and demonstrating good effects to conduct engineering demonstrations are selected. This stage aims to lay the foundation for the large-scale construction and deployment of the 5G-R system.

During the demonstration assessment stage, the following five tasks should be completed. The first one is refining the standard system of 5G-R. Then, trial assessments for the 5G-R system and its main application equipment are conducted, following the new product deployment process in the railway industry, including wireless networks, train control systems, and dispatch communication. The third step involves initiating the update of onboard terminals to facilitate the transition period from 450 MHz wireless train control and GSM-R to 5G-R. The fourth step involves initiating the construction of shared equipment and a centralized core network across the entire railway network. The final step is to initiate the construction of the bearer network.

During the application deployment phase, the following five tasks should be completed. The first task is to accomplish the centralized construction of shared equipment and the core network across the entire railway network. Then, the centralized transformation of onboard equipment is completed. The third task is the construction of the 5G-R system for new railway lines. For existing railways, the transformation of the 5G-R system is carried out based on maintenance cycles and prioritizing areas with urgent needs. Finally, the bearer network is constructed simultaneously. The ultimate goal is to comprehensively establish the 5G-R system, achieving a complete upgrade of the mobile communication systems for high-speed rail lines and major conventional mainline railways.

4.5.2 *Multimedia dispatching communication system*

The new requirements proposed by intelligent railways for dispatching communication include intelligent dispatching communications, intelligent maintenance and repairment, intelligent train crew communications, emergency intelligent communications, and multimedia integrated dispatching. The multimedia dispatching communication system is based on the mission-critical communication (MCX, which includes broadband push-to-talk (MCPTT), broadband data (MCData), and broadband video (MCVideo) services) technology specifications introduced by 3GPP. The system encompasses features such as group management, location management, group communications, and data and video communication capabilities.

The railway multimedia dispatching communication system is deployed at two levels: the group company and the railway bureaus. The inter-system communications and communications between terminals and systems are carried out by the railway data communication networks. The system architecture is shown in Figure 4.36.

The multimedia dispatching communication system provides various kinds of services, including the features of voice, video, and data communication capabilities. It also includes supplementary call services, railway emergency calls, function addressing and call range restrictions, service priority and preemption, visual location services, multimedia conferences, contact lists, notification sound playback, image/face/voice recognition, user management, monitoring, and data analysis.

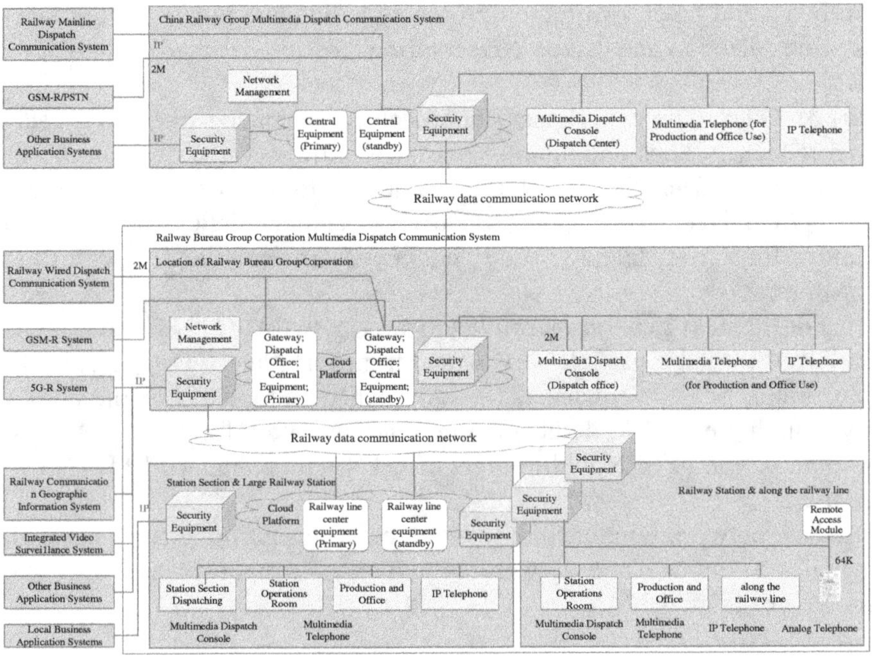

Figure 4.36. The architecture of the multimedia dispatching communication system.

4.5.3 *Next-generation railway bearer network*

The new generation of bearer network features an IP-based architecture, high bandwidth, low latency, wide connectivity, high-density access, high reliability, high security, intelligent control, high-precision time synchronization, and easy network maintenance.

There are three technical standards that comprise:

(i) **SPN**: In the same physical network, slicing is performed based on different traffic types, where each traffic type occupies separate network resources and has different management capabilities. This results in the formation of multiple parallel virtual networks.

(ii) **Enhanced IP RAN**: The core bearer technology is based on multiprotocol label switching (MPLS). The devices utilize various routing protocols and label distribution protocols to achieve functionalities such as route selection and resource reservation.

(iii) **M-OTN**: The evolutionary approach to enhancing traditional OTN packet technologies and simplifying OTN utilizes wave division multiplexing (WDM) as the underlying technology to organize the network at the optical layer. This approach provides characteristics such as high bandwidth, low latency, and enhanced routing and forwarding capabilities.

For the next-generation railway bearer network, OTN technology is preferred for large particle service and transport planes. For 5G-R backhaul links, SPN or enhanced IP RAN can be adopted. For the access layer, including 5G-R, dedicated line services, and information services, a combination of SPN with routers or OTN with enhanced IP RAN can be used.

The network architecture of the next-generation bearer network includes a control plane, forwarding plane, and support plane, which are shown in Figure 4.37.

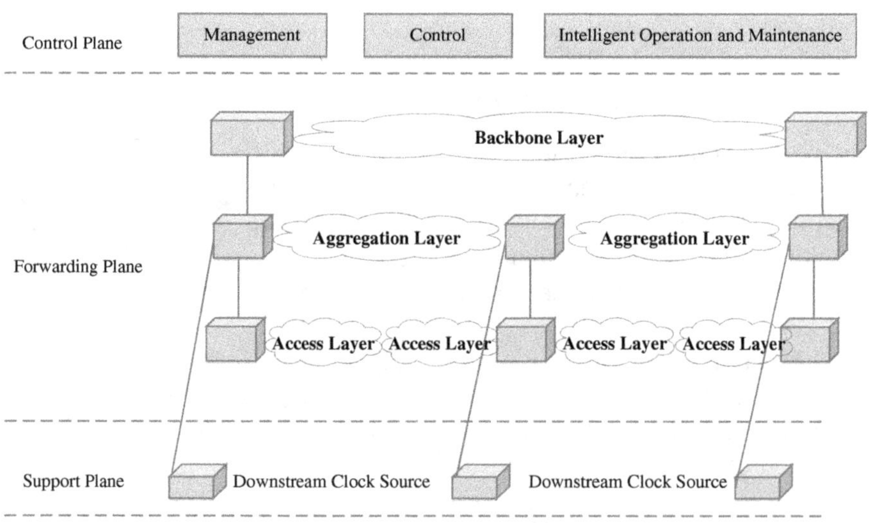

Figure 4.37. The architecture of the next-generation railway carrier network.

(i) **Control plane**: It possesses the capability for flexible configuration of services and network resources, as well as automated and intelligent network operation and maintenance capabilities.

(ii) **Forwarding plane**: The backbone layer is primarily responsible for establishing traffic channels between the group company and railway bureaus, between different railway bureaus, and between major data centers/business core nodes. The aggregation layer primarily handles traffic channels between various traffic aggregation nodes within a railway bureau and the dispatching center/core nodes located at the railway bureaus. The access layer directly serves users and mainly provides bearer access services within the jurisdiction of railway bureaus along the railway lines.

(iii) **Support plane**: It includes clock synchronization and time synchronization to achieve synchronous network deployment as needed, meeting the requirements of various types of services.

The evolution of the next-generation railway bearer network should adhere to the principles of fixed-mobile convergence and on-demand bearer. It should consider the high-speed, low-latency, high-reliability, and high-precision synchronization performance required by various applications. It should make full use of fiber optic resources along the railway lines to achieve resource sharing and rapid deployment at low cost. It is needed to build a comprehensive, technologically advanced, ubiquitous, high-bandwidth, intelligent, secure, and environmentally friendly bearer network to promote railway development.

4.6 Intelligent Detection and Monitoring of HSR

HSR equipment and facilities undergo performance degradation and accumulate damage during long-term service. It is essential to continuously monitor their operational status in complex and dynamic environments. To achieve this, China's HSR employs intelligent perception and information processing, resulting in a comprehensive monitoring system covering state acquisition, comprehensive assessment, early safety warnings, and maintenance guidance. This system monitors various elements, including mobile equipment, such as EMUs, as well as fixed infrastructure, such as industrial power supply, perimeter security, and environmental conditions.

4.6.1 *Intelligent detection and monitoring system*

The HSR intelligent detection and monitoring system encompasses infrastructure monitoring, mobile equipment surveillance, perimeter intrusion detection, environmental monitoring, natural disaster monitoring, and early warning systems. Various status information is obtained through self-inspection or mutual inspection between trains and the ground. Detection and monitoring methods primarily consist of vehicle-to-ground detection, ground-to-ground monitoring, vehicle-to-vehicle surveillance, and ground-to-vehicle detection, as illustrated in Figure 4.38.

4.6.1.1 *Vehicle-to-ground detection*

HSR infrastructure onboard dynamic monitoring can quickly obtain status information on public works, power supply, electrical infrastructure, and the line environment along the route section. This is of great significance

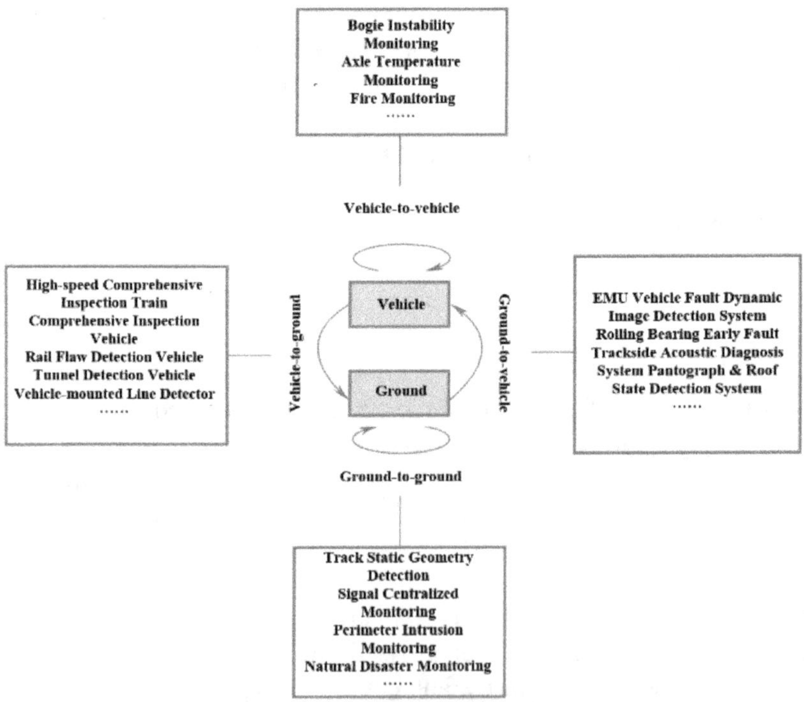

Figure 4.38. High-speed railway intelligent detection and monitoring method.

in ensuring operational safety, guiding maintenance and repair, and dynamic asset management. The detection equipment includes a professional detection vehicle with a single detection content, a multi-disciplinary comprehensive safety detection vehicle, and a high-speed comprehensive detection train. The maximum detection speed can be 10% higher than the train running speed. Since HSR is a complex system, there is a strong correlation between the subsystems of the infrastructure and between the infrastructure and mobile devices. Therefore, HSR mainly rely on high-speed comprehensive detection trains for comprehensive detection, and the detection results are used as an important basis for guiding infrastructure maintenance and repair. With high-speed EMUs as the carrier, the high-speed comprehensive detection train integrates advanced technologies such as modern sensing, intelligent identification, space-time positioning synchronization, large-capacity data exchange, and comprehensive data processing. It is the key technical equipment to realize periodic, high-speed, comprehensive detection.

4.6.1.2 *Ground-to-ground detection*

The static geometries of the track and contact net are an important basis for fault remediation and quality evaluation. In order to improve the timeliness of sensing the service status of infrastructure, ground monitoring devices are installed for line bridge and tunnel equipment, contact net and power supply equipment, electrical equipment, and the surrounding environment. Seamless rail and switch rail breakage, contact net vibration characteristics, cable temperature, compensation displacement, insulation status of power supply equipment, electrical characteristics of signaling systems, fault information, switch machines and natural disasters, foreign object intrusion, earthquakes, and other safety conditions that may directly endanger driving safety and comfort are monitored.

4.6.1.3 *Vehicle-to-vehicle monitoring*

High-speed trains are equipped with a train network control system, which integrates train operation control, status monitoring, fault diagnosis, and passenger information services. With the onboard computer as the core, information exchange between the subsystems of the train is realized through the network to achieve onboard equipment control and the monitoring, control, and management of the entire train. High-speed train onboard detection is based on the onboard network system to detect the status of key components on the train, including high-voltage systems, traction systems, braking systems, auxiliary power supply systems, fire monitoring, axle temperature monitoring, bogie instability monitoring, air-conditioning systems, door systems, and water supply and sanitation systems.

4.6.1.4 *Ground-to-vehicle detection*

In order to ensure the safe operation of high-speed trains, ground monitoring equipment is installed along the line and at the EMU maintenance base. This equipment is used to conduct status detection and monitoring of key components of the EMUs, including the EMU vehicle fault dynamic image detection system (TEDS), trackside acoustic diagnosis of early rolling bearing failures system (TADS), pantograph and roof status dynamic detection system (SJ), wheelset fault dynamic detection system (LY), mobile rim spoke flaw detection system (LU), and hollow axle ultrasonic flaw detection system.

4.6.2 *Intelligent detection and monitoring of infrastructure*

4.6.2.1 *Vehicle dynamic detection*

HSR trains have high speeds and short running intervals, so the maintenance organization method of "skylight repair" is generally adopted. In order to meet the application maintenance mode of "inspection and maintenance" separation and "precision repair," it is necessary to adopt efficient and high-quality detection methods. Dynamic detection is fast, frequent, efficient, takes up fewer resources, and facilitates centralized management and application of detection data. Therefore, HSR generally adopt a detection mode that mainly uses high-speed vehicle dynamic detection and is supplemented by manual ground inspection.

On the other hand, infrastructure such as HSR lines, bridges and tunnels, contact nets, communications, and signals are subject to dynamic reciprocating loads, such as train power, electricity energy consumption, and operating information. They are also subject to cold and heat, wind and sand, rain and snow, and freezing and thawing. Due to the combined effect of corrosion, other natural environments, and complex geological conditions, the operating state undergoes considerable changes compared with the no-load condition. High-speed vehicle dynamic detection is used to detect the operating status of infrastructure under load conditions, operating trains, and other working conditions and obtain dynamic parameters that cannot be obtained by manual static measurement, which can better reflect the real status.

Intelligent HSR use dynamic inspection equipment represented by high-speed comprehensive inspection trains, comprehensive inspection vehicles, mounted inspection equipment, and professional inspection vehicles to collect high-frequency and multi-dimensional inspection data, which can comprehensively and efficiently obtain the operation of the infrastructure status information with the help of new technologies such as big data and deep learning, mine the evolution rules of infrastructure service status, and provide data support for infrastructure design optimization, maintenance, and repair. At present, China's high-speed lines regularly run high-speed comprehensive inspection trains to detect and evaluate the train operation quality and infrastructure status; that is, a comprehensive "physical examination" of lines and other infrastructure is conducted to identify safety hazards and faults.

1. High-speed comprehensive inspection trains

High-speed comprehensive inspection trains have the capability to perform track geometry, wheel–rail force, vehicle dynamic response, contact net, communication, signaling, and integrated dynamic inspections. With the assistance of the onboard integrated system, achieving synchronization and information sharing across multiple professional domains becomes feasible, effectively averting potential disruptions to the normal transportation order resulting from single-profession inspections.

(1) *Track geometry detection*

Track geometric parameters include track gauge, track direction, height, level, and twist, as shown in Figure 4.39, which can scientifically evaluate the line quality status of different sections, test the effect of maintenance operations, and serve as a basis for management at all levels. It is an important basis for the department to conduct macro management and inspection and assessment of line quality. Through the periodic dynamic detection of track geometric parameters, track diseases that endanger driving safety can be discovered in a timely manner, guiding public workers to perform targeted maintenance and repairs, eliminating safety hazards, reducing blind spots in maintenance and repairs, and reducing maintenance costs.

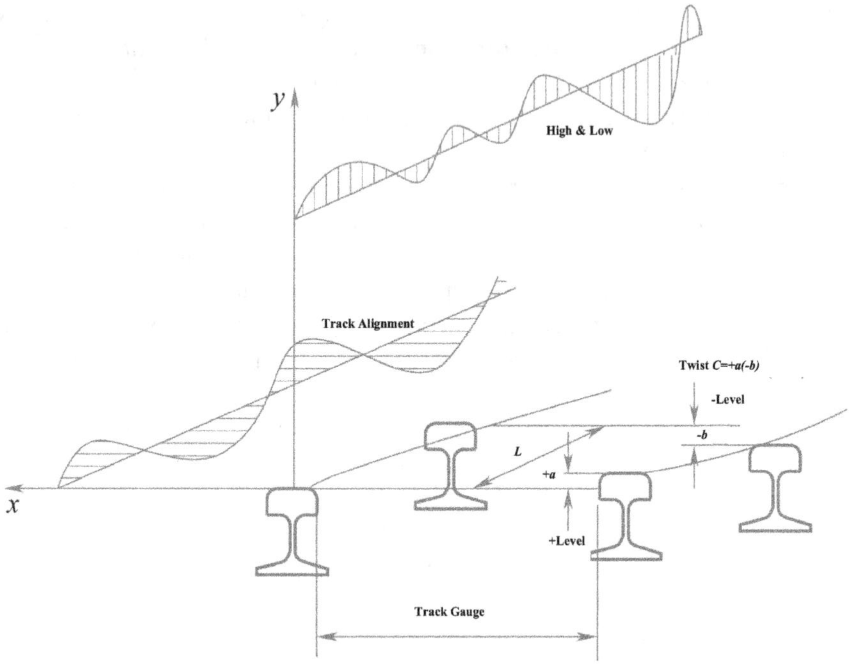

Figure 4.39. Gauge, track direction, height, level, and distortion items.

The track geometry detection system mainly consists of five parts: laser camera, inertial measurement, signal processing, data processing, and mileage correction and positioning synchronization, as shown in Figure 4.40. The track geometry detection adopts the inertial reference measurement principle and strapdown system structure. Inertial measurement elements and carrier phase difference Beidou/GPS are reasonably arranged in different parts of the train bogie and car body, and the gyro sensor and acceleration sensor compensation technology (CAS) are used. Analog and digital hybrid filtering technology and multi-body inertial measurement benchmark-integrated real-time processing technology, combined with high-speed laser cross-section photography and high-speed image processing, are used to synthetically calculate orbit geometric parameters.

(2) *Wheel–rail force and vehicle dynamic response detection*
The wheel–rail force detection system consists of a continuous force-measuring wheel pair, data collection, and processing equipment. In order

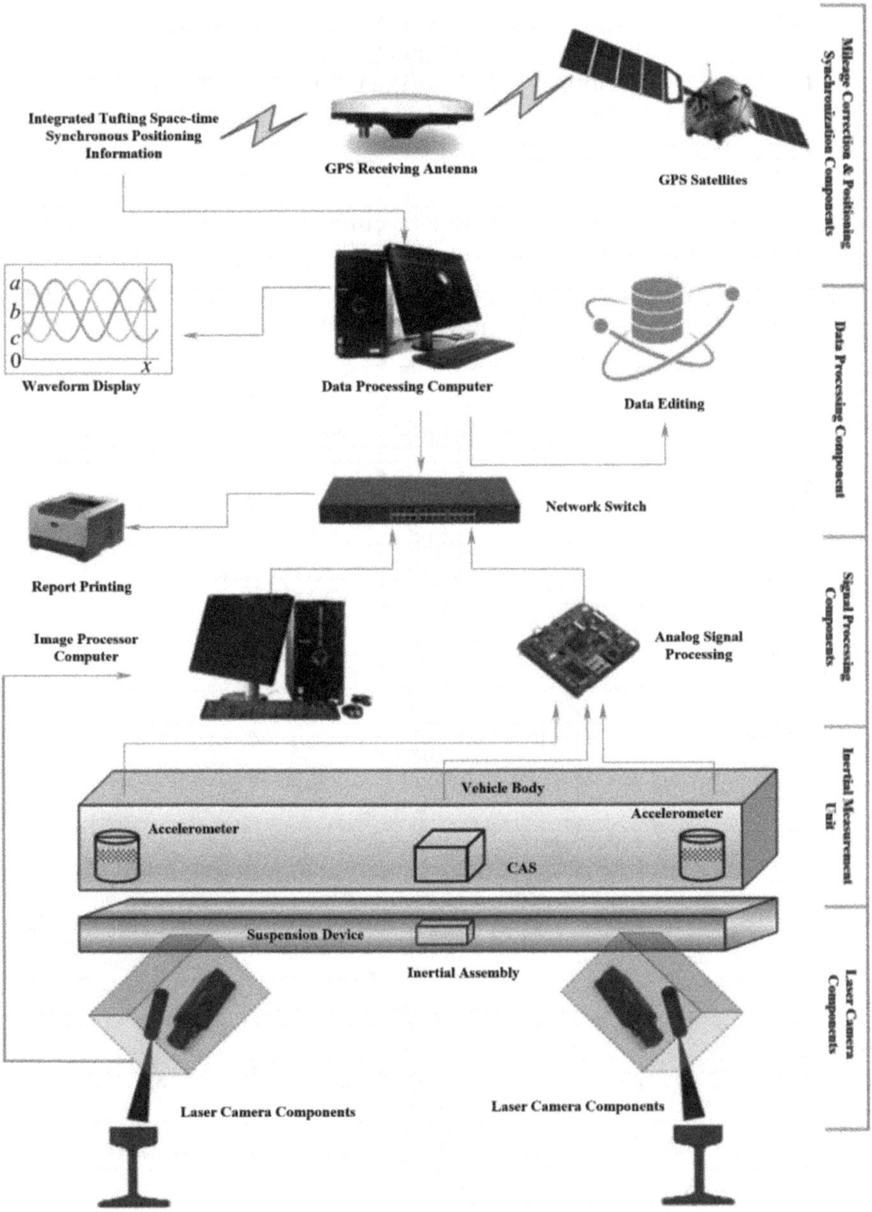

Figure 4.40. Schematic diagram of the track geometry detection system.

to obtain information such as line mileage, the wheel–rail force detection system generally also has auxiliary equipment such as gyroscopes and speed sensors. The continuous force measuring wheel pair adopts the strain measurement principle and includes two strain bridges, the vertical and lateral forces of the wheel and rail, as shown in Figure 4.41. The wheel–rail vertical and lateral forces test bridge uses a single-cycle double-bridge sine and cosine synthesis method, which can suppress and reduce the impact of the contact point position on the test bridge. The output signals of the bridge are all sinusoidal waves. Every time the wheel rotates, the waveform signal completes one cycle of output. The vehicle

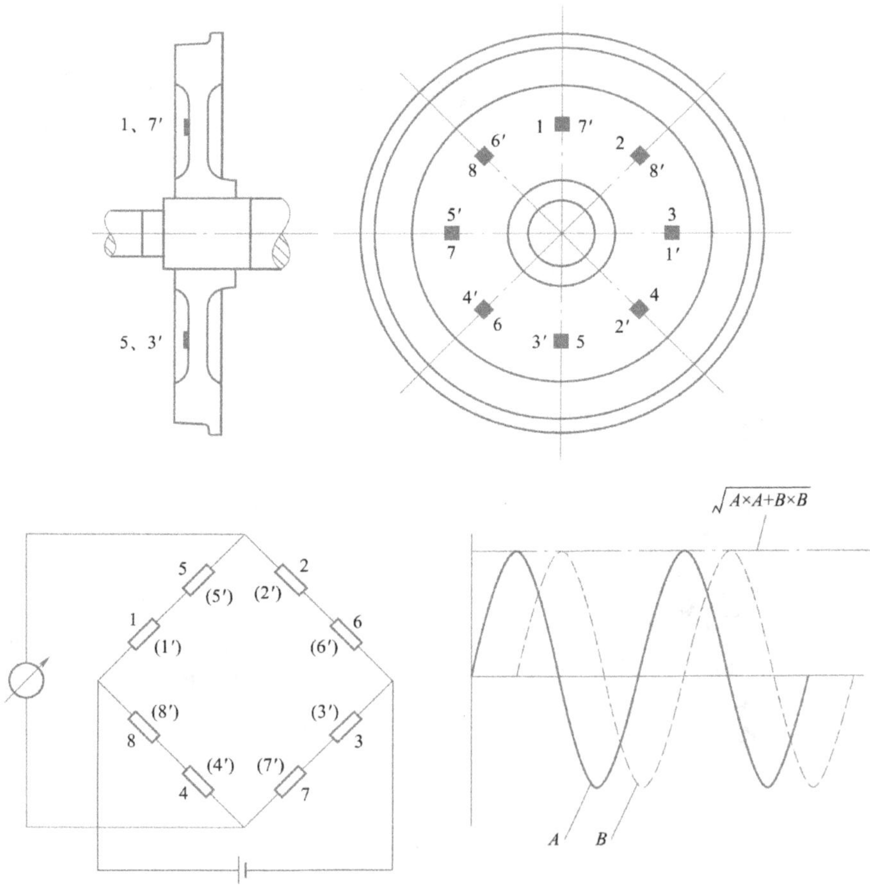

Figure 4.41. Test scheme and output waveform diagram of measuring wheel set.

operation stability (safety) index calculated by using the force measuring wheel set can also reflect the comprehensive impact of track status on vehicle operation safety and indirectly evaluate the line quality. The indicators for evaluating vehicle operation stability (safety) include derailment coefficient, wheel weight reduction rate, and wheel axle lateral force.

In order to reflect the actual vehicle response characteristics of train operation, the detection system selects two or three sections to arrange acceleration sensors (axle box, frame, or car body) at different positions of the vehicle. By analyzing the statistical characteristics and spectral distribution of the acceleration of the car body, axle box, and frame, we establish an evaluation method and analysis model for the vehicle operating status and use the high-speed detection of dynamic response indicators of train running acceleration to identify track faults. Combined with track geometry data, the impact of track status on vehicle operation safety is comprehensively analyzed to assist in the evaluation of track geometric irregularities and structural irregularities.

(3) *Contact net detection*

Contact net detection includes vehicle-mounted dynamic detection of three parts: contact net geometric parameters, pantograph dynamic parameters, and power supply parameters.

The geometric parameters of the contact net include the contact net pull-out value, conductance height, wear, and other geometric parameters. Dynamic detection is generally based on the principle of multi-eye stereo vision measurement using non-contact measurement methods to achieve automatic measurement of geometric parameters, as shown in Figure 4.42. This device can be used for both static acceptance testing and dynamic high-speed inspection. It has no impact on the current intake of the high-speed pantograph and is an important part of the contact net inspection.

The dynamic parameters of the pantograph and contact net mainly include pantograph and contact net contact pressure, hard point impact, roof acceleration, offline sparks, contact line dynamic height, and other parameters. The detection equipment mainly consists of a detection pantograph, contact pressure sensor assembly, hard point impact sensor assembly, conduction height sensor assembly, roof acceleration sensor assembly, high-voltage side data acquisition and processing unit, optical signal transmission device, power supply isolation transformer, low-voltage side

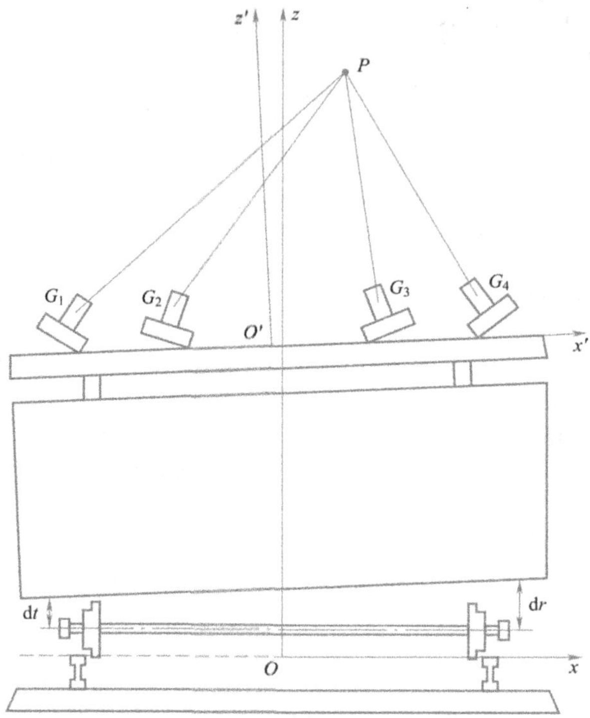

Figure 4.42. Contact net geometry parameter measurement principle.

signal processing unit, spark sensor component, spark signal processing unit, and software system, as shown in Figure 4.43.

The power supply parameter detection system divides the measured voltage using a high-precision voltage divider and then sends it to the input end of the conditioning circuit through a shielded cable, as shown in Figure 4.44. Because the transmission cable is shorter, it can effectively reduce the transmission path of analog signals and avoid losses and interference caused by signal transmission. After being conditioned by the voltage isolation sensor, the voltage signal is divided twice and is reduced to the range that the data acquisition unit can collect. It is first subjected to analog filtering and A/D conversion by the high-speed data acquisition unit and then processed by the data processing computer. After digital filtering and forwarding, the analysis software stores and displays the stored valid data.

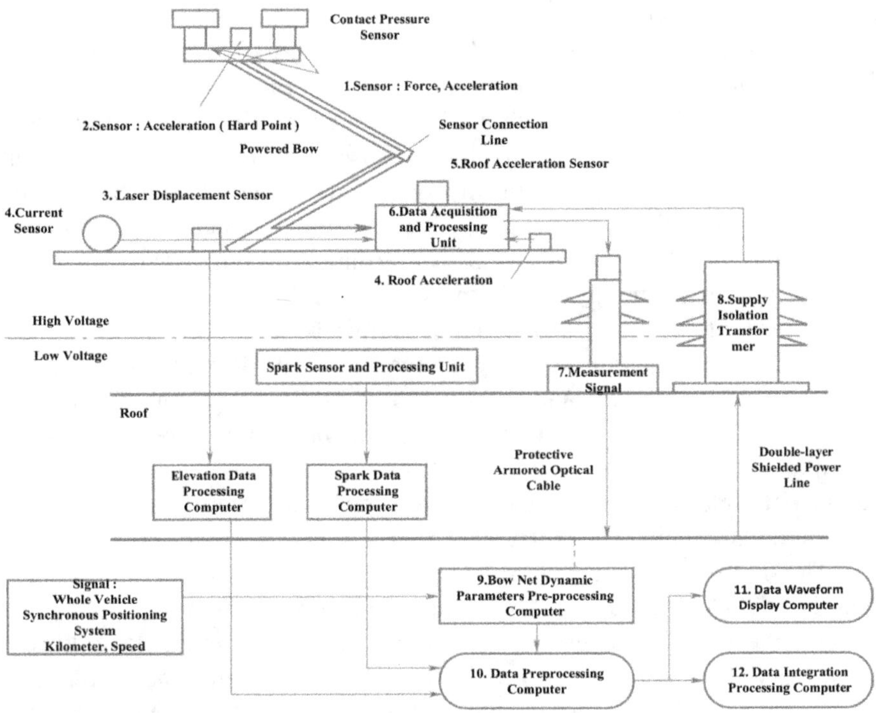

Figure 4.43. Bow net dynamic action parameter detection system architecture.

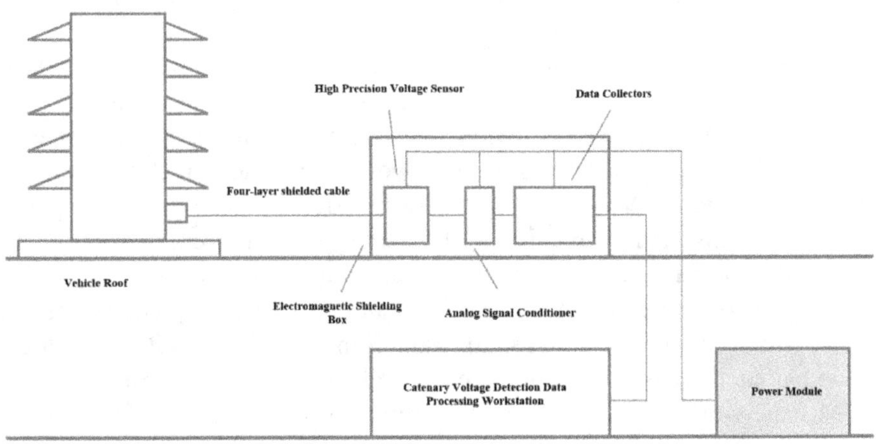

Figure 4.44. Power supply parameter detection system architecture diagram.

(4) *Communication detection*

For wireless communications, the basic condition to ensure reliable communication is to meet a certain intensity of radio waves. The intensity of radio waves is closely related to the transmitting device, the radio wave propagation environment, the propagation distance, and weather conditions. It is necessary to promptly discover the problems caused by equipment. Insufficient coverage requires timely system maintenance; weak field conditions caused by environmental impact require the design department to strengthen the wireless coverage system.

At present, China's railway wireless train operation communications mainly use two communication systems: the 450 MHz analog wireless train dispatching system and the GSM-R digital mobile communication system. The main purpose is to implement dispatching operation command and safety assurance. These two communication systems have different networking methods and different coverage requirements, so the test methods are also very different.

The high-speed inspection train communication detection project mainly includes GSM-R mobile communication system field strength coverage tests, GSM-R service quality tests, and electromagnetic environment tests along the railway line. The test system is equipped with four measurement receivers, as shown in Figure 4.45. The GSM-R vehicle communication module and the system test controller are configured. They use the unified time-space synchronization of the integrated test vehicle to complete the test positioning and the detection of various indicators in one round-trip crossing.

There are also different test requirements for different types of lines. On non-wireless train control lines, in order to meet the requirements of a round trip to complete the field strength coverage test, at least one measurement receiver is required for single network coverage, while at least two measurement receivers are required for interleaved coverage. Equipped with four measurement receivers, the wireless field strength test of an interleaved or dual-network redundant coverage network with the same site can be completed in one single pass, and one round trip can be tested twice, which enhances the test sample and improves the test efficiency. For the hub area, there is a cross-line section and a section where 450 MHz and GSM-R networks coexist. The configuration of four measurement receivers can be used to measure the wireless coverage of each adjacent cell.

(5) *Signal detection*

The signal detection system mainly realizes the detection of track circuit frequency parameters and transmission characteristics, compensation capacitor

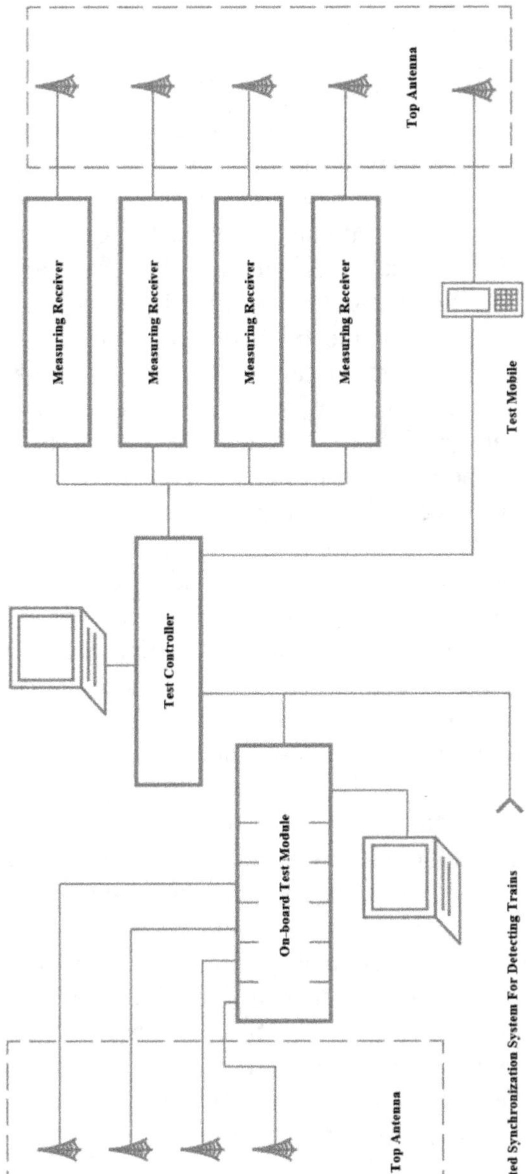

Figure 4.45. High-speed detection train communication detection system composition diagram.

working status, traction return current, etc.; transponder uplink signal characteristics, transponder messages, vehicle ATP operation data analysis, and Igsmr interface monitoring data analysis; UM wireless environmental monitoring data analysis, monitoring and comprehensive intelligent analysis, and processing functions.

The track circuit detection includes the test and analysis of the spectrum characteristic parameters of the track circuit, the detection and analysis of the transmission characteristic parameters such as the entrance and exit voltage and transmission attenuation of the track circuit, the testing of the position and characteristic parameters of the tuning area of the track circuit, and the interference detection functions such as the adjacent section and adjacent line of the track circuit. Compensation capacitor detection includes compensation capacitor position, step size parameter detection and analysis, compensation capacitor failure, and missing and other working state detection and analysis. Traction return current detection includes traction return total current detection, traction return unbalanced current detection, and traction return power frequency and harmonic component detection and analysis.

Transponder detection includes transponder position detection, transponder message analysis, comparison, analysis, transponder message bit error rate, bit error detection and analysis, transponder uplink signal strength, power, range of action, and spectrum characteristics detection. Vehicle ATP monitoring and data analysis includes vehicle ATP human–computer interaction information, input data, key output data, and operating status data record analysis functions. Train-controlled onboard Igsmr interface detection includes the Igsmr interface between the GCD and MT of the onboard ATP device and the real-time, noise-free monitoring and recording of message exchanges between the ATP and RBC layers.

The vehicle-controlled vehicle UM wireless environment detection includes CTCS 3-level train control system application data, the business data of vehicle module interaction with the wireless network, the transmission and reception record of CTCS 3-level train control and system train control data at the wireless interface when the communication timeout fault occurs, fault location, and the CTCS 3-level train control system vehicle wireless communication terminal working situation.

(6) *Integrated system*

The comprehensive inspection train is generally composed of 4–6 inspection vehicles, loaded with various special inspection equipment. The special detection equipment can use the onboard integrated system of the vehicle when collecting data and maintain strict synchronization in speed, time, and mileage position with the help of its data network. Each special detection data can also use the data network to realize the data sharing of the whole train. It not only reduces the repeated equipment of multiple special inspection vehicles but also provides the possibility for rapid and accurate, comprehensive analysis and processing of data.

The data network is the vehicle local area network of the detection vehicle, as shown in Figure 4.46, which provides a high-speed, reliable, and large-capacity data transmission channel for each detection system, and it has the function of real-time transmission of data, video, and other information about each system. The data network backbone adopts optical fiber as the transmission medium and establishes a network management center in the data integrated processing carriage, equipped with core network switching equipment to monitor and manage the transmission of data information.

In order to carry out a comprehensive analysis of multi-disciplinary detection data, the integrated system has the functions of space-time calibration and mileage positioning calibration, as shown in Figure 4.47. Time calibration provides a timing function based on a high-precision ethernet timing protocol (PTP) hardware card. The PTP time synchronization card is deployed in a computer that needs to perform high-precision time synchronization, so that the time synchronization between the time synchronization card and the time server is based on the PTP protocol, which can achieve microsecond-level time synchronization accuracy. Spatial synchronization uses the main and standby photoelectric encoders installed on the shaft head to distribute the distance pulse signal generated by it to each detection system as the signal source for the sampling trigger and distance calculation of the detection system.

The onboard data integrated processing system comprehensively manages the large value alarm data in the detection process. It can automatically receive and transmit the train operation positioning data, the large value alarm and attachment data of each detection specialty, and the large value alarm point video surveillance data of track inspection dynamics to the ground center. It also provides SMS alarms, equipment account queries, detection daily report generation, GIS comprehensive display, and detection data waveform comprehensive display, and other functions. The system can provide real-time detection and application of the line and provide a variety of effective means and bases for the on-site evaluation and comprehensive analysis of the inspectors.

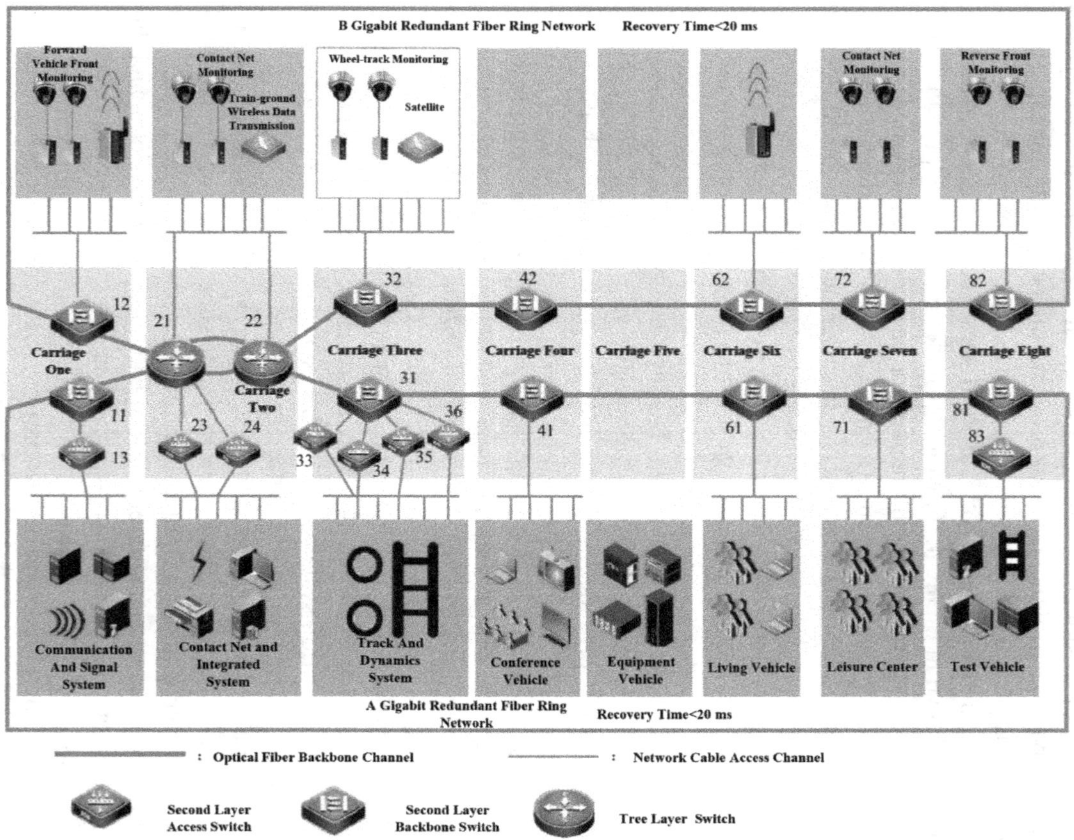

Figure 4.46. Data network architecture.

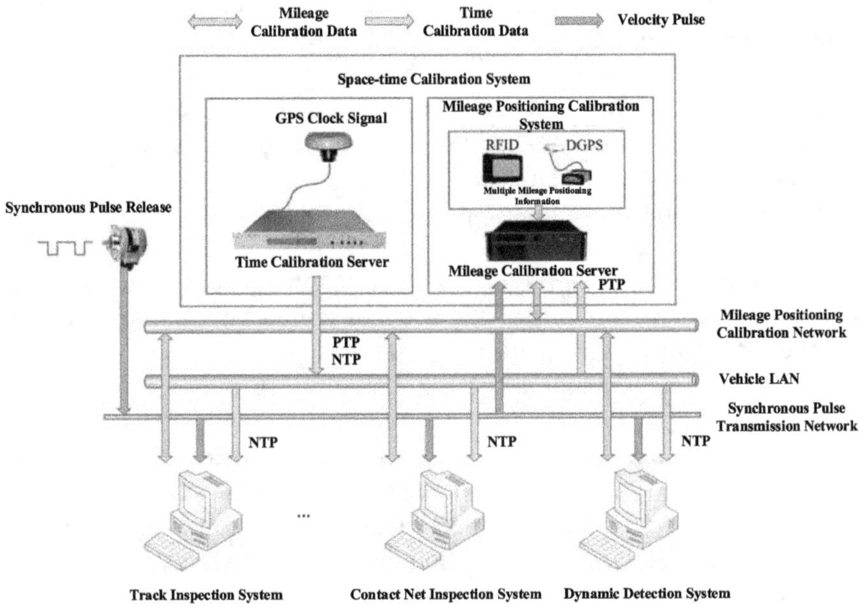

Figure 4.47. Space-time synchronization and positioning.

2. Integrated inspection vehicle

A high-speed integrated inspection vehicle is a patrol vehicle equipped with track surface status, rail profile, line limit, contact net component suspension status, trackside signal equipment status, and other detection equipment, mainly using high-speed cameras, LIDAR, intelligent information processing, and other technologies to achieve integrated comprehensive inspection of public works, power supply, electricity, and other major aspects, replacing manual patrol operations.

The rail state inspection system uses a high-speed dynamic image acquisition device installed on the inspection vehicle to scan the track and correlate it with the line mileage information to obtain complete track image data. Then, based on image processing and pattern recognition technology, the analysis and processing are carried out to realize the intelligent detection of rail surface, sleeper (track plate), and fastener state.

The rail profile detection system is a rail profile measurement system based on laser camera structured light measurement and image processing. It can measure the entire rail profile from the bottom to the top of the rail and calculate the vertical wear, side wear, total wear, and other items of the rail.

Line limit detection is based on the rail top surface and the track center line (track coordinate system) as the measurement standard and adopts the measurement method of laser scanning or camera to accurately detect the position and approaching state of the buildings on both sides of the line in real time, so as to realize rapid automatic detection and dynamic analysis of the building clearance.

Based on intelligent image processing technology, the inspection system of contact net collects high-definition images of key equipment and parts of contact net contact suspension system in real time, analyzes the characteristics of the state of contact net parts, and realizes the intelligent identification and state evaluation of common contact net parts defects, such as clamp shedding, insulator damage, and dropper deflection, through image information mining, so as to identify the state change, damage, and foreign matter of contact net parts. The detection of contact net geometric parameters, such as pull-out value, contact wire height, and the slope of the locator, is realized. The contact net suspension state detection and monitoring system can realize efficient inspection of the contact net, comprehensively grasp the contact net state, and guide the elimination of the hidden dangers of the contact net.

The appearance inspection system of trackside equipment is a vehicle-mounted visual inspection system for detecting the appearance state of trackside equipment of HSR. The system is mainly composed of an image data acquisition subsystem, an image data analysis subsystem, an image data management subsystem, and a synchronous positioning device. The image data acquisition subsystem uses linear array or area array image sensing to continuously scan or shoot at equal intervals to complete the digital imaging of the trackside equipment of the electric signal in the running state of the inspection vehicle and associates the image data with the line mileage information in real time to obtain the complete image data of the trackside equipment of the electric signal. Based on machine learning theory and pattern recognition technology, the image data analysis subsystem filters the image data of the trackside equipment from a large number of background image data. The image data management subsystem constructs a data management mechanism, combining real-time synchronous information with a ground ledger, and realizes a closed-loop management system for image acquisition, analysis, and report release and review verification of trackside equipment inspection.

3. Onboard inspection equipment

The onboard detection equipment can be mounted on the operating EMU, and an increase in detection frequency as a supplement to dynamic detection can effectively improve the safety early warning ability.

The vehicle-mounted line inspection instrument uses embedded intelligent sensors to obtain horizontal and vertical vibration signals during train operation, and it integrates with the date, time, vehicle type, vehicle number, train number, mileage, speed, line number, and other information in the dynamic monitoring system (DMS) of train control equipment to generate a comprehensive data package that can reflect the track state, detect the bad locations of line shaking in time, and transmit them wirelessly to the client.

The contact net safety inspection device (referred to as a 2C device) can be installed in the driver's room of an operating EMU, electric locomotive, or other rail vehicle to inspect the technical status and external environment of the contact net and guide the operation and maintenance of the contact net. The onboard contact net operation state detection device (referred to as a 3C device) is installed on the operating EMU to realize the dynamic detection of the contact net, and the detection results are used to guide the operation and maintenance of the contact net.

4. Professional inspection vehicle

(1) *Vehicle-mounted rail flaw detection*

Vehicle-mounted rail flaw detection is the main form of large-scale rail flaw detection, and a large rail flaw detection vehicle is a rapid detection equipment that integrates acoustics, electronics, computers, machinery, railway vehicles, and other multi-disciplinary technologies, mainly detecting internal damage to in-service rails. The latest flaw detection vehicles are also equipped with a patrol inspection system, which can simultaneously detect rail scratches, rail wear, rail profile, fastener status, etc., and can be gradually developed into multi-project composite detection equipment.

(2) *Tunnel inspection vehicle*

The tunnel inspection vehicle can realize the detection of lining surface state, lining interior, and tunnel bottom (roadbed) state.

When the lining surface state detection equipment continuously measures the running speed of 0.9 km/h, it can identify cracks with a width of more than 0.3 mm. Using the software provided by the system, combined with manual identification, most of the surface faults, such as lining cracking, lining falling, rock spalling, and water leakage, can be identified, and the fault information can be stored in the database for statistical analysis. The historical data can also be compared and analyzed to grasp the development speed and severity of the faults.

The internal state detection of lining is to realize the rapid non-destructive detection of tunnel lining and tunnel bottom by means of vehicle-mounted geological radar. The system is mainly composed of a hydraulic manipulator support system, a tunnel bottom and lining detection system and a data recording and processing system. The effective detection depth of lining state detection is 1.5 m, which can explore the inside of a lining and its back and meet the requirements of detection and evaluation of lining arch and wall.

(3) *Roadbed radar detection*

The roadbed radar detection system is composed of an antenna, a geological radar control unit, a positioning device, a road surface image acquisition device, and data processing software. The antenna of the vehicle-mounted geological radar is generally installed at the bottom of the passenger car, and the roadbed detection is carried out under the traction of the locomotive or the rail car. The antenna is divided into three columns (parallel to the line, respectively arranged in the center of the line, and outside the sleeper). In order to meet the requirements of detection depth and accuracy, three lower frequency antennas and one higher frequency antenna are generally selected to complete the detection of roadbed, and the detection results can obtain three longitudinal sections.

The premise of parameter classification and statistical analysis is to pick up the stratum and determine the reflection coefficient or dielectric constant of each layer. According to the dielectric constant of each layer, the thickness of each layer is calculated. According to the reflection coefficient or dielectric constant of the ballast and road base layer, the degree of dirt on the road bed is derived, the water content of the roadbed is evaluated, etc., and then line faults such as ballast thickness and foundation bed settlement, slurry and mud turning, and foundation deformation are counted.

4.6.2.2 *Ground monitoring*

1. Line infrastructure detection and monitoring

Static track inspection is used to assess the quality of tracks, promptly identify track anomalies, eliminate safety hazards, and provide technical guidance for track anomaly remediation and quality assessment.

The static detection of the track generally includes track geometry, rail straightness and profile, seamless line displacement, and rail temperature and stress.

2. Contact net detection and monitoring

The measurement of contact net geometric parameters adopts laser precise ranging technology, and the supporting software platform realizes the full digital measurement of real-time storage. Through the laser ranging module and angle measurement module, the distance and angle are measured, respectively. Using trigonometric function relationships, the geometric parameters of the contact net, such as the height of the contact net, the pull-out value, and the anchor section joint, can be quickly measured.

A ground monitoring device for contact net and power supply equipment (referred to as 6C) is installed in the specific position of the contact net and the traction substation. It is used to monitor the vibration characteristics of the contact net, the temperature of the clue, the compensation displacement, the insulation state of the power supply equipment, and other parameters, or the technical state of the contact net at a specific position. The monitoring results are used to guide the operation and maintenance of the contact net and power supply equipment.

3. Electrical equipment monitoring

A centralized signal monitoring system with the integrated application of computer technology, microelectronics technology, network technology, fieldbus technology, automatic control technology, sensor detection technology, interface technology, software programming technology, and other group technologies is used to achieve the railway signal indoor and outdoor target system monitoring.

The centralized signal monitoring system can monitor, record, and replay the electrical characteristics, equipment operation process, equipment operation status, operator's operation process, equipment failure, or abnormal situation of signal equipment, such as railway signal interlocking, block, train control, hump, TDCS/CTC, and power screen, in real time and raise the alarm in the case of monitored excessive and fault information.

4.6.3 *Intelligent detection and monitoring of mobile devices*

Intelligent detection and monitoring of mobile devices include onboard detection and monitoring based on the train network control system and ground equipment detection and monitoring. The main monitoring objects of the high-speed train network monitoring and control system include: high-voltage system, traction system, braking system, auxiliary power supply system, fire alarm system, axle temperature monitoring system, bogie instability system, and air-conditioning system. This section mainly introduces mobile device detection and monitoring technology based on ground monitoring equipment, including the trouble of moving EMU detection system (TEDS), the trackside acoustic diagnosis system (TADS) for early rolling bearing faults, and the dynamic detection system for pantograph and roof conditions.

1. Trouble of moving EMU detection system

The TEDS is a system that uses trackside camera devices to collect images of visible parts, such as the bottom of the train body, side skirts, connecting devices, and bogies. It uses linear array image acquisition, 3D imaging, image recognition, and other technologies to automatically compare and analyze faults and alarms, achieving monitoring of the status of visible components at the bottom and side of the train. The implementation of this system has improved the detection rate of problems during the operation of high-speed trains, ensured driving safety, and achieved assistance in the inspection of high-speed trains, thereby improving work efficiency, reducing labor intensity, and improving the overall management level of high-speed train operation. TEDS includes detection station equipment, high-speed train depot (rolling stock depot) monitoring center equipment, railway bureau group monitoring center equipment, national railway group query center equipment, and network transmission equipment, as shown in Figure 4.48.

The TEDS detection station system mainly consists of two parts: trackside equipment and computer room equipment. The trackside equipment includes a linear array image acquisition module, a 3D imaging module, a laser light source module, an AEI antenna, a wheel sensor, and trackside installation protection devices (inner bottom box, outer bottom box, side box, and junction box), air blowing, dust removal and drainage devices, video monitoring devices, cables, and optical fibers. The equipment in the computer room includes vehicle information acquisition

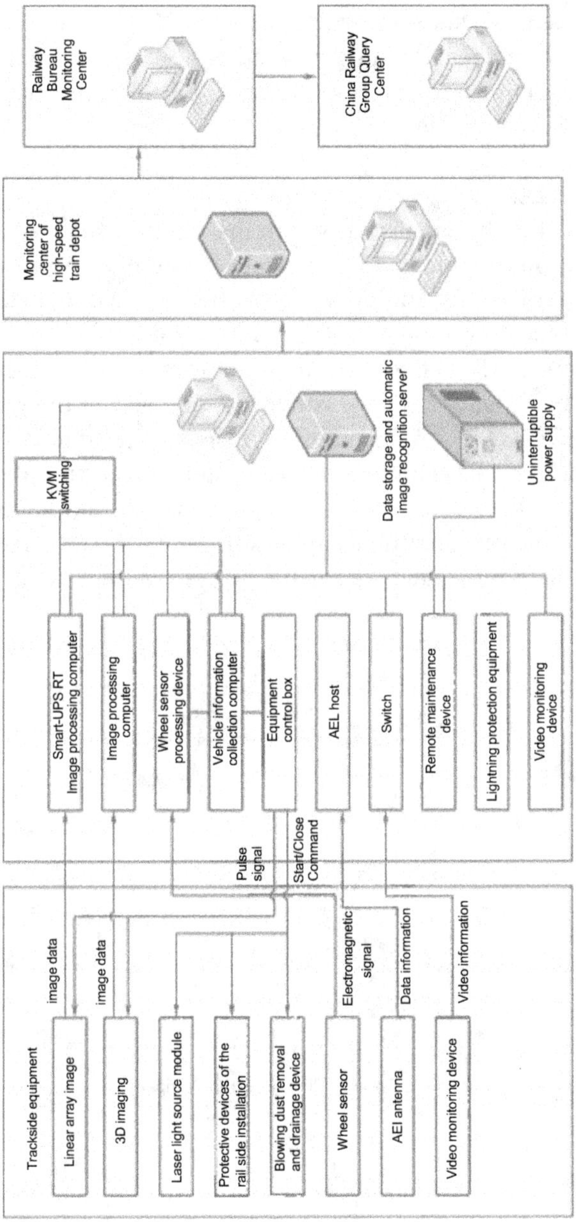

Figure 4.48. Schematic diagram of TEDS composition.

computers, image processing computers, data storage servers, and image automatic recognition servers, as well as AEI hosts, video monitoring devices, wheel sensor processing devices, KVM, signal lightning protection devices, power lightning protection devices, equipment control boxes, remote maintenance devices, switches, UPS uninterruptible power supplies, and air conditioning.

2. Trackside acoustic diagnosis system

The TADS for rolling bearings is a system that utilizes an array of acoustic sensors installed on the trackside to collect real-time sound from the rolling bearings of railway vehicles. It uses signal processing technology, acoustic diagnosis technology, and computer network technology to achieve online and early diagnosis and prediction of cracks, damages, and other faults in the inner, outer, and roller components of the rolling bearings of running vehicles. TADS mainly consists of ground detection stations and network systems (Figure 4.49). Ground detection stations are the core of TADS, used for data collection, axle counting, automatic identification of train numbers, fault discrimination, fault location, and real-time transmission of data messages. The network system consists of servers of

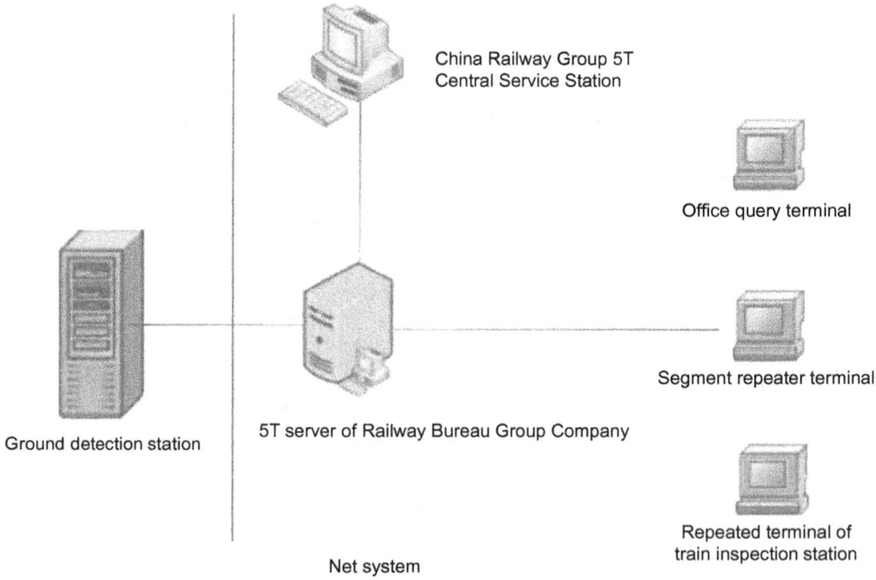

Figure 4.49. TADS composition.

the Railway Bureau Group Company, the Central Service Station of China Railway Group, and terminals, used to achieve data upload, receive messages, establish databases, reproduce alarm information, and real-time monitoring equipment.

The TADS ground detection station can be divided into two parts: outdoor and indoor, with the outdoor part mainly composed of the TADS trackside equipment and AEI equipment and the indoor part mainly consisting of cabinets and UPS power supply. The outdoor equipment of TADS detection station mainly includes protective boxes (acoustic sensor arrays), wheel sensors, and AEI equipment.

3. Dynamic detection system for pantograph and roof status

Then, the personnel uses AI methods to intelligently monitor and track obstacles and contact foreign objects in the network. It adopts high-speed, high-resolution, non-contact image analysis and measurement technology to achieve dynamic automatic detection of key characteristic parameters, such as pantograph sliding plate wear, centerline offset, bow head tilt, working pressure, and indoor visual observation of roof foreign objects and key component states.

The pantograph wear and centerline detection subsystem is mainly composed of on-site detection equipment for wear, centerline, and bow head tilt modules, as well as control, support, and processing equipment located in the equipment room. The on-site detection equipment for wear, centerline, and bow head tilt modules mainly includes triggering and control, image acquisition, and light compensation equipment.

The roof status monitoring subsystem uses high-speed and high-resolution cameras to monitor and record the entire process of the roof when the high-speed train passes through the equipment detection area. The playback video uses a high-definition and high-resolution large screen, which allows for indoor observation of the comprehensive status of the roof, such as foreign objects, pantographs, porcelain bottles, and main circuit breakers. Any abnormal situations found are recorded, and fault reports are generated, which can serve as a basis for maintenance and repair.

4.6.4 *Intelligent monitoring of perimeter intrusion*

The perimeter of a HSR refers to the enclosed area boundaries that require physical or electronic protection. Due to the high speeds of trains on HSR, incidents involving unauthorized access by individuals or large livestock into the perimeter pose a serious threat to operational safety. Perimeter intrusion alarm systems constructed along HSR lines utilize sensor and electronic information technologies to intelligently monitor perimeter intrusion incidents and issue alarms, making them crucial for ensuring the safety of HSR operations.

1. Single sensor-based perimeter intrusion intelligent monitoring technology

Common technology techniques for perimeter intrusion monitoring include pulse electronic fences, infrared thermal imaging, vibration cables/fiber optics, millimeter-wave/laser radar, and intelligent video analysis. These methods using single types of sensors are prevalent. Due to their low cost and simplicity, especially with the rapid development of deep learning technology in intelligent video analysis, various methods are still in use today, as shown in Table 4.10.

Due to the extremely complex railway environment, as well as various harsh weather conditions such as wind, frost, snow, rain, thunder, fog, sand, haze, and lighting conditions such as darkness and strong light, all types of single-sensor intrusion monitoring have certain limitations. For example, vibration fiber optics may have a high false alarm rate in windy conditions and cannot detect non-contact intrusion events, such as objects falling from the sky. Visible light-based intelligent video analysis offers high levels of intelligence, flexibility, and visual results but may have reduced accuracy in low-visibility conditions, such as nighttime, fog, haze, or strong light interference. Thus, single-sensor technologies struggle to meet the requirements of zero false positives and low false negatives for perimeter intrusion intelligent monitoring in complex HSR scenarios under all weather conditions. The trend is toward multi-sensor fusion for HSR perimeter intrusion intelligence.

2. Multi-sensor fusion-based railway perimeter intrusion intelligent monitoring technology

Currently, most railway perimeter intrusion monitoring systems based on multi-sensor fusion combine visible light cameras with one or more other sensor technologies. Common combinations include visible light + vibration fiber optics, visible light + infrared thermal imaging, and visible light + radar methods.

Vibration fiber optics + Visible light intelligent video analysis: This approach uses vibration fiber optics for large-area intrusion detection and

Table 4.10. Comparison of characteristics of various perimeter intrusion detection technologies.

Sensing technology	False alarm rate	Missed detection rate	Environmental impact	Weather impact	Software complexity	Intuitiveness	Comprehensive cost	Primary advantages	Primary disadvantages
Vibration fiber optics	Higher	Lower	Small	Small	Lower	Low	Common	Good environmental and weather adaptability, not affected by obstruction	High false alarm rate
Infrared thermal imaging	Common	Higher	Less	Common	Higher	Higher	Higher	Personnel intrusion monitoring has good performance, especially at night	High cost
Laser radar	Lower	Common	Less	Common	High	Higher	High	Precise position and size can be obtained, and the ambient weather adaptability is good	High cost, data processing is more complicated.
Millimeter-wave radar	Lower	Higher	More	Small	Higher	Low	Higher	Good weather applicability	Difficult to detect low-speed targets
Intelligent video analysis	Common	Common	Less	More	High	High	Common	The image is rich in information, more intuitive, and has strong detection functions	Greatly affected by light, and difficult to detect at night

uses video analysis for local confirmation after an initial intrusion is detected. It offers advantages in terms of wide coverage and low false alarms but cannot detect targets entering from the air across the guardrail, mainly due to shrub obstruction and light interference.

Infrared thermal imaging + visible light intelligent video analysis: This method relies on infrared thermal imaging for large-area intrusion detection and uses video analysis for local confirmation after initial identification of the area where the intrusion occurred. Visible light can be equipped with a laser-fill light source, enabling clear visible light images even at night. This approach offers advantages in broad coverage, sensitivity to thermal radiation from people and animals, and low false alarms. However, it may struggle to capture intruders in areas with obstructions, and it lacks clarity for targets with minimal temperature differences from the background on account of the high complexity of the scene along the HSR.

Millimeter-wave/laser radar + visible light fusion monitoring: Millimeter-wave radar offers high accuracy in detecting moving objects, providing information on target direction, distance, and speed. However, it cannot provide details about target shape and size and performs poorly at detecting slow-moving targets in complex backgrounds. Laser radar can provide specific information about target position, shape, and size through 3D point clouds but involves substantial data processing. Combining millimeter-wave/laser radar data with visible light imagery at the data, target, or decision levels can improve detection performance.

In scenarios with minimal obstruction and relatively simple backgrounds, such as throat areas and channel gates, millimeter-wave radar can detect moving targets, and the target or decision level fusion with the intrusion target obtained by visible light image intelligent recognition realizes perimeter intrusion intelligent monitoring. In complex scenes with significant reflective background objects, such as tunnel mouths, using millimeter-wave radar may result in strong background clutter and a reduced signal-to-noise ratio, greatly impacting monitoring effectiveness. In such cases, a fusion monitoring approach using 3D laser radar and visible-light imagery is more suitable. After the installation of laser radar and cameras, a joint calibration is performed to establish correspondence between 3D laser point clouds and 2D images, allowing data-, target-, and decision-level fusion using the recognized targets.

3. Development directions for railway perimeter intrusion intelligent monitoring technology

While multi-sensor fusion technology significantly reduces false alarms in perimeter intrusion systems, fixed combinations may not fully apply to various typical HSR scenarios and weather conditions. Investigating different types of sensor fusion modes for different scenarios is a development direction for HSR perimeter intrusion intelligent monitoring.

Furthermore, current multi-sensor fusion monitoring solutions often involve simple combinations of different devices. To fully leverage the advantages of deep integration of various sensor technologies, research into multi-sensor deep fusion algorithms based on deep neural networks and integrated development of hardware and software for fusion perception devices, along with the application of edge computing technology, is also an important direction.

4.6.5 *Intelligent monitoring of environment*

Environmental intelligent monitoring plays an important role in monitoring the factors in the local environment of vehicles which threaten the safety of train operations, such as foreign object encroachment and obstacles in the track. As for the classification of monitoring methods, environmental intelligent monitoring of HSR can be divided into two categories: ground equipment or vehicle-mounted equipment. The two types of monitoring methods complement each other and play significant roles in guaranteeing the operational safety of HSR.

1. Intelligent monitoring of HSR environment based on ground equipment

At present, the intelligent monitoring system of a HSR environment based on ground equipment mainly includes systems that monitor foreign object encroachment or obstacle encroachment along the track.

(1) *Intelligent monitoring system for foreign object encroachment*

Foreign object encroachment monitoring is mainly focused on monitoring foreign objects falling on the bridge of railway. The on-site monitoring equipment for foreign object encroachment of the public span railway is used to monitor the power grid, which is generally set on both sides of the railway bridge of the upper span railway. The monitoring network for foreign object encroachment is shown in Figure 4.50.

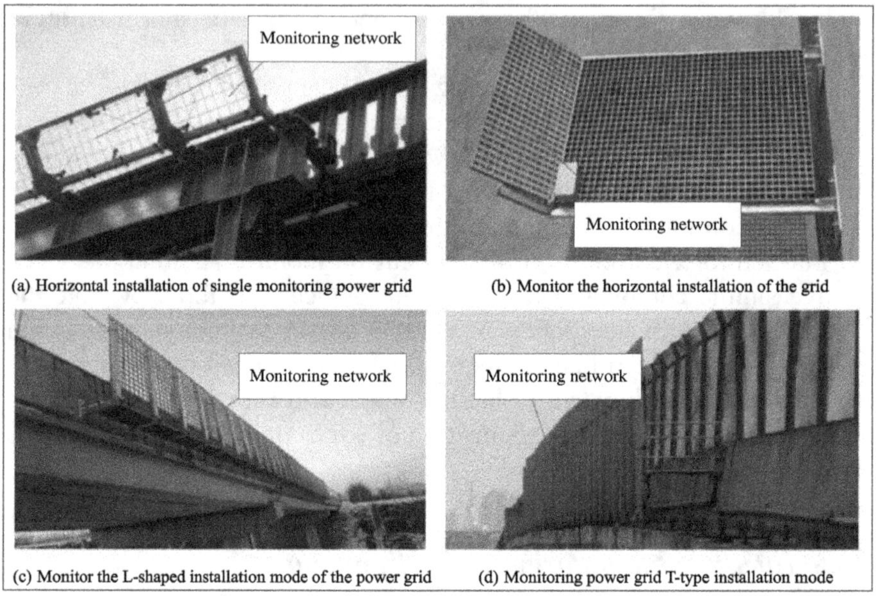

Figure 4.50. Monitor grids for foreign object encroachment, (a) horizontal installation mode for single monitor grid; (b) horizontal installation mode for monitor grids; (c) L-shape installation mode for monitor grids; (d) T-shape installation mode for monitor grids.

When a foreign object falls into the monitoring equipment at the site, the short circuit of the monitoring grid triggers an alarm, and the alarm information is sent to the dispatching center through the communication network. The dispatching center is equipped with a foreign object encroachment monitoring and alarm terminal, which can display encroachment alarm information and respond when foreign object intrusion occurs. In addition to intelligent monitoring of foreign object encroachment, the related device of the public cross-railway bridge also has a certain blocking effect on foreign objects, which further avoids encroachment by such objects.

(2) *Intelligent monitoring system for obstacle encroachment along the track*

Intelligent monitoring of obstacle encroachment along the track is used to monitor and warn of obstacles on the railway track and hazardous sources along both sides of the track. On-site sensors include LiDAR sensors, video acquisition equipment, displacement sensors, fiber-optic grating sensors, and single-grid sensors. The monitoring method adopts a displacement meter for real-time monitoring of slopes and dangerous rocks; fiber-optic grating monitoring and single-grid monitoring technologies for real-time monitoring of collapse and rockfall within the monitoring range and providing early warning when collapse or rockfall occurs; and laser scanning and video monitoring technologies for real-time monitoring of the monitoring range along the track and provides early warning when collapse or rockfall is detected within the line boundary. Laser scanning and video monitoring technologies are used to monitor the monitoring range along the track in real time, and an alarm is automatically raised when collapse and rockfall or other obstacles are found within the track boundary, as shown in Figure 4.51.

2. Intelligent monitoring of HSR environment based on vehicle-mounted equipment

Environmental intelligent monitoring of HSR based on vehicle-mounted equipment is mainly employed to shoot track environment video using a vehicle-mounted camera, make use of intelligent video analysis methods to analyze and process track environment video, and carry out automatic identification and monitoring of environmental abnormal factors. At present, the environment intelligent monitoring system based on vehicle-mounted equipment can mainly monitor the track environment, such as track obstacles, contact network foreign objects, and guardrail integrity.

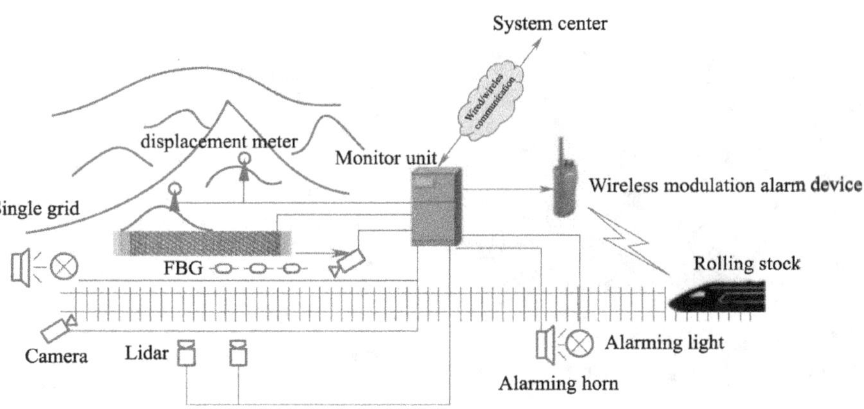

Figure 4.51. Foreign body detection equipment along track.

(1) *Intelligent monitoring of foreign objects on vehicle tracks*
The vehicle-mounted track foreign object intelligent monitoring system collects track environment information through vehicle-mounted cameras, LIDAR, and other sensing devices. Then, the personnel uses AI methods to intelligently monitor and track obstacles and contact foreign objects in the network. Its system composition is shown in Figure 4.52. The system consists of one wide-angle camera and one telephoto camera, two LIDARs, as well as a data processing host and a display unit. The wide-angle and telephoto cameras are used to collect near- and long-range line video images, respectively, while the LIDAR is used to collect three-dimensional information about the track. The data processing host processes the LiDAR and image data, then detects track obstacles and foreign objects on the contact network, and finally sends out sound and light alarms when abnormal conditions occur.

At present, the data processing mainly adopts the intelligent detection method based on the fusion of image and LiDAR processing. The main steps of the commonly used processing methods include railway scene area segmentation, object detection and classification, and foreign object encroachment detection. First of all, the railway scene is segmented into areas. The track area, contact network area, and other areas of concern for foreign object encroachment are segmented out. Then, the images in the segmented area are intelligently recognized; in this case, the foreign objects are detected and classified using deep learning and other methods to extract suspected intruding objects. Finally, LIDAR and image fusion are used to determine whether there is an intruding foreign object based

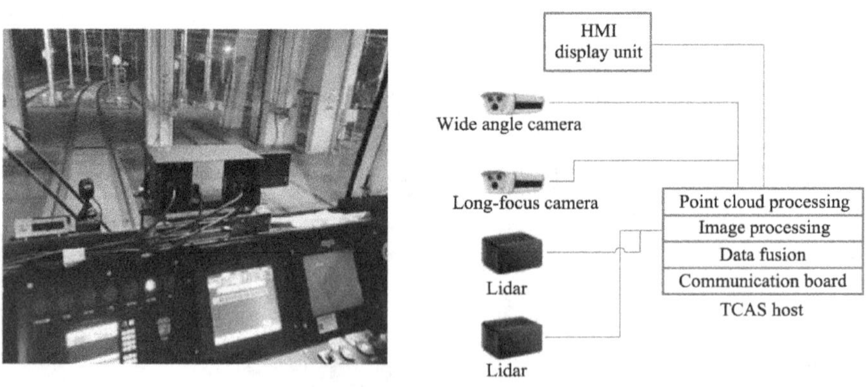

Figure 4.52. On-board foreign body detection equipment.

on the location and size of the foreign object. The alarm is triggered when the foreign object is considered to be intruding.

(2) *Intelligent monitoring of HSR guardrail integrity*
The composition of the HSR guardrail integrity intelligent monitoring system is shown in Figure 4.53. It mainly includes a guardrail integrity testing computer, two digital cameras, a rotational speed sensor, a testing result management computer, a wireless data transmission module, a set of lighting devices, as well as an image acquisition card and a data acquisition card installed in the computer.

Two cameras are installed behind the windows on both sides of the integrated train to capture the guardrails on the left and right sides of the track. The system adopts high-performance digital cameras with a short exposure time and high gain to ensure that the images remain clear when the train is traveling at high speeds. At the same time, the camera also has automatic exposure, automatic white balance, automatic aperture, automatic gain, and other functions, which offer a wide range of automatic adaptation to external light changes. The digital camera lens can be fitted with a polarizing filter to reduce the effect of reflected light from the windows if necessary. It transmits the image signal via optical fiber to the image capture card and saves it in the computer memory.

The guardrail integrity detection computer processes and recognizes the captured image to determine if the guardrail in the image is missing or

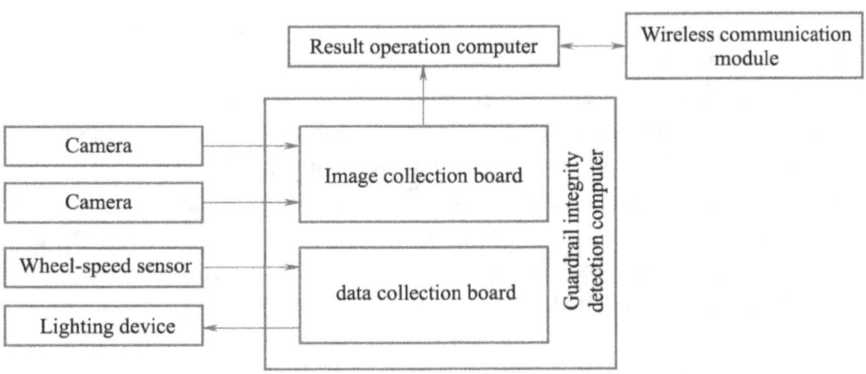

Figure 4.53. Schematic diagram of intelligent monitoring system for integrity of high-speed railway guardrail.

broken. When an abnormality is detected, the computer will automatically record information such as the location of the current train and save the current image, making it easier for track maintenance, rapid identification of the condition and location of the guardrail abnormality, and repair in a short time.

4.6.6 *Intelligent monitoring of and early warning for natural disasters*

The natural disaster monitoring and early warning system is a complex integrated system that combines sensors, computers, communication, networks, data processing, and other technologies, as well as meteorological engineering, aerodynamics, safety science, prediction theory, and reliability theory. It uses information technology to monitor natural disaster events that may pose a safety risk in real time and issue graded speed limit instructions based on the prescribed operational management measures to ensure both safety and transportation efficiency. The HSR natural disaster monitoring and early warning system mainly includes the monitoring, alarming, and emergency handling of disasters, such as wind, rain, snow, and earthquakes.

4.6.6.1 *Wind monitoring system*

As the speed of high-speed trains continues to increase, the impact of strong winds on their operation is receiving more attention. The use of lightweight materials in train manufacturing can increase their speed, as there is a relationship between train speed and train load. However, when a train is operating at high speeds, it generates lift and pitching moments, which increase with the speed, causing the train to be in a "floating" state. Trains operating under such conditions are easily affected by strong winds, leading to traffic accidents. In some special sections, such as large bridges, viaducts, embankments, hills, and mountains, the flow field around the train undergoes substantial changes, resulting in significant changes in aerodynamics. This increases the possibility of train derailment and overturning.

(1) **Strong wind monitoring**

(a) *Installation of anemometers*: The anemometer calculates the average wind speed during a certain period by converting the number of rotations of the wind cups. The railway implements train operation control based on this average wind speed. Anemometers are usually installed in train operation control sections that frequently experience strong winds, such as embankments and elevated bridges, where the aerodynamic effects on the train body are significant.

(b) *Installation of windbreaks*: The installation of windbreaks in key areas such as bridges and cuttings, where gusts and strong winds are common, is an effective measure to mitigate strong wind disasters. The optimal position for windbreaks is 5.7 m away from the track center on the windward side. At this position, the aerodynamic forces on high-speed trains are minimized, resulting in better aerodynamic performance, and the crosswind speed at the overhead contact line is relatively lower. For example, the Xinjiang section of the Lanzhou–Xinjiang High-Speed Railway, spanning 710 km, has windbreak projects along 462 km of its length, accounting for 65% of the total length of the railway. The application of these windbreak technologies is a first in China's HSR construction and represents the largest scale worldwide.

(c) *Enhanced study of localized gusts*: Conducting research on localized gusts is essential in order to promptly forecast and monitor gusts that may affect the safety of train operations and adjust train control strategies accordingly. Currently, the prevailing approach involves employing a comprehensive train control method that takes into account both aerodynamic forces and tipping resistance. This method relies on wind speed sensors to predict the aerodynamic impact on the train body. Multiple wind speed sensors are strategically placed at uniform intervals equivalent to the length of the train. When the monitoring value of any sensor exceeds the alarm threshold, a control command is issued.

(d) *Implementation of a high wind warning system*: In the high-wind warning system, the upper limit of short-term wind speed is predicted based on the monitoring values from wind speed sensors. When the predicted upper limit or the monitoring value from any specific monitoring point exceeds the alarm threshold, a train speed restriction command is issued. The speed restrictions are lifted when the predicted upper limit or monitoring value falls below the alarm threshold.

(2) **Short-term localized wind speed and direction forecasting**

Wind speed and direction monitoring sensors have already reached a mature stage. Current research on railway wind disasters mainly focuses on short-term localized wind speed and direction forecasting. Various methods, such as numerical forecasting, time series analysis, Kalman filtering, wavelet analysis, neural networks, and deep learning, are utilized for wind speed prediction. Each method has its own advantages, disadvantages, and applicable scenarios. Among these methods, neural networks and deep learning have experienced rapid development in recent years and have been widely applied in numerous engineering fields.

In key areas prone to frequent wind disasters, increasing the number of wind speed and wind direction monitoring points, enhancing research on local gusts, and conducting short-term local wind speed and wind direction forecasting studies are essential steps in constructing a comprehensive HSR wind warning and monitoring system. This system plays a crucial role in reducing wind disaster losses and ensuring the safety of railway transportation.

4.6.6.2 *Rain monitoring system*

Rain disasters refer to various disasters caused directly or indirectly by rainfall. Concentrated rainfall in a short period often leads to meteorological and geological disasters, such as floods and landslides, which pose a significant threat to railway safety.

Rainfall monitoring equipment can be divided into two types: mechanical and non-mechanical. Mechanical devices include tipping bucket rain gauges, siphon rain gauges, and double-valve capacity rain gauges, while non-mechanical devices include radar rain gauges, pressure-sensing rain gauges, and laser rain gauges.

Rainfall monitoring points should be established based on the meteorological characteristics of different regions, especially in railway sections located in vulnerable areas prone to landslides, mudslides, hazardous rocks, rockfalls, or collapses. Each rain gauge should be installed at a height of approximately 4 m above the track surface, either on contact wire poles without obstructions or within maintenance bases, maintenance work areas, and maintenance workshops.

4.6.6.3 *Snow monitoring system*

Snowfall refers to the depth of new snow falling over a period of time (generally 24 h) but does not include drifting snow and blowing snow. Snow monitoring also typically includes other forms of precipitation, such as ice pellets, glaze, hail, and sheet ice. Snow depth usually refers to the total thickness of snow on the ground at the time of observation. Precipitation is often used by meteorological departments to indicate the amount of snowfall, and the water equivalent of snow is the vertical depth of water obtained by melting snow. At present, there are three main methods of snow depth monitoring:

(i) **Measurement method based on image processing**: The principle involves replacing the scale of the human eye observation scale by the camera. This method uses the camera to capture the scale of the established scale and uses an image processing method to identify the measured value. This method is greatly affected by the weather and light conditions, and its robustness is not good.

(ii) **Snow depth measurement method based on ultrasonic sensors**: The accuracy of snow depth measurement by ultrasonic sensors is greatly affected by temperature and wind speed. In practical applications, the monitoring point should be set as far as possible to avoid wind, so as to improve the stability of the sensors in low-temperature environments. Meanwhile, it is recommended to take the average value of a multi-point measurement.

(iii) **The method of measuring snow depth based on the principle of phase-based laser ranging**: This method is used in non-contact industrial measurements due to its mature technology. Furthermore, the method of embedding polytetrafluoroethylene material into the snow measuring board effectively reduces the error caused by laser penetration of accumulated snow. The laser measurement method can meet the observation needs in severe winter and blizzard weather, and its measurement stability and accuracy are superior to those of the ultrasonic measurement method. However, the fixed laser ranging method can generally only obtain the depth of accumulated snow at a single point, which is easily affected by the uneven surface of the snow. In practical applications, the propagation of lasers is interfered with by large-diameter particles in the air (haze weather, sandstorm weather, etc.), and people or objects entering the measurement range may block the measurement line, which can easily lead to erroneous measurement results.

Considering the advantages and disadvantages of existing snow depth measurement technologies and methods, in the actual application environment of HSR, it is necessary to collect the snow thickness on the surface of the rail sleepers. According to the installation conditions and requirements of the measurement object, measurement devices based on laser snow depth sensors are widely used. However, pure laser measurement is easily interfered with by other external factors. Therefore, it needs to be used in conjunction with other auxiliary methods to obtain high-quality measurement data.

From the perspective of practical applications in high-speed rail, the snow depth monitoring subsystem needs to continuously obtain real-time information about the extent of snow cover on and around the railway tracks. When the snow accumulation reaches a certain depth, the system issues alarms to the train scheduling, maintenance, power supply, and signaling departments. In sensitive areas such as deep cuttings and tunnel entrances, where snow accumulation is prone to occur, snow monitoring is particularly essential.

4.6.6.4 *Seismic monitoring system*

China experiences frequent, strong, shallow-focus earthquakes with widespread distribution, making it one of the countries with the most intense seismic activity and severe earthquake-related disasters in the world. As the high-speed rail network expands, the mileage of HSR located in earthquake-prone areas gradually increases. China's mainland faces a wide distribution of earthquakes with scattered and complex seismic sources, significantly impacting the safe operation of HSR. Establishing earthquake monitoring and early warning systems for HSR located in earthquake-prone areas is of utmost importance.

During an earthquake, two types of seismic waves are generated: P-waves (longitudinal waves) and S-waves (transverse waves). P-waves travel faster and cause less damage, while S-waves have slower propagation and cause more significant destruction. High-speed rail earthquake early warning involves placing seismometers along the high-speed rail line or in proximity to potential seismic sources. When an earthquake occurs and reaches an alert level, the system utilizes the speed differential between electromagnetic waves and seismic waves, as well as the speed differential between P-waves and S-waves. This provides several seconds to several tens of seconds of warning time for trains traveling within the affected earthquake zone before the destructive seismic waves arrive. This allows the trains to slow down or come to a halt, thereby preventing safety accidents.

The high-speed railway earthquake monitoring and early warning system includes two modes: threshold alarm and P-wave alarm.

Threshold alarm: After an earthquake occurs, if the ground motion parameters monitored by seismic stations along the HSR reach or exceed a predetermined threshold, it sends out an earthquake alarm message. At this time, an earthquake disaster may have occurred in the region, and the threshold alarm can only prevent adjacent trains from entering the disaster area and prevent the occurrence of secondary disasters.

P-wave warning: Through the deployment of a seismic network or strong seismic network and real-time communication transmission lines along the HSR, real-time earthquake monitoring, using the early P-wave information of the earthquake, quickly analyzes the distance between the earthquake epicenter and the railway line, earthquake magnitude, focal depth, and other parameters, or directly estimates the ground motion parameters along the HSR, when the predicted ground motion parameters reach or exceed the predetermined threshold. Earthquake warning messages are issued, and trains running on potentially damaged lines are controlled so that high-speed trains have more time to slow down or stop before the main shock hits the line to prevent derailment and eliminate potential threats.

(1) **Determination of earthquake magnitude:** The magnitude of an earthquake, denoted as M, can be estimated based on the dominant period of the P-wave, denoted as T. Previously, for earthquakes with a magnitude of 6 or higher, it was necessary to consider the rupture process of the fault (the time required for the fault to rupture completely from its initial breaking point) to determine the dominant period, which took around 3 s. With advancements in technology, it is now possible to reliably determine the dominant period before the fault ruptures completely, and earthquakes with a magnitude of 7 or higher can be calculated in approximately 1 s. The calculation process can also be adjusted based on continuously updated data.

(2) **Determination of the epicentral direction:** The core component of seismic monitoring used in engineering is the three-axis force-balanced accelerometer. It monitors the seismic acceleration components in three directions: x, y, and z (typically set as east-west (EW), north-south (NS), and vertical (z)). By performing vector operations on the vibrations in

these three directions, the direction of the incoming seismic waves can be determined in real time, and the development of fault rupture can be tracked in real time. Additionally, the focal depth can also be determined statistically based on the incident angle of seismic waves entering the surface.

(3) **Determination of the source-to-station distance:** In deriving the earthquake magnitude, it is necessary to determine the epicentral distance (the distance from the starting point of fault rupture on the fault to the monitoring station) or the P-wave amplitude. Conversely, based on the magnitude M and the P-wave amplitude, the source-to-station distance can be derived, taking the incident angle into account to calculate the epicentral distance and depth.

4.6.7 *Intelligent video analytics*

The video surveillance of HSR is divided into two categories: comprehensive video surveillance system and professional video surveillance system. The comprehensive video monitoring system mainly covers HSR stations and intervals along the track and is constructed simultaneously with a new railway project; the professional video monitoring system mainly covers production places such as each train station, part of the storage yard, transformer (distribution) substation, and passageway gate, and the professional video monitoring system is constructed, operated, maintained, and used by each profession.

The railway integrated video monitoring system is one of the significant technical methods to assist train command, production organization, passenger and freight transport services, operation monitoring, rescue and relief, and security prevention. It has already been built as a mode of access, storage, distribution/forwarding, application, and management, including access regional nodes and core nodes. In terms of video resource management, it tends to adopt an integrated mode to unify the management of video resources, such as stations, inter-district tracks, and premises along the track, and the Beijing–Shanghai High-Speed Railway and the Beijing–Jiangshan Intercity Railway have taken the lead in completing the construction of an integrated video monitoring system. In terms of video resource utilization, "video +" technology, combining "cloud + edge + end" and fusion of multi-sensor technology (such as radar and vibration optical fiber), is adopted, and Intelligent analysis algorithms are established for typical scenarios of the track safety environment, so as to provide timely and accurate early warning information for users. This is the urgent demand of the current business application, which is the professional coordination and efficient disposal of the production command department.

4.6.7.1 *Main functions of the system*

In full consideration of network bandwidth resources, application timeliness, and other factors, based on the "cloud" as the main, and taking into account the "side + end" combination of modes, intelligent video analysis of typical railway scenes is achieved. Applications that are stronger in real time, are closer to the scene application, and occupy large bandwidth are placed on the side (end). Cross-specialty, deep, and less real-time applications are placed on the cloud side. This solution maximizes the utilization of network bandwidth resources. Based on the principles of focusing, precise control, professional coordination, and efficient disposal, the application scenes are classified, and the high-risk lots adopt a case-by-case strategy.

(1) **Intruder monitoring of key areas**
This function is based on the deep learning target detection algorithm to establish a model, and under the activation function and data enhancement processing, it can quickly and accurately detect intruders in key areas. Through the real-time analysis of the video of the track interval and the premises along the track, it provides alarms for the detected intruders and provides the abnormal status and image information to the duty personnel to assist in monitoring the key areas, as shown in Figure 4.54.

(2) **Video comparison of key areas**
Based on the image similarity algorithm, different image templates are established. The peak signal-to-noise ratio, structural similarity, and hash value algorithms are used to compare the image similarity. By comparing the cyclic video images of the slope at the tunnel entrance, the slope of the roadbed, the bridge, and other upper-span buildings, it can timely discover the abnormal situation and track the change pattern of its status. By comparing the cyclic video images of the slope of the tunnel entrance, the slope of the roadbed, the bridge above the bridge, and other buildings above the bridge, it can discover abnormal situations in time and track the

Figure 4.54. Intruder monitoring of key areas.

Figure 4.55. Video comparison of key areas.

pattern of change in their status. The video comparison of key areas is shown in Figure 4.55.

(3) **Identification of worker access**

The convolutional neural network algorithm is used to extract the features of the door opening and closing state, eliminating the influence of external conditions, such as light and darkness changes, on the judgment. According to the operation plan information, the video stream of the operating door camera to be analyzed is obtained, and through the intelligent analysis algorithm, the attribute features of the operating personnel are extracted and access identification is carried out. The intelligent analysis algorithm extracts the attribute characteristics of the operator and carries out in/out identification. The entry/exit identification of operators is shown in Figure 4.56.

(4) **Foreign object monitoring of contact network**

Based on the image similarity algorithm, an image template is first established. The peak signal-to-noise ratio, structural similarity, and hash value algorithms are used to compare the image similarity as a judgment condition for image changes. The similarity of two images is calculated by extracting neural network features. If there is an anomaly, then a structured alarm message is generated, which is secondarily confirmed by the user. It can detect foreign objects, such as bird nests, kites, and plastic bags, in the image. With an increase in the sample base, deep learning-based contact network foreign object monitoring will be gradually realized. Contact network foreign object monitoring is shown in Figure 4.57.

(5) **Imaging monitoring of night-vision camera**

The image histogram algorithm is used to calculate the distribution of pixel intensities and the number of pixels per intensity value in the image to determine whether the night vision status of the laser camera is normal. It adopts the principle of time-sharing multiplexing to monitor the status of the night vision camera at night. The monitoring of the imaging effect of the night vision camera is shown in Figure 5.58.

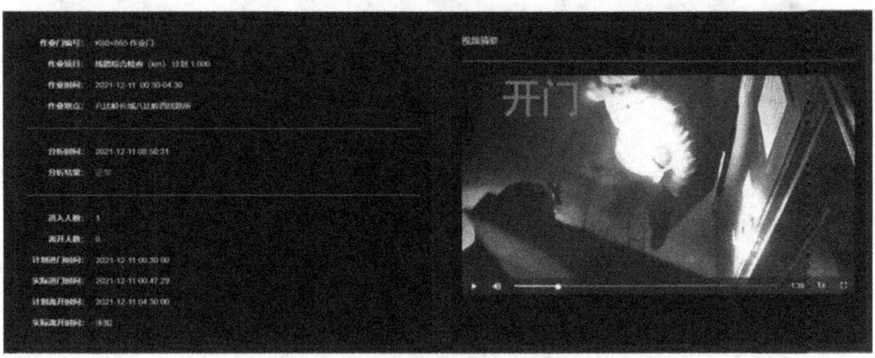

Figure 4.56. Identification of worker access.

Figure 4.57. Foreign object monitoring of contact network.

Figure 4.58. Imaging monitoring of night-vision camera.

(6) **Angle shift of camera surveillance**

Based on the algorithm model of the optical flow method, the correlation of pixels in the image sequence in the time domain is used to find the correspondence between the previous frame and the current frame, so as to calculate the motion information of the object between adjacent frames and to monitor the angular shift of the camera. Because of this, the camera

Figure 4.59. Angle shift of camera surveillance

angle shift status is monitored. The camera view range monitoring is shown in Figure 4.59.

Typical intelligent analyses in stations also include application scenarios such as personnel retrogression, personnel density, and legacy monitoring. In terms of track safety environment, we are carrying out the research and development of video analysis scenarios such as mudslides and water-flooded tracks and falling rocks at tunnel entrances and will gradually improve the application functions along with sample collection, algorithm research and development, and training.

4.6.7.2 *Key technologies of the system*

(1) **Cloud storage technology:** Cloud storage is one of the core technologies of cloud computing. With video data storage and management as the center of the transparent mode for the user, any authorized user can connect to the cloud through the network for data access. Video cloud storage uses storage virtualization and cloud computing management technologies to build an easy-to-manage, dynamic, efficient, flexible, scalable, stable, and reliable video storage center. Among them, storage virtualization provides unified storage resource services for the video surveillance system. The platform mainly consists of a video front-end collection area, a video storage area, network switches, video cloud storage management nodes, and video cloud storage nodes.

(2) **"Cloud + Edge + End" convergence technology:** Through the "cloud + edge + end" integration mode, the system can carry out the intelligent video analysis architecture design through the unified standard interface, provide the basic general video analysis capability for each professional user, and provide intelligent empowerment for each profession. It can avoid repeated investment and construction, save resources, reduce costs, and improve the application effect. Intelligent analysis can be divided into four levels: front-end analysis, basic analysis, professional analysis, and high-level analysis. Through the division of labor at the four levels, it can complete the task of intelligent video analysis, reflecting the ability of intelligent video analysis and improving business intelligence.

4.6.7.3 *Application effects*

Through carrying out intelligent video analysis, unified access, centralized storage, standardized management, and efficient use of integrated video data resources of HSR are realized, which is conducive to promoting the sharing and use of railway video data and can significantly enhance the comprehensive application of video and the ability of intelligent analysis. It can significantly improve the comprehensive application of video and intelligent analysis capability and support the automation and intelligence of equipment operation and maintenance, production organization, track safety, and environmental management.

(i) "Intelligent Analysis + Data Mining" improves professional and technical safety protection capability. On the one hand, through data mining, it provides decision support in the form of analysis reports for managers and decision makers at different levels, such as the State Railway Group, the Railway Bureau Group, stations, and sections, so as to realize decision-making data. On the other hand, through intelligent analysis, it solves the actual problems and pain points of field staff in the process of work, improves production efficiency, and ensures operational safety. From the aspects of intelligent operation and maintenance, security, business visualization, work process recording, data integration and sharing, etc., the system can be used to provide support for decision-making.
(ii) The mode of "platform + application" is adopted to promote the transformation of traditional applications. By making full use of the

railway data service platform to create an open video big data service platform, it provides intelligent video analysis and sharing services for users of various systems and empowers the relevant application systems. On the one hand, it fully enhances the value of video data; on the other hand, it fully enhances the application effect, forming the positive effect of "1 + 1 > 2". Through sharing and integration, it is conducive to the integration of technical strength and achieves a good interactive situation of "concentration of expertise and sharing of results." The adoption of "cloud + edge + end" combination mode promotes improvements in the railway video application mode and the innovative development of railway video monitoring technology and equipment.

(iii) The company has been able to control the hidden dangers of track safety and effectively improve the level of risk prevention and control. On the basis of making full use of the resources of the integrated video monitoring system of the reinforced railway, comprehensively applying the means of intelligent video analysis can reduce the number of on-site inspectors and guards, significantly increase the proportion of technical safety protection of video inspection and intelligent video analysis, firmly control the track problems and hidden dangers, and improve the efficiency of the production organization, so as to achieve the purpose of reducing costs and increasing efficiency.

(iv) By helping China's HSR improve its intelligent operation and management level, video big data analysis series technology is an important part of the intelligent operation of China's HSR. Through the construction of video big data analysis, it can not only promote professional business processes but also improve the automation and intelligence level of operation and maintenance work. What's more, it provides intelligent means of support for the safe operation of railway.

Review Questions

1. What are the main methods for detecting the status of high-speed railway vehicles and infrastructures?
2. What are the main functions of the ATO system for high-speed railway?

3. What are the main methods for detecting the status of high-speed railway vehicles and infrastructures?
4. What are the technologies of the new generation of carrier networks for railway?
5. What are the main types of high-speed railway mobile equipment and infrastructure condition inspection methods?

Chapter 5

Intelligent Operation of High-Speed Railway

Intelligent operation of high-speed railway widely adopts cutting-edge technologies, such as cloud computing, the Internet of Things, mobile communications, big data, and artificial intelligence, and integrates them with advanced high-speed railway operating technologies. Through comprehensive sensing, ubiquitous interconnection, fusion processing, active learning, scientific decision-making, and other means, the intelligent operation of high-speed railway realizes convenient and intelligent passenger transportation services, transportation organization, infrastructure maintenance, and EMU maintenance. The key technologies for intelligent operation mainly include intelligent passenger stations of high-speed railway, intelligent ticketing, intelligent train diagram compilation, intelligent traffic control, intelligent comprehensive dispatching, intelligent infrastructure maintenance, EMU PHM, and intelligent maintenance.

Intelligent operation is one of the three major parts of the intelligent high-speed railway system framework, closely coordinated with intelligent construction and intelligent equipment. The intelligent operation can be horizontally divided into three fields: passenger service, transportation organization, and maintenance. Passenger service includes intelligent passenger transportation and intelligent ticketing. Transportation organization includes intelligent comprehensive dispatching and intelligent traffic control. Maintenance includes integrated maintenance of railway lines, railway communication and signaling, power supply, and intelligent maintenance of EMUs. The intelligent operation part can be summarized into 3 areas, 6 directions, and 25 innovations. In the future, the innovation content can be expanded as needed.

This chapter opens up the multi-business collaboration of passenger flow — vehicle flow — maintenance flow based on the concept of "module-data driving, axis-surface collaboration." Intelligent ticketing and intelligent passenger transportation directly interact with passengers, providing them with multichannel electronic and convenient ticketing services. Travel services include the entire process of entry, waiting, boarding, and exit, achieving accurate collection and prediction of passenger flow information. Intelligent comprehensive dispatching and intelligent traffic control provide vehicle flow dispatch that matches passenger flow and converts precise passenger flow demands into vehicle flow allocation. It realizes integrated dispatch management of traffic, passenger transportation, vehicles, maintenance, etc., and ensures automatic, safe, and on-time operation of trains. Integrated maintenance of railway lines, railway communication and signaling, power supply, and intelligent maintenance of EMU provide reliable infrastructure and mobile equipment support for high-speed railway transportation. By grasping the deterioration principles and evolution rules of infrastructure and mobile equipment, predictive maintenance can be realized, maintenance efficiency can be improved, and maintenance costs can be reduced.

5.1 Intelligent HSR Passenger Station

Intelligent passenger stations of high-speed railway are intelligent complexes that extensively apply new generations of information, control, and communication technologies to realize interactive and coordinated development of passengers, stations, trains, and cities. They are oriented toward goals of convenience, safety, efficiency, and energy saving. They are controlled through automated information services and intelligent management. Their main functions include providing service information for passengers and operation information for passenger stations, intelligent management of passengers and station personnel, and intelligent fault diagnosis and maintenance management of station equipment.

A high-speed railway station has the characteristics of a complex spatial structure, numerous equipment, dense personnel, high mobility, and diverse businesses. It is not only a passenger service and railway passenger operation base but also an urban complex integrating comprehensive transportation, road-land coordination, and commercial properties. Building intelligent passenger stations to achieve convenient passenger services, system interconnection, overall collaborative management, and station–city integration has become an important development direction.

Intelligent passenger station data mainly include operation plan data of travel service, production command data, management data of passenger equipment, emergency management data, image information of luggage X-ray, image information of passenger bag placement/retrieval, passenger luggage-related information, contraband alarm and disposal information, status information of security equipment, personnel on-duty information, route planning data, navigation data, positioning information data, site information data, voice data, and portrait data. A business analysis model mainly includes a passenger transport management warning analysis model, a passenger credit reporting model, and an equipment fault prediction model. Integrated fusion applications based on models and data provide support for improving passenger transportation management and safe operation.

5.1.1 *Overall framework*

The overall framework of an intelligent passenger station is built around passenger services, operation production, and station management. It is mainly composed of information collection, information processing, application modules, and application interfaces, as shown in Figure 5.1. Following the mainstream "platform + application" framework, the big data platform and optimized computing platform are the brain and core of the intelligent passenger station, respectively. The platform collects basic information through station operation intelligent terminals and front-end passenger services and is connected to the superior dispatching and command platform and other station platforms. It analyzes and processes basic information to provide support for passenger services, passenger station operations and operation management, and other applications. It achieves the organic integration and transmission of passenger flow, business flow, and information flow, and it improves passenger station passenger services and operation and management levels.

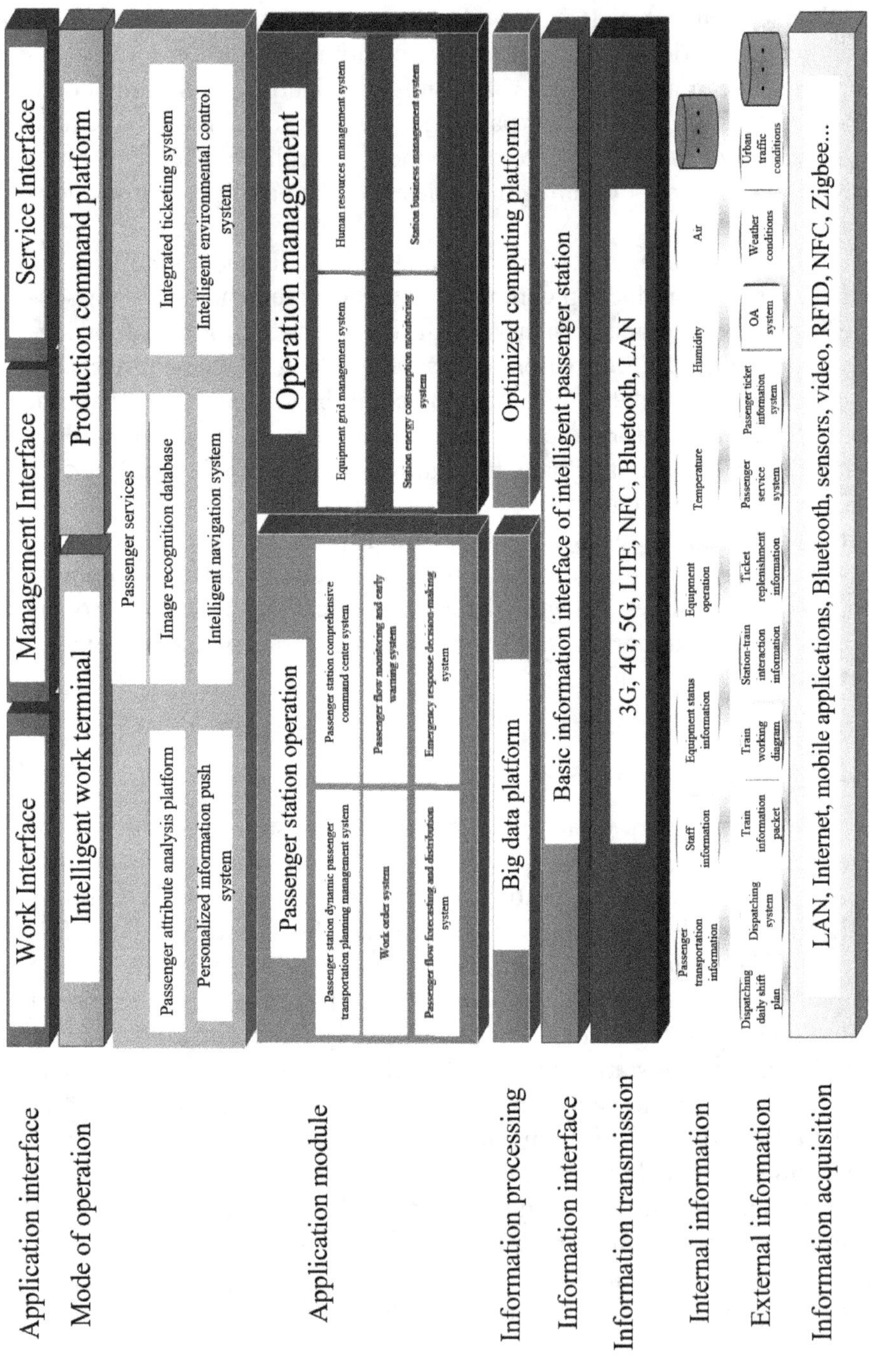

Figure 5.1. Intelligent passenger station: Overall framework diagram.

(1) **Information collection:** The basic data of a passenger station originate from external related systems and internal monitoring information. The external information includes dispatching daily shift plans, train operation diagrams, passenger service system information, station–train interaction information, weather, and urban traffic conditions. Internal information includes passenger information, staff information, equipment status information, and temperature. The above information is obtained through LAN, Bluetooth, sensors, video, Zigbee, or other technical means. The information is then transmitted to the basic information interface through 3G/4G/5G, NFC, or other means at the information transmission layer and finally collected into the big data platform and optimized computing platform.

(2) **Information processing:** The passenger station big data platform and optimized computing platform store and manage the collected multi-source information. They apply data analysis algorithms, such as feature extraction and cluster analysis, and optimization algorithms, such as deep learning and graphical models, to extract multi-dimensional features of data and mine data patterns to provide public information and algorithm support for each application module of the station.

(3) **Application module:** The application module layer is composed of passenger service, passenger station operation, and operation management.

The passenger service application module consists of a passenger attribute analysis platform, an image recognition database, an integrated ticketing system, a personalized information push system, an intelligent navigation system, and an intelligent environmental control system. The main functions include the following: intelligent identification and personalized services for different travel groups, such as key passengers and frequent travelers; smooth services throughout the entire process of passenger transportation connection, in-station inspection, in-station travel, boarding verification, and out-station transfer; intelligent somatosensory environment control in the station; passenger positioning and guidance in the station; visualization of station status and facility; and equipment utilization visualization.

The application module of a passenger station operation organization consists of the station's comprehensive command center and dynamic passenger transportation planning management, work order, passenger flow monitoring and early warning, passenger flow forecasting and distribution, emergency response decision-making and other systems. Its main functions include optimization and adjustment of the dynamic passenger transportation plan of the station, intelligent compilation of the dispatching daily shift plan, comprehensive intelligent command of the station management, and circulation of service orders.

The operation and management application module consists of equipment grid management, human resources management, station energy consumption monitoring, station business management, and other systems. Its main functions include equipment history update, status monitoring, operation situation assessment and maintenance plan compilation, staff position management and work arrangement, station energy monitoring, station operation point and advertising space leasing management, and station business information integration and release.

(4) **Application interface:** Through intelligent operating terminals and production command platforms, the application interface provides:

- station managers with comprehensive operational information, such as station passenger flow distribution, train arrivals and departures, personnel distribution, and environmental monitoring;
- staff with auxiliary tools, such as quick contact, coordination support, and timely feedback;
- passengers with personalized functions, such as in-site navigation, information inquiry and display, and travel plan adjustment recommendations.

5.1.2 *Key technologies*

The key technologies of an intelligent passenger station mainly include the following:

(1) **Theoretical technology for all-weather and full-region monitoring of passenger station operating environments based on the Internet of Things:** Based on the principle of "safety, energy saving, and comfort," the Internet of Things technology is used to monitor and sense the flow density and queue length, passing speed, moving direction, environmental comfort, and other station operating environment information of key areas such as station waiting rooms, ticket gates, entrances, exits, and transfer channels.

(2) **Collaborative linkage technology based on information, personnel, equipment, and environment:** Through a unified data interface, real-time collection and sharing of station operations, trains, personnel, equipment, environment, and other information are realized. Relying on the real-time flow and sharing of information, the collaborative linkage of personnel, equipment, and environment is realized.

(3) **Passenger station safety emergency decision-making and handling technology based on artificial intelligence and big data:** In view of typical scenarios of various emergencies at the station, such as large-scale train delays, large numbers of stranded passengers, fires, and violent terrorism, scientific decision-making suggestions are generated based on the actual operating conditions and emergency response plans of the station. This technology can rapidly deploy various emergency resources in the station, monitor and transmit back on-site conditions to ensure accurate and timely issuance of instructions, and realize integrated emergency technology that runs through emergency monitoring, planning, organization, response, disposal, recovery, and assessment.

(4) **Intelligent audio and video analysis technology based on deep learning and edge computing:** Through artificial intelligence technologies, such as neural networks, deep learning, and edge computing, video feeds of key areas such as the station entrance identity verification office, ticket gates, waiting halls, ticket halls, automatic ticket collection halls, transfer channels, and areas near elevators, along with images and audio, are used for deep learning training and intelligent recognition. Face detection, ID card information comparison, passenger flow trend monitoring, passenger flow analysis and early warning, and other operations can be implemented, thus reducing the burden on station staff and optimizing the passenger transport organization process.

(5) **Virtual gate technology based on biometric recognition:** Through a variety of detection methods, the passengers' fingerprints, face shape, iris, retina, hand shape, gait, voice, and other biometric characteristics are collected. Technologies such as image recognition, voice processing, and intelligent video analysis are used to identify the passengers' biometric characteristics, and the results are compared with passenger identity and ticket purchase information to ensure normal and convenient travel for

ticketed passengers and provide early warning and tracking of passengers exhibiting unusual behaviors.

(6) **Multilingual interactive passenger service technology based on speech recognition:** Based on technologies such as signal processing, pattern recognition, sound generation and hearing mechanisms, and artificial intelligence, this technology intelligently analyzes passengers' query words and provides voice answers. The voice interactive service provides intelligent recognition functions in multiple languages and accents, enabling barrier-free query services for travelers from different countries and regions.

5.2 Intelligent HSR Ticketing

With the development of the digital era of interconnected everything, data-driven processes, and intelligent decision-making, the application of "Internet + travel" has become deeply integrated, giving rise to high-speed rail intelligent ticketing. By building a self-aware and adaptive comprehensive railway passenger transport service model, a multi-dimensional, multi-level perception is achieved, which can enable proactive and in-depth identification of users' explicit and implicit needs. This also provides personalized, one-stop passenger transportation services for passengers during the entire travel process. The construction of intelligent ticketing systems has extensively improved passenger travel satisfaction and happiness while achieving comprehensive optimization and reconstruction of passenger transportation production organizations and passenger transportation service processes.

The intelligent ticketing business involves massive internal and external dynamic and static data and models. The data mainly include those related to ticket transaction, management, seats, pricing, basic ticket checking, ticket checking transaction, railway membership, passengers, e-cards, waiting list, subsequent transfer, verification, train stops, remaining tickets, travel service products, merchant information, orders, transactions, settlements, user reviews, invoice information, train capacity, dispatch volume, attendance and passenger occupancy rates, ticket revenue, and passenger flow forecast. The data also includes information of external civil aviation, highways, and tourism. The mechanism models involved mainly include: dynamic monitoring, early warning and adjustment of ticketing organization strategies, passenger user portrait, optimization model of train operation plan, internet ticketing risk prevention and control model, sales revenue model, and passenger flow forecast by time period, region, line, and direction. Intelligent ticketing carries out accurate passenger flow prediction, one-picture-a-day application, and internet risk control based on the integrated application of models and data. It provides passengers with better and more efficient travel services.

5.2.1 *Overall framework*

After more than 20 years of development, the China Railway Ticketing System has gradually developed from a single ticket sales business system in the station to the world's largest comprehensive ticketing system. It covers more than 3,000 passenger stations online (the "12306" online ticketing system, including the website and mobile app of the same name) and offline and provides ticket sales, extended services, marketing decisions, and other services. With the continuous emergence of new functions, such as independent seat selection, continuous transfer, 12306 meal ordering, online car hailing, railway tourism, new ticket systems, and high-speed railway Wi-Fi coverage, ticketing services are becoming intelligent and humanized. The overall structure of high-speed rail intelligent ticketing mainly consists of electronic tickets, intelligent ticketing organization, customized services, and passenger portraits, as shown in Figure 5.2.

5.2.1.1 *Electronic tickets*

Railway electronic ticket is a kind of electronic data stored in the railway ticket system and a railway passenger transportation contract voucher embodied in the form of electronic data. It is an electronic form of the attribute of the traditional paper ticket transportation contract and carries all the related feature information of railway passenger transportation services. Intelligent ticketing has achieved effective separation of the three attributes of paper tickets through research on key technologies of electronic tickets and the upgrade of the ticket system framework. The three attributes of a paper ticket consist of transportation contract vouchers, ride vouchers, and reimbursement vouchers. While completely solving the existing problems of paper tickets, it also

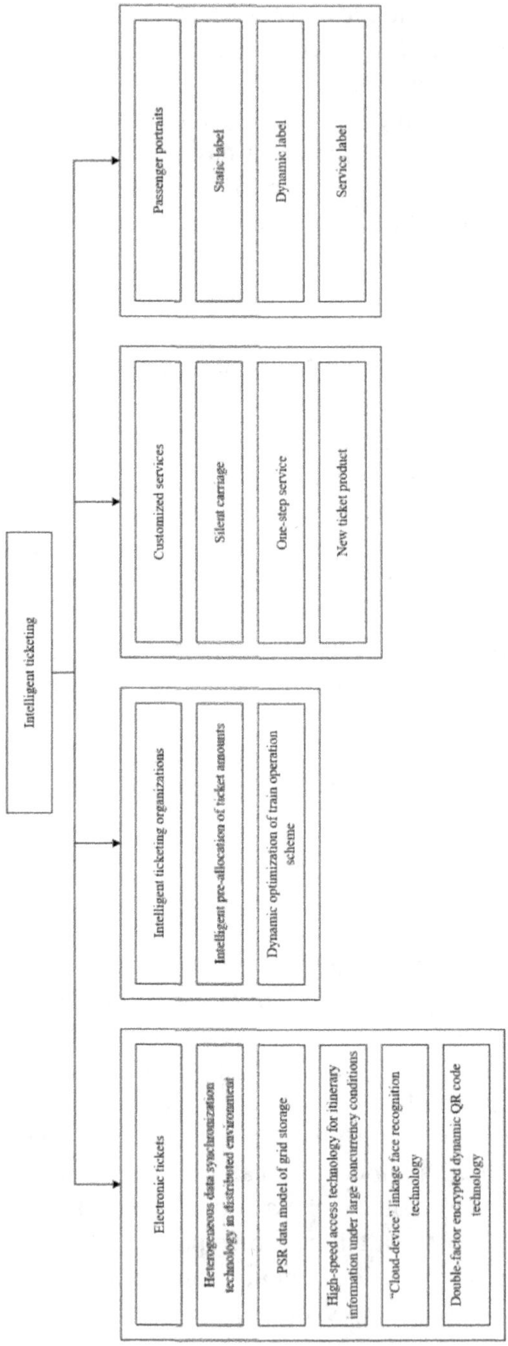

Figure 5.2. Overall framework diagram of high-speed railway intelligent ticketing.

greatly improves passenger travel satisfaction. Its main connotations include:

(1) **Full service informatization:** It realizes information-based services for passengers throughout the entire process, comprising ticket purchase, entry, waiting, boarding, exit, transfer, etc. After purchasing electronic tickets, passengers do not need to exchange them for paper tickets. Valid ID documents used to purchase tickets are again used to handle real-name verification at the station, boarding, ticket inspection at the exit, and train inspection. The passenger ticket system identifies the validity of the electronic ticket corresponding to the document based on the document information. The electronicization of tickets and the informatization of full-service services can effectively eliminate the risks of passenger tickets being lost or damaged and counterfeit tickets being purchased, and they completely solve the problems of paper tickets.

(2) **Self-service ticketing services:** For passengers who hold readable ticket purchase documents, such as second-generation ID cards, residence permits for Hong Kong, Macao, and Taiwan residents, or electronic passports, whether they are entering the station for real-name and ticket verification, checking for tickets to board a train, exiting the station, obtaining travel information, changing tickets, or printing reimbursement vouchers, they can follow self-service procedures through the station equipment. For passengers who have completed identity verification on the 12306 mobile app, they can directly use the "Ticket Check QR Code" in the app to check for tickets. For passengers with unreadable ticket purchase documents, they can also fill out the ticket purchase information form through a semi-self-service equipment and complete the procedure after their certificates are manually verified.

(3) **Integration of online and offline services:** It also aims to establish connectivity between online and offline channels, break down service barriers between different service channels, and realize "buy in one place, seen everywhere" for electronic tickets across all channels. Electronic tickets purchased by passengers on the 12306 online ticketing system can be changed or refunded at the station window. Ticket changes or refunds of electronic tickets purchased by passengers at the station window or at a ticket vending machine (TVM) can also be processed on the 12306 online ticketing system.

As a new ticketing system for railway passenger transportation, electronic passenger tickets improve business processing efficiency and reduce system complexity in the process of reshaping the passenger service process. Technological breakthroughs have been made in various aspects, such as heterogeneous data synchronization technology in a distributed environment, passenger service record (PSR) data model of grid storage, high-speed access technology of itinerary information under large concurrency conditions, face recognition technology with "cloud–device" linkage, and double-factor encrypted dynamic QR code technology:

(i) *Heterogeneous data synchronization technology in a distributed environment*: Different application scenarios in a distributed environment use different types of databases and file systems, forming a variety of hybrid deployment states of databases and file systems. Heterogeneous data synchronization in a distributed environment offers an efficient solution to the centralized storage of key business data and meets the needs of cross-channel changes and rapid ticket verification.

(ii) *PSR data model of grid storage*: Currently, passenger travel needs are developing in a diversified manner, and more and more extended service contents will appear in passenger travel bookings in the future. E-tickets construct a passenger-centered data structure covering passenger travel services and other content — a PSR. The PSR data model effectively supports passengers' personalized services and centralized business management needs.

(iii) *High-speed access technology for itinerary information under large concurrency conditions*: Ordinary relational databases cannot effectively support the huge access pressure on massive PSR data. Faced with the demands for high-concurrency data queries by tens of millions of passengers per day, e-tickets have innovated high-speed access technology for trip information under concurrent conditions.

(iv) *Face recognition technology with "cloud–device" linkage*: Railway face recognition technology solves the problem of automatic verification of "person–certificate" consistency during the self-service real-name system entry process based on "cloud–device" linkage. It overcomes the problems in facial recognition of passengers with electronic tickets in complex railway scenes, such as camera backlighting caused by the placement of gates, large age discrepancies between passengers and their certificates, and the low performance of existing gate hardware equipment.

(v) *Double-factor encrypted dynamic QR code technology*: The electronic ticket is designed with a QR code as one of the information carriers, and the QR code is dynamically updated according to time after encryption as its identification code to ensure security.

5.2.1.2 *Intelligent ticketing organization*

Intelligent ticketing organization refers to a seat management method that combines refined seat management and seatless/free seat inventory management on the basis of the existing ticketing organization. It realizes the automation and intelligence of seat adjustment through automatic generation of ticketing organization strategies by marketing and manual-assisted adjustments. At the same time, it provides convenient and fast operation methods to improve the efficiency of business operations.

Ticketing management must follow the principles of fairness, justice, and openness while also standardizing authority and equipment management. It improves business volume statistics and assessment mechanisms, strengthens the control and monitoring of key businesses, standardizes business operations, and improves business management efficiency. In addition, it establishes information communication mechanisms and a platform to realize information sharing and provide data support for relevant departments. It performs intelligent pre-allocation of tickets and dynamic optimization of the train operation plan through data processing.

(1) **Intelligent pre-allocation of ticket amounts:** Based on passenger flow analysis and prediction, seats will be reasonably allocated outside the pre-sale period, and dynamic adjustments will be made during the pre-sale process based on sales progress and passenger flow conditions to achieve the best match between seat resources and passenger flow demand. The core functions of intelligent ticket pre-allocation include passenger flow prediction between train stations and automatic ticket allocation. In terms of data, it uses pre-sale data combined with historical data to build a massive data sample training model to enhance the real-time accuracy of predictions. In terms of technology, it uses big data to analyze passenger information and ticket sales data, and it integrates the offline sand table exercise data and the online prediction algorithm to build a passenger flow

forecasting system. By establishing prediction models suitable for different scenarios, such as time series, linear regression, and exponential smoothing, it can achieve train passenger flow OD prediction, station passenger flow OD prediction, and ticket amount intelligent pre-segmentation to provide scientific predictions of the national railway dispatch volume.

(2) **Dynamic optimization of train operation diagrams:** Based on passenger flow forecasting combined with the allocation of transportation capacity resources, the passenger trains' operating sections, operating routes, train types, running pairs, marshaling content, stops, and EMU operation plans are dynamically optimized and adjusted.

5.2.1.3 *Customized services*

Customized services refer to providing suitable high-speed rail passenger service solutions according to the needs of different groups of people. Passengers can independently choose the service they are most satisfied with. As a higher level of service, this kind of customized service allows intangible services to provide greater value to gain passengers' satisfaction. The "Silent Carriage" is tailor-made for passengers seeking a quiet environment to get a better travel experience. The "one-step" service focuses on the entire journey of key passengers and makes them feel warm. Services such as "high-speed rail + tourism," "high-speed rail + hotel," and "high-speed rail + catering" enable passengers to enjoy "one-stop" dining, accommodation, travel, shopping, entertainment, fun, and extended services at high-speed rail stations.

5.2.1.4 *Passenger portrait*

The railway passenger portrait system conducts data modeling of existing and potential passengers who use railways as a mode of travel and depicts them in different data dimensions. By modeling and analyzing critical information such as passengers' static demographic attributes and dynamic behavioral preferences, easy-to-understand semantic tags are abstracted, and a complete picture of the user's information is formed. The passenger portrait system can help business managers distinguish different customer groups, analyze the differences in demand preferences and behavioral characteristics between different groups, and formulate more targeted marketing strategies for different customer groups. Combined with the business status of the high-speed rail passenger transportation department, the passenger portrait system can be used to provide technology assistance in the analysis of passenger group behavior characteristics, daily and holiday passenger transportation marketing strategy research, and precision marketing strategy research.

5.2.1.5 *Intermodal transportation services across transportation modes*

(1) **Rail–water combined transport:** It supports the sales of "bus tickets + boat tickets" and "train tickets + bus tickets + boat tickets" to achieve one-time ticket purchase, one-time payment, and seamless transfer.

(2) **Air–rail combined transport:** It supports the sales of air tickets and "aircraft + high-speed rail" air–rail combined transport tickets, allowing passengers to experience seamless travel by airplane and high-speed rail. China Railway and China Eastern Airlines jointly launched an innovative initiative to establish air–rail combined transport service centers at airports and train stations. It provides passengers with one-stop ticket purchasing services for air tickets and train tickets, meets their multimodal transfer needs, and provides conveniences such as waiting and security checks. It not only improves the efficiency of the comprehensive transportation hub but also brings a more convenient travel experience to the majority of passengers.

(3) **Bus ticket service:** By aggregating multi-supplier resources covering 30 provinces, autonomous regions, and municipalities across the country, it saves passengers the trouble of finding online ticket purchase channels at local passenger terminals and booking across multiple platforms.

(4) **Online car-hailing service:** This service provides online car-hailing as well as airport and station pickup and drop-off services. It introduces regional transportation capacity suppliers, solves the problem of large-scale station passengers arriving at night, and provides a more convenient and favorable car-hailing travel experience.

(5) **Intermodal transportation information platform:** This platform carries out the correlation and integration of multi-source data originating from urban transportation, railways, civil aviation, highways, and waterways. It provides railway passengers with various inquiry functions, namely comprehensive traffic information, itinerary traffic information summary, convenient transfer guidance and recommendations, and travel announcement information to support the better development of interline transport services.

5.2.2 *Key technologies*

The vision and goal of intelligent ticketing are to achieve inertia-free passenger travel, automated business processing, resource sharing, and service integration; improve the service quality of the entire journey; and achieve "seamless connection" throughout the journey, relying on a wide range of perception systems and focusing on the passenger service "MaaS + full journey." It includes providing passengers with full-process closed-loop automated sensing passenger transportation services; integrating the information of railways, combined transport parties, social parties and passengers, conducting centralization and integration, and providing passengers with comprehensive network services relying on the new generation of network communication and information security technologies; relying on intelligent perception to obtain passenger needs throughout the entire travel journey through multiple channels, combining augmented reality, virtual reality, and artificial intelligence technologies to enhance service decision-making capabilities through deep learning, genetic algorithms, and other methods to achieve intelligent travel for passengers. The key technologies of intelligent ticketing include ticketing service perception automation, service networking, execution autonomy, and other related technologies:

(1) **Perception automation:** Through the use of diversified automatic sensing technologies, comprehensive sensing of the status of each production factor is achieved. Through research on biometric identification technology, visual interaction technology, and other technologies, we can achieve the acquisition and analysis of passenger identity and service needs. Based on railway products, through intelligent perception and execution, a comprehensive resource integration model is established. By using intelligent analysis and processing to form ubiquitous interconnection and intelligent perception of passengers, operators, equipment, and the environment, it creates a convenient, smooth, warm, and comfortable

travel environment for passengers, promotes the innovative development of railway passenger services, and realizes personalized and differentiated service models.

(2) **Service networking:** A widely interconnected information sharing mechanism is built through research on technologies such as intermodal information fusion, human–computer interaction, shared fusion, and collaborative linkage safety services. Based on railway passenger transport information, it integrates and processes comprehensive information on domestic aviation, highways, water transport, and foreign transportation industries. Through extensive, safe, and reliable interaction of various types of information, it forms a railway electronic ticket as the core, covering diversified passenger transport for different travel purposes. Through extensive, safe, and trustworthy interaction of various types of information, a networked system of diversified passenger transportation extension services covering different travel purposes has been formed, with railway electronic tickets as the core. It realizes the deep integration of multiple types of service information, such as "people–equipment–environment–trains–places." It forms a comprehensive transportation integrated service model for passenger travel and improves the passenger travel experience and the quality of intelligent ticketing.

(3) **Execution autonomy:** Mainly through research on technologies related to human–machine collaboration, using intelligent analysis and processing to comprehensively integrate railway operation management data, railway passenger transportation data, combined transportation data, other social data, user portraits, and other information resources to provide a basic source for intelligent ticketing. It deeply combines the new generation of artificial intelligence with railway passenger service needs to form a full self-service intelligent passenger service model based on intelligent identification and artificial intelligence. It realizes the active execution of intelligent decisions and forms an integrated service system for intelligent perception, decision-making, and execution.

5.3 Intelligent HSR Train Diagram Compilation

Intelligent train diagram compilation of high-speed railway fully utilizes next-generation information technologies, such as cloud computing, big data, and the Internet of Things. It adopts multi-department collaboration and centralized deployment with a distributed access model to achieve remote, geographically dispersed, collaborative, and automatic railway diagram compilation for large-scale railway networks. This system aims to meet the requirements of "One Network, One Diagram" for high-speed railway, homogeneity of stations and sections, "operationalization of train diagrams, and integration of train diagram compilation," and train diagram intelligent compilation.

5.3.1 *Overall framework*

The overall framework of the train diagram intelligent compilation system of high-speed railway mainly includes the infrastructure, data, service, application, and security management layers, as shown in Figure 5.3.

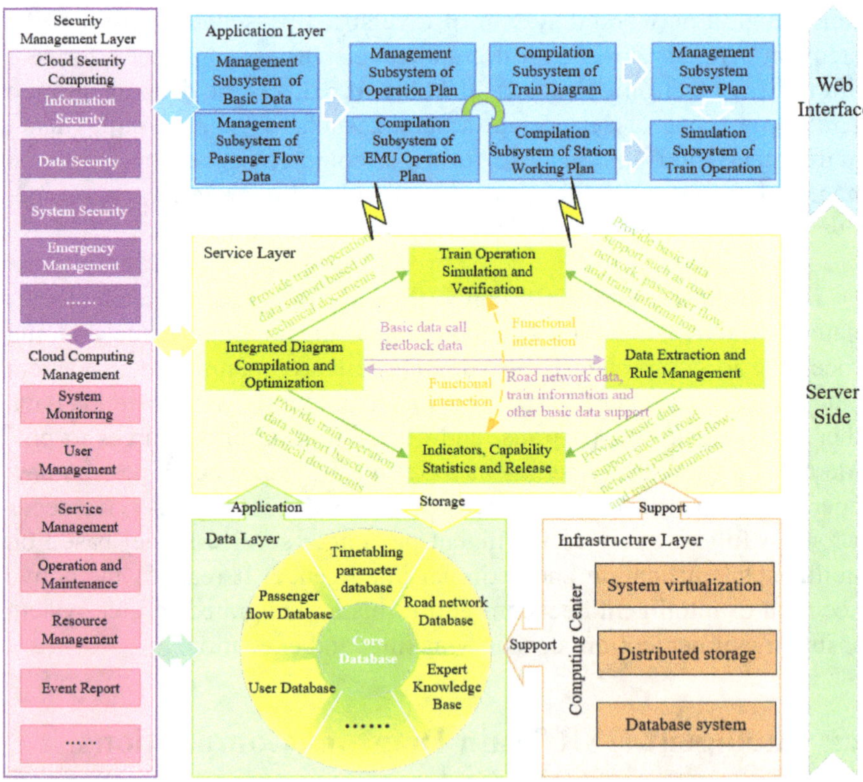

Figure 5.3. Overall framework of intelligent train diagram compilation system of high-speed railway.

(1) **Infrastructure layer:** In the cloud computing framework, the infrastructure layer serves as the lowest physical resource pool and is responsible for providing data storage and computing resources to the upper layer. The infrastructure layer mainly applies computer cluster technology to build a cloud computing center with high performance and reliability. It integrates resources such as servers through virtualization technology to form a virtualized resource pool for computation, storage, networking, and other dynamically manageable resources. This effectively improves resource utilization and saves costs. Additionally, it employs scalable distributed storage technology to overcome storage performance bottlenecks, enhance system security, and improve storage efficiency.

(2) **Data layer:** The data layer is mainly responsible for data perception, storage, exchange, and management in the cloud, providing multi-granularity data support for system operation. The data layer mainly applies data management technology to classify, organize, store, and manage massive amounts of data. It realizes the extraction and integration of key data, such as the basic road network, passenger flow, and transportation resource allocation. Combined with distributed storage technology, it builds distributed databases for various types of data to improve data exchange efficiency.

(3) **Service layer:** The service layer mainly focuses on building basic functions and automation algorithms for the application layer. It provides users with various automated programming and human–computer interaction functions through embedded algorithms. By reconstructing the optimization model of the refined train diagram, it designs a solution algorithm that can be implemented in parallel to achieve rapid solution of the train diagram optimization model.

(4) **Application layer:** The application layer based on the web interface provides services for users with different permissions, provides core business functions for users with graphical interfaces, and provides convenient human-computer interaction tools. By calling service layer algorithms in the cloud, automatic optimization functions and intelligent decision-making assistance are achieved, such as conflict detection, train diagram compilation, station working plan compilation, EMU operation plan compilation, and train operation simulation.

(5) **Security management layer:** The cloud computing security and management layer runs through the overall framework, monitoring and safely maintaining resources, tasks, data, functions, etc., at all levels of the cloud computing system to ensure the normal operation of the system. In the infrastructure layer, access monitoring and hardware reinforcement of virtualization systems are strengthened. In the data layer, data monitoring, protection, and encryption processing are implemented.

The intelligent compilation system for high-speed railway train diagrams is centrally deployed in the main data center of China Railway Group. With the strong peak computing power and massive data storage capacity of the cloud computing center, model solutions for optimizing train diagram compilation in the railway network are achieved, including the optimal calculation of timetables, station track utilization plans, and EMU crew routing. The system provides different users with corresponding system functions and data resources based on their set permissions. It also offers web-based applications and services using the B/S model to customers who are unable to access the private cloud of China Railway Group, as shown in Figure 5.4.

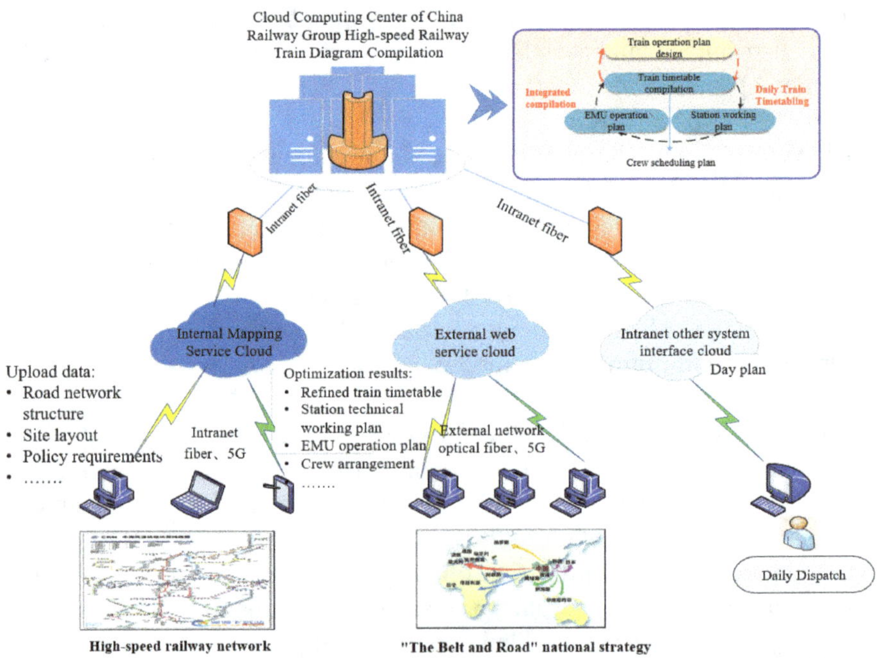

Figure 5.4. Deployment of intelligent compilation system for high-speed railway train diagram.

5.3.2 *Key technologies*

Train diagram intelligent compilation is a complex multi-objective optimization problem involving multiple departments, such as planning, passenger transportation, train maintenance, maintenance of railway lines, railway communication and signaling, and power supply. The key technologies mainly consist of two categories. The first category is intelligent compilation, including data management technology, integrated intelligent compilation technology of train diagrams, and simulation technology of train operation traction. The second category covers models and algorithms that support intelligent compilation, including those for the optimization and compilation of train diagrams, station working plans, EMU operation plans, and crew routing.

5.3.2.1 *Intelligent compilation service of high-speed train diagram*

(1) **Data management technology**

Infrastructure data: Based on the geographic information system (GIS), the road network data are associated with the actual geographical coordinates and then standardized and digitized. The basic road network and mapping parameters are refined, visualized, and managed. Service functions automatically display multi-level and multi-granular road network models on three levels: macro, meso, and micro.

Passenger flow data: By reading or inputting passenger flow OD data and combining it with the macroscopic road network, it can perform statistical analysis of section passenger flow under road network conditions, analyze passenger flow in and out of hub stations, and transfer passenger flow. The results can be visualized in various forms.

Train operation plan: By integrating with the compilation system interface of a train operation plan, the data can be read to achieve a graphical display of the plan. It provides human–computer interaction tools for compilation and adjustment, and it performs relevant indicator statistics, serving as one of the important inputs for train diagram compilation.

(2) **Integrated intelligent compilation technology of train diagram:** Based on optimization models and intelligent solving algorithms, it analyzes data related to, for example, infrastructure, real-time passenger flow, and train operation plans and fully considers factors such as station

operations, stop plans, and maintenance time. It also realizes the "integrated" compilation and adjustment of train diagrams, station working plans, EMU routing plans, crew plans, etc. It automatically completes train running time calculation, train operation conflict detection/resolution, train operation plan matching, station working plan dynamic simulation, and other services. The technology uses a graphical display interface to provide human–computer interaction compilation, adjustment, and statistical analysis tools which display the operation indicators of the adjusted plan in real time, intelligently assist staff in making decisions, and optimize the efficiency of the high-speed railway system at the macro level.

(3) **Train operation traction simulation technology:** In order to verify the train operation plan and identify real capacity bottlenecks, microscopic train operation simulations are performed. On the basis of the microscopic road network (line ramps, curves, bridges, tunnels, blocks, and station signals), conditional constraints such as traction power supply information, train control level, and simulation goals (e.g., shortest time first, lowest energy consumption first) are considered to conduct train operation simulation with high reliability. The simulation results can not only evaluate the accuracy of the diagram parameters (running hours, train intervals) but also verify the feasibility of the diagram, identify capacity bottlenecks, and provide a decision-making basis for the adjustment and optimization of diagrams.

5.3.2.2 *Core service models and algorithms*

(1) **Model and algorithm for train diagram optimization:** The intelligent compilation function of train diagrams is mainly realized by calling mathematical optimization models and automatic solving algorithms on the cloud. There are many models for optimizing train diagrams, and one relative mature model is the resource allocation model based on a spatiotemporal network. Taking the operation process of a train in a local road network as an example, the station arrival, stopping, departure, and section operations as objects, and the stations and sections as occupied resources, the train operation process can be regarded as a collection of operations with temporal and spatial continuity. A spatiotemporal network diagram of train operation is shown in Figure 5.5. The vertical axis in the

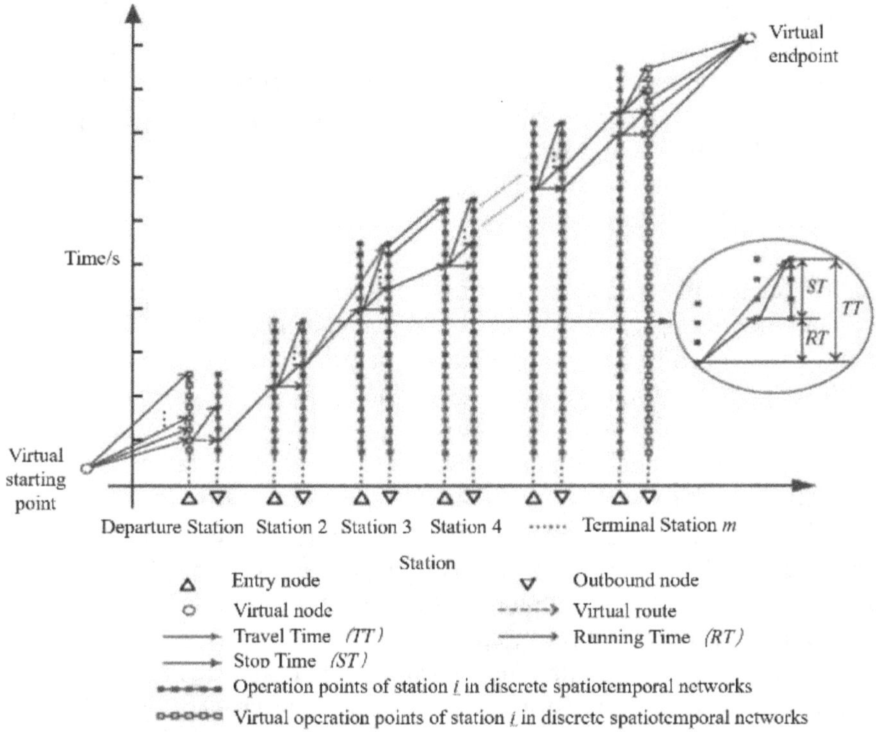

Figure 5.5. Spatiotemporal network diagram of high-speed railway train operation.

diagram represents continuous time, discretized into time points based on actual needs and in units of seconds or minutes. The horizontal axis represents continuous space, discretized into operation points based on the spatial positions of stations and the arrival and departure operations. The trajectory of a train on a two-dimensional spatiotemporal plane forms a spatiotemporal network diagram. Under the premise of simply considering the blocks and route conflicts of the microscopic road network as constraints, when more and more trains need to run in the road network and occupy the entry and exit routes, tracks, and sections, conflicts in resource occupation (station routes, tracks, and sections) will inevitably arise. Effective resources should also be allocated.

The train diagram optimization model aims to maximize capacity or meet passenger transportation needs. Based on optimization methods such as column generation, the Lagrangian relaxation algorithm, and branch and bound, a parallel computing program module that combines exact and heuristic algorithms is constructed to perform computing tasks. Innovations are made in computing task decomposition, computing resource allocation, progress coordination, and other aspects. The goal is to achieve collaborative compilation for large-scale railway networks with multiple levels and users.

(2) **Model and algorithm for station working plan optimization:** When the station working plan only takes into account the departure line occupancy plan, it can be integrated with the diagram compilation model into an overall optimization model. If we further consider the coordination of station train and shunting operations (such as entry and exit operations and intra-station transfer operations), the optimal compilation model of the station working plan is equivalent to the optimization model for the train diagram of a local road network, which must be based on the microscopic basis of the road network. Turnouts, crossovers, and tracks (i.e., branched sections and non-branched sections in the station) are allocated to each train as occupied resources.

Essentially, the compilation, adjustment, and optimization of station working plans can be transformed into a job-shop problem in industrial scheduling. The trains are considered jobs, and the tracks are machines. Suitable processing machines and times are arranged for each train operation, usually based on the premise of meeting the timetable calculated from the train diagram. The optimization objectives include maximizing station capacity or balancing the utilization of station resources. Various operation time standards, route assignments, and track occupancy conflicts serve as constraints. The automated process completes the working plan for each train at the station and feeds the results back to the diagram compilation module.

Small-scale intermediate stations can usually quickly generate corresponding station working plans based on simple rules. Station working plans for connecting stations with multiple lines or large-scale passenger stations need to consider the interference of train and shunting operations, special operation rules (such as the suction and water supply), parallel and hostile approaches, and other factors. A specialized subsystem needs to be developed to optimize the compilation of working plans for the hub station.

The automatic compilation of station working plans is completed by calling algorithms on the cloud. Mature algorithms for solving job shop problems in industrial scheduling fields, such as the Lagrangian relaxation algorithm and the alternating vector multiplier method, can be modified to solve the automatic compilation model of station working plans.

(3) **Optimization models and algorithms for EMU operation plan and crew routing:** The automatic compilation method of the EMU routing plan mainly meets the requirements of train operating line responsibility. It takes the minimum number of EMUs as the optimization goal and the maintenance mileage and time of the first-level maintenance as the main constraints. Then, it transforms the problem into a vehicle routing problem with time windows (VRPTW). The conversion of the EMU connection network diagram based on the train diagram is shown in Figure 5.6. Each train can be regarded as a customer in the VRP. The train running distance and time can be regarded as the customer's needs. Each EMU is regarded as a vehicle in the VRP. The compilation of EMU operation plans has more complex constraints and a larger computational scale compared to the classical VRPTW. When designing single-line EMU routes, the focus is mainly on the connectivity and continuity of the train operating lines on that specific line. For network-wide EMU operation compilation, consideration is extended to include the coupling of train operating lines on different routes. Regulations such as EMU allocation and maintenance

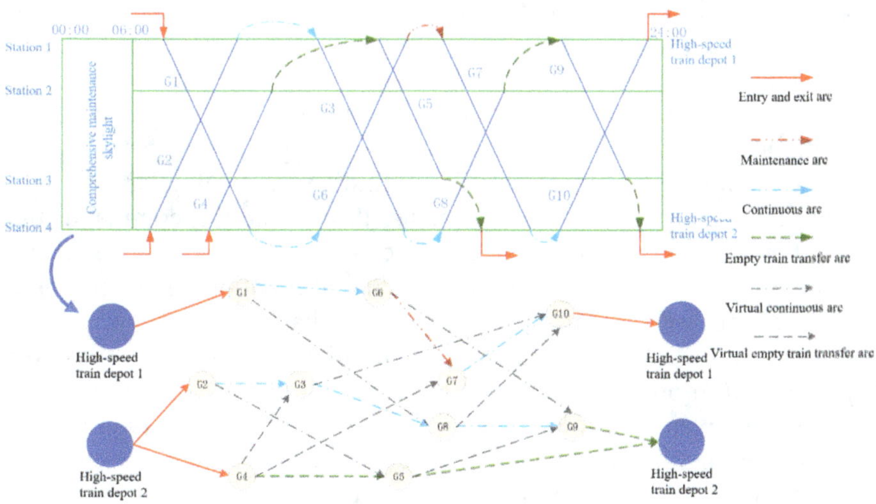

Figure 5.6. Conversion of EMU connection network diagram based on train diagram.

dispatching are not only technical problems but also management problems.

In order to further improve the utilization efficiency of EMUs, under the current management system, generally more mature algorithms for solving VRPs can be improved and used to solve the optimization model for EMU routing, such as the Lagrangian algorithm in the exact algorithm, the alternating vector multiplier method, and bionic evolutionary algorithms in modern optimization algorithms.

The model and solution algorithm for the automatic compilation of crew routing plans can refer to the optimization model and algorithm for EMU routing. The crew segments are regarded as customers in the VRP, and the crew groups are regarded as vehicles in the VRP. The model for crew routing plans can also be transformed into a VRPTW. The improved algorithm for solving the optimization model of EMU routing can also be used for solving the optimization model of crew routing.

5.4 Intelligent Centralized HSR Traffic Control

Intelligent centralized control of high-speed railway introduces artificial intelligence technology and algorithms on the basis of the existing centralized traffic control (CTC) system to realize the intelligent adjustment of train operation plans and provide train dispatchers with advanced safety inspections and more efficient technical means for assistance. Intelligent CTC aims to realize the intelligent assistance of train dispatching command, intelligent collaborative control, and emergency decision-making, so as to improve the efficiency of train dispatching command and operation efficiency holistically.

The intelligent CTC system utilizes different types of data, including the travel, routing, and three-way plans, the scheduling command, the EMU bottom application plan, on time/delay information, the actual operation data of the train, the route warning information, the fixed route, the station operation control, the automatic change information of the train number, the ATO platform door information, and the catenary power supply status information. The business analysis model mainly includes the automatic plan adjustment model, the traffic flow optimization model, and the overall situation deduction model of transportation and production. Based on the integration and application of models and data, the transportation production law can be accurately controlled to improve the efficiency and benefit of high-speed railway transportation.

5.4.1 *Overall framework*

High-speed railway dispatching can be divided into three levels: strategic, tactical, and operational, including operation plan formulation and adjustment. Compared with the existing dispatching, intelligent CTC focuses on the transformation from "artificial experience-based passive reaction" to "scientific, efficient and active regulation," which aims to realize "one map a day" and "optimization and adjustment", be able to quickly compile operation maps when transportation resources and constraints change dynamically, and restore the traffic order as soon as possible when emergencies cause traffic disorder. Specifically, at the strategic and tactical levels, the iteration of the hierarchical distribution of the operation plan and train diagram is transformed into the dynamic integration of "the operation plan — train diagram". At the operational level, it changes from manual experience scheduling into data-driven predictive intelligent dispatching to realize the optimal allocation of resources and the improvement of emergency disposal capacity.

The intelligent CTC system of high-speed railway is based on the existing CTC system and takes the automatic adjustment function of stage plan as the core, taking into account heterogeneous information sharing and intelligent safety control functions, and optimizing the existing CTC system. The intelligent CTC system is mainly optimized and improved in the intelligent adjustment of train operation plans, the safety control of train routes and order, and the information sharing of train information data platform. Its main functions include the following:

(1) **Intelligent adjustment of train operation plan:** In the case of emergencies, such as severe weather or equipment failure, the key information databases, such as train routing, minimum return time, and arrival–departure line utilization, are used, and the basic strategy of not changing the train running order and stopping stations is followed. By establishing the intelligent adjustment strategy of a comprehensive train operation plan related to speed limits, such as delayed trains, total delayed time, and arrival–departure line utilization, the intelligent and rapid adjustment of the train operation plan can be realized. This improves the dispatcher's emergency response efficiency. At the same time, by constructing an expert knowledge base of train operation plan adjustment, the delays and adjustment schemes caused by different factors are classified so as to realize the accumulation of adjustment cases and experience and solve problems such as dispatcher's emergency scheme learning and adjustment scheme optimization.

(2) **Safety control of train route and order:** By integrating CTC-related driving and signal logic relations, it can expand self-disciplined control conditions and improve driving safety. Intelligent control includes one-key single locking and unlocking of routes and switches to realize intelligent focus control of key train approaches, control of inconsistent scheduling plans and execution paths to prevent human errors in the dispatcher stage, and effective control of non-single lock operation of bad branching switches.

(3) **Traffic information big data platform:** The CTC system and railway transportation information integration platform are deeply combined to realize the interconnection and real-time sharing of CTC and multi-professional information, such as passenger transport, power supply, construction, and disaster prevention. It also provides the emergency handling process, a comprehensive display of train operation, passenger ticket (passenger number and seat) information, conductor information display, automatic control of line power supply, automatic mapping of construction order symbols, automatic extraction of disaster prevention, speed limit information, etc.

The overall architecture of the intelligent CTC system is shown in Table 5.1. At present, the Railway Bureau divides the jurisdiction of

Table 5.1. The overall architecture of the intelligent CTC system.

<table>
<tr><td rowspan="4">dispatching section 1</td><td>Train diagram operation terminal</td></tr>
<tr><td>Train diagram adjusting terminal</td></tr>
<tr><td>Application server dispatching console 1</td></tr>
<tr><td>Intelligent adjusting server</td></tr>
<tr><td rowspan="4">dispatching section 2</td><td>Train diagram operation terminal</td></tr>
<tr><td>Train diagram adjusting terminal</td></tr>
<tr><td>Application server dispatching console 2</td></tr>
<tr><td>Intelligent adjusting server</td></tr>
<tr><td colspan="2">Database server</td></tr>
<tr><td colspan="2">Train diagram server</td></tr>
<tr><td colspan="2">Bureau general adjustment server</td></tr>
<tr><td colspan="2">Interface platform</td></tr>
<tr><td colspan="2">External interface</td></tr>
</table>

train management according to the dispatch control section, and the intelligent CTC design takes a single dispatch station as the minimum adjustment unit.

(1) **Intelligent adjustment:** For the relatively routine adjustment planning involving the station of the dispatching platform (such as track blocking, section blocking, and slowing down between sections), combined with the characteristics of the dispatch section, some suboptimal paths are given and added to the retrieval results to provide users with choices without deep adjustment planning. Through the accumulation in the process of use, an expert database and a history learning library are gradually established to meet the paths of user needs.

(2) **Data fusion and secure control:** The design of the intelligent CTC system security control mainly consists of two aspects: control in the operational process and control during emergencies. When integrating and processing with other systems, it can be classified and handled according to these two aspects, as shown in Table 5.2.

The above data are broken down into broad categories. In practice, the fused data should first be selected, cleaned, and filtered. Then, the data should be finely categorized and introduced into the intelligent CTC's safety control module for classification and processing.

(3) **External interface:** The existing CTC system has already exchanged data with interlocking, train control, TSRS, RBC, TDMS, Global System for

Table 5.2. Data fusion processing methods for interface systems.

Interface system	Data processing methods	Main uses
TDMS system	Operational processes and support	Day shift schedule, passenger handling, etc.
C3+ATO	Operational processes	Automatic train stop, platform door knocking, etc.
PDMS	Operational processes	Power arm power-off operation
PSCADA	Emergency	Line power status
Disaster monitoring	Emergency	Line disasters and speed limitations
DMS	Operational processes	Train attributes and auxiliary information

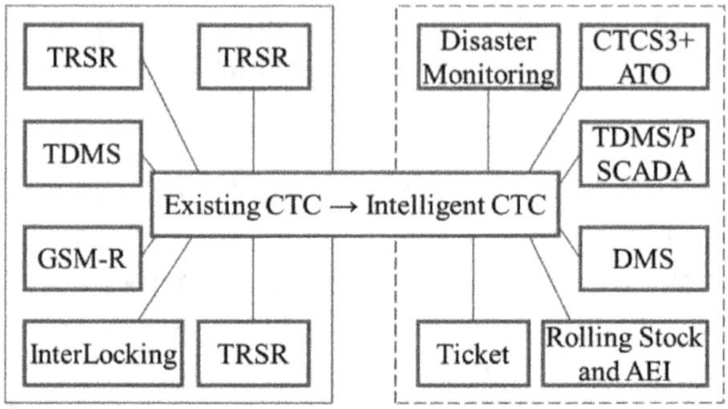

Figure 5.7. Newly introduced interfaces for the intelligent CTC system of the Beijing–Zhangjiakou high-speed railway.

Mobile Communications-Railway (GSM-R) systems, etc. In order to enhance the precision of centralized control, the CTC system needs to interface with more external systems, integrate and share data, and achieve a higher level of safety monitoring and intelligent adjustment. The interfaces newly introduced for the intelligent CTC system of the Beijing–Zhangjiakou High-Speed Railway consist of six aspects, as shown in Figure 5.7:

(i) *CTCS3 + ATO Interface*: The CTC system and the CTCS3 + ATO system communicate through the RSSP-II safety protocol. The CTC sends real-time adjustment plans from the train dispatcher to the onboard equipment. After boarding, the onboard ATO system can accurately achieve functions such as automatic train operation according to the plan and automatic platform stopping.

(ii) *Disaster monitoring system interface*: The CTC system communicates with the disaster monitoring system to obtain fault alarms related to weather abnormalities (such as wind, rain, and snow) and foreign object intrusion on the railway line. This information effectively guides the dispatcher in adjusting the train operation plan.

(iii) *Power supply and dispatch management system (PDMS/PSCADA) interface*: This interface connects the CTC system with the power supply system, breaking down the barriers between train dispatch and power dispatch. It facilitates the management of processes such as power supply interruptions in track sections and ensures electrical safety protection.

(iv) *Dynamic monitoring system (DMS) interface*: The DMS interface enables the CTC system to retrieve real-time onboard status information from trains, allowing dispatchers to have detailed information about train assignments, train operation levels (CTCS-2/CTCS-3), and other relevant details.

(v) *Ticketing system interface*: The CTC system obtains information about the seating occupancy of high-speed train passengers. This information serves as a basis for intelligent adjustments of train schedules in case of special delays, providing strategies to minimize the impact on passengers. It also provides guidance for extreme situations, such as modifying the timetable or canceling train services.

(vi) *Rolling stock and AEI system interface*: The CTC interacts with the rolling stock system to obtain information about shunting operations and automatic equipment identification (AEI) train numbers. This interface enables the tracking, identification, and processing of the positions of EMUs.

5.4.2 *Key technologies*

In response to the needs of "improving the capacity of line transportation and emergency response capabilities, reducing system operating costs, and increasing passenger satisfaction," China's high-speed rail intelligent scheduling system will gradually develop toward optimizing the allocation of human-train-railway resources, comprehensive perception of multi-disciplinary information across regions, intelligent real-time decision-making, and intelligent coordinated scheduling based on passenger demands. The key technological directions include breakthroughs in big data storage and processing platforms, methods for modeling complex operational scenarios, research and development of intelligent decision-making technologies, human–machine interaction design, integration of train scheduling command and operation control, and cross-regional collaborative scheduling command for multiple disciplines. The key technologies of intelligent scheduling are shown in Figure 5.8:

(1) **Integrated monitoring and data processing platform based on CPS technology:** Intelligent perception and big data processing are the

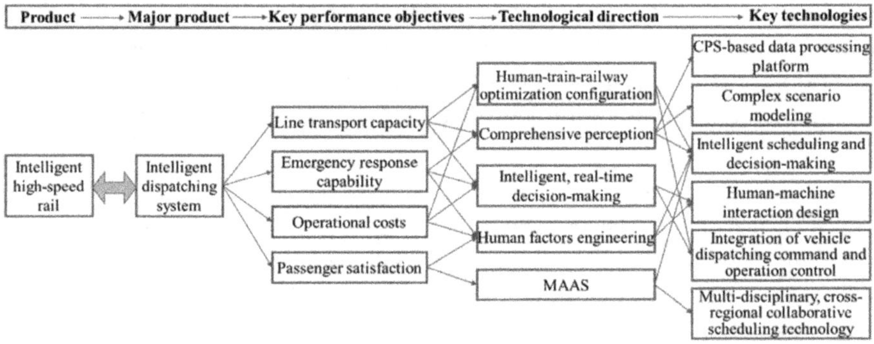

Figure 5.8. Composition of key technologies for intelligent scheduling.

foundation and prerequisite for intelligent decision-making. As the "brain" of high-speed rail operation management and train operation control, the intelligent scheduling command system needs to research an integrated monitoring and data processing platform based on CPS technology. This platform should comprehensively perceive accurate information about the entire process of high-speed rail train operation, including train operation plans, train operation status, equipment operation, passenger demands, environmental monitoring, and equipment maintenance within specific areas. Research should also be conducted on CPS network fusion technology and network structure optimization to build a secure, reliable, high-capacity, and low-latency information transmission and sharing network that covers the "train service, mechanical service, civil service, electrical service, and trains" aspects of China's high-speed rail system. By fully utilizing diverse sources and heterogeneous sensor data resources in different times and spaces, research should be conducted on data storage, processing, and analysis methods for the high-speed rail scheduling command monitoring data processing platform, overcoming the problem of "data silos" across disciplines. Furthermore, valuable information should be extracted and analyzed from a large amount of historical and real-time data, especially relevant information critical to dispatcher decision-making, such as evaluating the train operation status and predicting train delay time. This information should provide a sufficient basis for the formulation and execution of comprehensive decision-making, as shown in Figure 5.9.

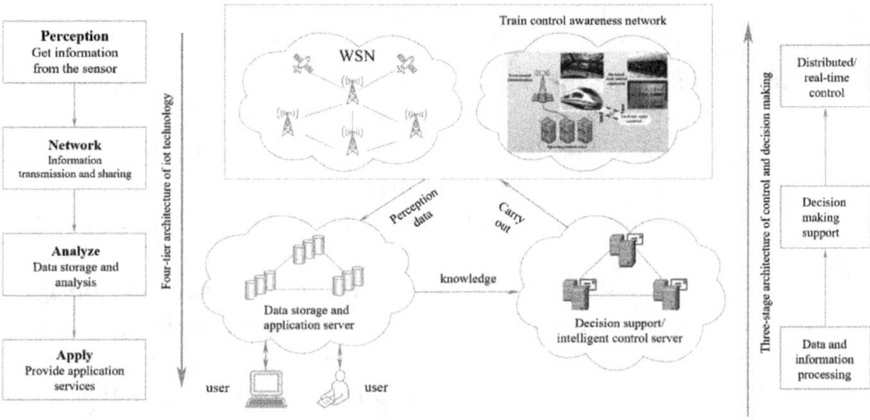

Figure 5.9. Integrated monitoring data processing platform of high-speed railway based on CPS technology.

(2) **Intelligent perception, precise recognition, and matching in complex scenarios:** Based on the analysis of coupling relationships between historical multi-domain data, the reconstruction of missing data and accurate prediction of future states can be achieved. This involves integrating the foreign object intrusion system with the high-speed railway train control system and incorporating a network of heterogeneous sensors, such as static cameras and infrared radars strategically placed along the high-speed railway line, along with an onboard dynamic sensor monitoring network. Further, technologies such as big data analytics and deep learning are leveraged to achieve rapid and precise collaborative sensing and accurate identification of unforeseen events.

(3) **Intelligent decision technology based on artificial intelligence and large-scale distributed optimization:** Intelligent decision-making technology is an interdisciplinary field that combines operations research, management science, computer science, artificial intelligence, and the internet. Key technologies for integrating intelligent decision-making technology with scheduling applications include research into the networked operation-oriented high-speed railway system's state evolution deduction method, establishment of a state-driven train delay propagation model, development of artificial intelligence-based methods for assessing

the impact of train delays on network operations and intelligent prediction, an artificial intelligence-based network train delay propagation impact assessment and intelligent forecasting methods with automatic grading based on the severity of different delays, breakthroughs in real-time solution methods for large-scale distributed optimization models, rapid emergency response during unforeseen events.

This ensures standardized, automated, and intelligent train scheduling.

(4) **Intelligent scheduling human–computer interaction design:** Utilizing engineering analysis methods that consider human factors and are based on cognitive structure theory, this research examines the dispatcher's ability to collect environmental information from digital devices using sensory inputs, such as sight and hearing, in emergency scenarios, such as unforeseen events. It aims to construct models for high-speed railway dispatchers' perception, attention, operational response, and dispatching behavior. The goal is to achieve optimal human–machine interaction effects, reduce the dispatcher's workload, prevent errors, and effectively enhance human–machine interaction efficiency.

(5) **High-speed train dispatching command and operation control integration:** Incorporating the integrated model of dispatching and train operation control (Figure 5.10) and leveraging advanced sensing,

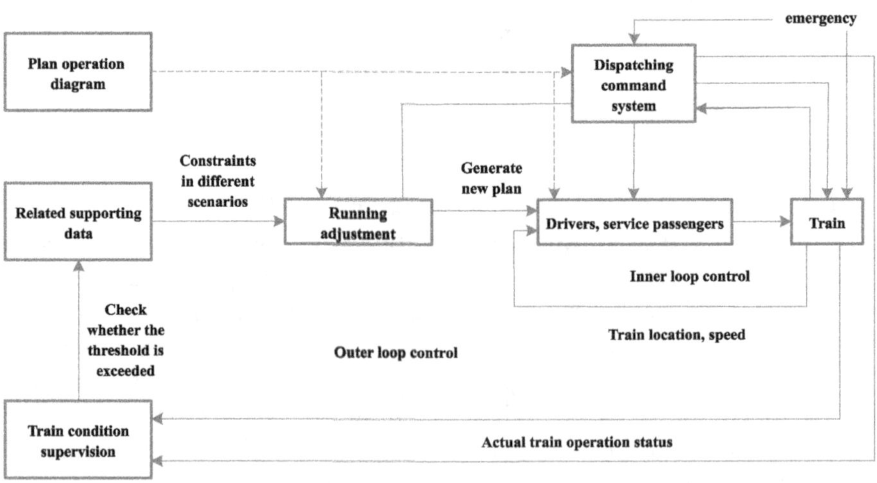

Figure 5.10. Integration of high-speed train dispatching command and operation control.

transmission, and control methods and technologies, we aim to transition from manual dispatching and human driving to intelligent dispatching and autonomous driving. This transition will ultimately lead to the integration of dispatching and train control, achieving globally optimal operational efficiency for the entire rail network and significantly enhancing our ability to respond to unforeseen events. It is proposed based on big data and a machine learning evolution model of emergency train autonomous driving states to evaluate and predict the impact of train delay propagation. Based on the analysis of the correlation between capacity resources and feasible routes of network passenger flow, a "cloud computing" model of dynamic matching between capacity resources and network passenger flow is constructed. A high-speed railway operation control modeling method oriented toward train–track networks is proposed. In the integrated train operation control system, an integrated model of high-speed train operation control (inner-loop control) and dynamic scheduling (outer-loop control) is proposed.

(6) **Cross-regional cooperative dispatch and command technology of high-speed railway:** The high-speed railway dispatching and command system is a comprehensive system composed of multiple interconnected and interdependent units. This system is characterized by a multitude of components, strong coupling relationships between elements, and evident dynamic characteristics. The main components of the high-speed railway's inter-regional collaborative dispatching and command technology include:

Regional collaborative dispatching within the high-speed railway network: This primarily involves collaborative dispatching among different dispatch offices and divisions within the high-speed railway network, spanning multiple specialties.

Collaboration between high-speed railway and existing mainline railway dispatching systems: This ensures coordination between trains crossing different lines, maintains cross-line regional capacity, and enables seamless, accurate, and rapid transitions between the two systems.

Effective coordination between high-speed railways and intercity railways, regional railways, and urban mass transit: This synergy promotes the overall development of regional transportation networks and drives regional economic growth. The integrated dispatching and command system with multi-specialty collaboration is illustrated in Figure 5.11.

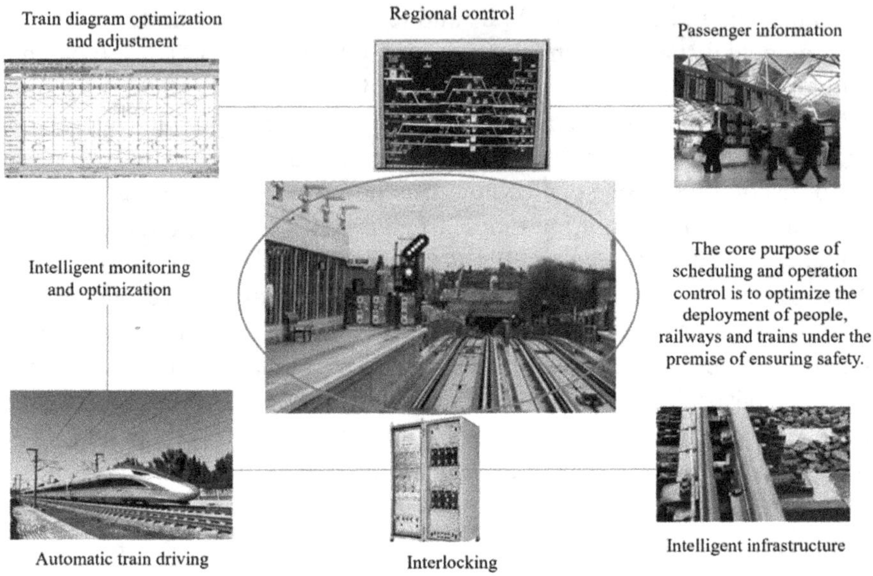

Figure 5.11. Integrated dispatching command system with multi-specialty coordination.

5.5 Intelligent and Comprehensive HSR Dispatching System

As the core work of high-speed railway transportation production, transportation dispatching and commanding involve the whole process of production organization. Timely and accurate collection of relevant information is necessary in order to make accurate dispatching and command decisions. In 2017, the transportation dispatch management system 5.0 (TDMS) was fully implemented. This is an important information system for the operation and management personnel of the National Railway Group, the Railway Bureau, and related stations to carry out transportation production dispatching and command. The existing dispatching management information system has certain problems, such as poor data sharing among subsystems, inconvenience in cross-post use, and a low degree of automation, so it is urgent to carry out digitalization and intelligent upgrades.

Intelligent comprehensive dispatching is aimed at ensuring transportation production safety, improving transportation efficiency, enhancing transportation benefits, and optimizing transportation service quality.

At the same time, it uses model data integration drive as a technical means to improve the perception, decision-making, execution, and evaluation abilities of the transportation dispatching command department. Intelligent comprehensive dispatching data mainly include construction and maintenance daily plans, operation execution feedback data, emergency plans, earthquake information, professional basic information, regulations and telegrams, emergency events, disposal processes, emergency cases, emergency instruction notices, and emergency drill information. The business analysis model mainly incorporates the traffic flow prediction model, the general situation prediction model of transportation production, and the auxiliary decision model of emergency response.

5.5.1 *Overall framework*

Intelligent comprehensive dispatching uses operations research, big data, artificial intelligence, computer simulation, and other technologies. Based on the automatic acquisition of information such as transportation demand, facility and equipment status, environment status, and production dynamics, it can realize the integrated preparation of plans for high-speed rail transportation, train operation, rolling stock operation, crew, and production operation. At the same time, it can realize the dynamic monitoring of the plan's automatic delivery and execution process and the automatic evaluation of production efficiency and transportation benefits.

The intelligent comprehensive dispatching system provides multilevel applications for the National Railway Group, the Railway Bureau, station sections, and workshops. The system adopts the "platform + application" mode, and its functions cover core applications such as cooperative preparation of the comprehensive transportation plan, passenger dispatching management, train operation management, vehicle management, construction and maintenance management, dispatching command management, and emergency auxiliary decision-making. The system provides functional integration interfaces for related application systems such as lines, stations, vehicles, maintenance, railway lines, railway communication and signaling, and vehicle services, as well as integrated display channels for integrated video and audio video analysis. The overall framework of the intelligent comprehensive dispatching system is shown in Figure 5.12.

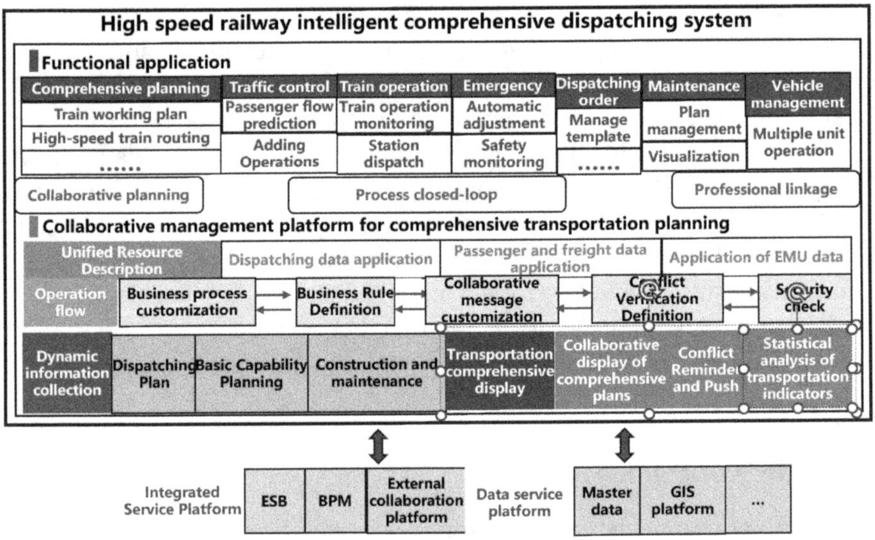

Figure 5.12. Technical framework of intelligent comprehensive dispatching system.

(1) **Transportation comprehensive plan collaborative management platform:** The transportation comprehensive plan collaborative management platform module is divided into four parts: unified resource description, dynamic information collection, business process control, and transportation comprehensive display. On top of the unified resource description and dynamic information collection, the transportation comprehensive plan is managed and compiled collaboratively through an integrated platform. Through mutual control and information interaction with the business processes of various business fields, the unified transportation plan, the real-time release of transportation resources, and the unified coordination and management of transportation operations are realized.

(2) **Comprehensive planning:** The comprehensive planning module forms an organic whole of the comprehensive plan through structured processing of the basic map data, taking the train work plan as the guide, and running through the plans for EMU operation, station operation, construction, and maintenance. It includes three functions: train

operation comprehensive management, daily shift plan comprehensive preparation adjustment, and comprehensive operation diagram display. Train operation comprehensive management is mainly aimed at the scenes of temporary passenger shutdown and extra operation, involving the coordination process of the National Railway Group, the Railway Bureau, and station sections. Daily shift plan comprehensive preparation adjustment is the daily work of comprehensive plan dispatching, involving two levels of the Railway Bureau and station sections. All maintenance, trains, crew, travel clothing, and station operations revolve around a plan. When the plan changes, all the work needs to be adjusted synchronously so as to realize intellectualization and automatization of planning and adjustment, preparation and issuing of dispatching commands, and the collaboration of various work types. Finally, the closed-loop monitoring from plan issuing to actual feedback is formed.

(3) **Passenger transport dispatching management:** Passenger transport dispatching management includes the planning management of passenger transport operation, the adjustment of additional operations and shutdowns, and the supervision of passenger service. By reading the data of the basic map of passenger transport, the "one map a day" of passenger transport is formed. Then, the basic information, such as the locomotive bound on the basic running line and the undercarriage routing, is comprehensively displayed, and the relevant functions of conflict detection and auxiliary adjustment of the basic running line are provided. Additional shutdown adjustments include additional shutdown application format management, additional shutdown auxiliary approval, open path selection, basic map running line selection, vehicle underpass check, safety management, additional shutdown, and shutdown command preparation. Passenger service supervision includes mastering the information of ticket pre-sale, sale, and replacement, as well as the actual passenger flow. Further more, train operation status is monitored to receive real-time train operation adjustment plans, train operation performance information, and train operation delay and forecast information. It also predicts passenger train delays caused by abnormal passenger flow or passenger detention, as well as passenger transfer difficulties caused by early warnings and alarms. All kinds of information affecting passenger transportation in real time is received, including accidents, weather, disasters, train speed limits,

and equipment failure alarms, and the conductor and stations along the way are informed of the affected situation.

(4) **Train operation management:** Train operation management needs to closely monitor the train operation and various real-time information sent back from the site in order to organize the normal operation of the train according to the issued plan. The train operation adjustment plan is prepared in time when the train is out of order, and relevant personnel are actively organized to implement it to ensure that the normal train operation order is restored as soon as possible.

By uploading the actual operation information, the train running position, train number, train speed, train delay information, train entry and exit section status, and operation status information in the train base can be displayed in a variety of ways in real time.

(5) **Vehicle management:** The basic information about EMU involved in an intelligent comprehensive dispatching system comes from the EMU management information system (EMIS). It includes information such as vehicle type, train group number configuration, formation, quota, maintenance status, history, maintenance, and operation plan. This function can be viewed using the system or directly queried in the EMIS system.

(6) **Construction and maintenance dispatching management:** It realizes the functions of plan management, comprehensive maintenance visual management, construction maintenance dispatching command and safety management, disaster prevention safety monitoring, and so on. It receives the plans for train operation, construction and maintenance, road trains, train operation adjustment, and power supply. Then, according to these plans and the preparation of each maintenance department, the dispatcher will adjust and confirm the comprehensive maintenance plan and send it to the relevant department.

(7) **Dispatching command management:** Based on the unified basic data management, the intelligent control system of dispatching command integrated by China State Railway Group, the Railway Bureau and station sections are constructed. Diversified input, formatting, intelligent card control, automatic transmission, and whole-process integrated closed-loop management of dispatching commands are realized.

(8) **Emergency assisted decision:** This function provides auxiliary decision-making for the dispatching emergency operators of the National Railway Group and Railway Bureau levels in dealing with emergency incidents. It includes train delay monitoring and control, in-transit vehicle distribution, abnormal stop monitoring and control, inter-office information sharing, automatic train adjustment, intelligent safety detection, and other functions.

The automatic train adjustment function means that by obtaining the planning information and abnormal information on the transportation plan collaborative management platform, it makes full use of the redundant time existing in the original drawing (such as the time reserved for construction and adjustment and the natural gap during the laying of the drawing) and changes the running line of the on-time train without increasing the delay time. Through the automatic adjustment operation, the adjustment scheme is generated.

5.5.2 *Key technologies*

Intelligent comprehensive dispatching mainly consists of key technologies such as whole-process management technology, comprehensive preparation and adjustment technology, and construction technologies, which are described as follows:

(1) **Whole-process management technology for transportation production:** In order to meet the demands of passenger transport, with the rational use of transport capacity resources as the goal, the information distributed in the transport production information system is integrated through the construction of a transportation comprehensive plan collaborative management platform, to realize the full flow of passenger information. Instead of manual work and telephone communication, it ensures effective collaboration of relevant professional systems and provides guarantees for transportation safety and management quality improvement.

(2) **Technology of comprehensive preparation and adjustment of transportation comprehensive plan:** The integrated construction technology of "platform + system" is used to realize the integration of China State Railway Group, the Railway Bureau, station sections, and

workshops. With the train operation plan as the core, the plans for construction and maintenance, car bottom operation, and station operation, along with other professional plans, are adjusted from decentralized to collaborative preparation. All maintenance, passenger transport, and station operations are carried out around a plan. So, the efficiency of the overall plan optimization preparation is improved. When plans change, all work is adjusted simultaneously. The cooperation of the whole process of passenger transport production is realized through information support.

(3) **The closed-loop management technique of traffic control from planning to execution:** The system is based on the basic train operating diagram, combined with passenger demands and the establishment of the operation plan. Based on the opening plan and daily messages, a train work plan is initially prepared. Then, the work plan for the station section is made. The implementation of station operations is supervised to collect feedback on station operation performance. On this basis, statistical and stage trend analyses are performed, and follow-up plans are adjusted. The coordinated closed-loop management of the basic plan, opening plan, and station operation plan can be realized to improve the quality of transportation dispatching decisions.

(4) **The technology of transportation integrated plan collaborative management platform construction:** The module function and information flow of the transportation comprehensive plan collaborative management platform are designed. Based on a unified resource description and dynamic information collection, it can realize a unified transportation plan, real-time release of transportation resources, mutual control of business processes, information exchange, and unified coordination and management of transportation operations.

(5) **Business subsystem construction technology based on microservice framework:** By building the microservice framework of an intelligent comprehensive dispatching system, the high availability of business functions is guaranteed, and the support ability for business innovation is enhanced. The processes of development, testing, and maintenance are unified to reduce the cost and risk of building new business applications.

(6) **Unified PaaS platform building technology:** Through the unified PaaS platform, an intelligent comprehensive dispatching system application program and operation environment are created. It has several platform security-related capabilities, such as PaaS platform user security authentication, multi-tenant management, automatic security patch upgrade, and secure transmission. At the same time, it provides key core technology components such as distributed service management, containerization technology, distributed service monitoring, and platform health and application performance monitoring.

5.6 Intelligent HSR Infrastructure Maintenance

High-speed railway infrastructure maintenance can be divided into infrastructure management, inspection, and maintenance according to the work content. In terms of maintenance, it can be divided into three parts by profession: railway line, railway communication and signaling, and power supply. The railway civil engineering is mainly responsible for the maintenance of track, roadbed, bridge and culvert tunnels, and other facilities; the railway telecommunication and signaling is responsible for communication and signal systems; and the power supply engineering is mainly responsible for traction and electric power supply systems. In the maintenance process, the maintenance management that integrates the three parts of railway line, railway communication and signaling, and power supply is called a comprehensive maintenance mode.

Intelligent maintenance of high-speed railway infrastructure, which is aimed at equipment such as high-speed railway tracks, subgrades, tunnel bridges and culverts, communication signals, contact networks, and substations, utilizes networking, intelligent systems, and other technical means to realize the functions of intergrated management of train line, communication and power supply, comprehensive data perception, real-time fault diagnosis, intelligent state analysis, early accident safety warning, autonomous planning decision-making, and the whole-life-cycle management of equipment assets, and improves efficiency of infrastructure maintenance management.

Infrastructure intelligent maintenance data are primarily composed of data related to equipment assets, mobile detection, fixed monitoring, on-site inspection, status analysis and decision-making, production dispatching, and material and tool implementation. The business models involved are those of equipment fault patterns, fault diagnosis, predictive maintenance decision support, etc. In addition, it also includes characterization models, such as infrastructure BIM design models and completion models, provided by the intelligent construction part. The intelligent maintenance of infrastructure integrates the mechanism model, characterization model, and data of the entire lifecycle of infrastructure, providing support for the visualization, intelligence, and predictive maintenance of infrastructure.

5.6.1 *Overall framework*

With the development of the production value of high-speed railway infrastructure operations, maintaining the good condition of infrastructure equipment becomes the main content of professional equipment maintenance work for railway lines, railway communication and signaling, and power supply during the operation period. In recent years, with the continuous development of railway infrastructure informatization construction, a large number of new technologies and equipment have been put into use. The railway civil engineering, railway communication and signaling, and power supply engineering have synchronously built relevant detection and monitoring platforms and maintenance systems, generated and stored massive amounts of unstructured and structured data. It has become an important development direction that integrates new information technologies, such as mobile communication, cloud computing, the Internet of Things, big data, artificial intelligence, carrying out comprehensive digitization, networking, and intelligentization for the entire process of infrastructure equipment maintenance. This realizes the intelligent maintenance of equipment asset concentration and the testing, inspection, fault diagnosis, and fusion analysis of equipment. In addition, it implements production plan management, quality evaluation, comprehensive decision-making, health status prediction, and other functions, thus improving safety service level and management efficiency and reducing labor intensity, skill requirements, and the costs of maintenance. The overall framework of intelligent maintenance of infrastructure can be divided into infrastructure, service, application, and user access layers, as shown in Figure 5.13.

(1) **Infrastructure layer:** It consists of servers, storage, network equipment, security equipment, a virtualization platform, basic software, etc., and provides infrastructure resources, such as storage, computing,

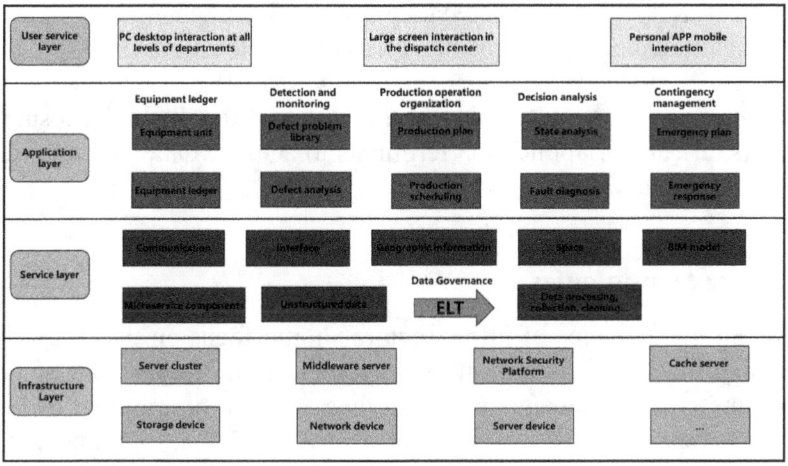

Figure 5.13. Overall framework of intelligent maintenance of infrastructure.

databases, operating systems, and geographic information software, for upper-layer applications.

(2) **Service layer:** It is composed of microservice integrated development environments, data governance, and common service components. Among others, data governance mainly realizes the collection, cleaning, management, and storage of structured and unstructured data and provides communication, interface, geographic information, and spatial services, as well as public algorithms, BIM models, and other public service components on the basis of complete data governance.

(3) **Application layer:** It aims at safe production process management, production maintenance decision analysis, active maintenance in-depth application, etc. To achieve these, it focuses on infrastructure equipment and facilities, applies dynamic and static detection technologies, and utilizes the data aggregation and processing capabilities of the service layer, the support of the basic platform, BIM, and spatial data services and algorithms. It mainly includes the following functions: equipment ledger (including equipment unit, equipment ledger, etc.), detection and monitoring (including defect problem library and defect problem analysis), production operation organization (including operation plan, operation monitoring, production dispatching, and evaluation system), decision analysis (including state analysis, safety early warning, fault diagnosis,

production analysis, and maintenance decision), and emergency management (including emergency plan and department locations).

(4) **User access layer:** Users access the system through channels such as browsers, clients, or application terminals to achieve data interaction and functional application.

5.6.2 *Key technologies*

Intelligent maintenance of infrastructure should focus on the research of multi-major collaborative business processes, online adjustment of facility parameters, maintenance robots/drones, a space-time model of infrastructure, online non-destructive repair of facilities, optimization of maintenance resources, and other technologies so as to provide matching management means for efficient and intelligent railway operation. The key technologies of intelligent maintenance of infrastructure are shown in Figure 5.14.

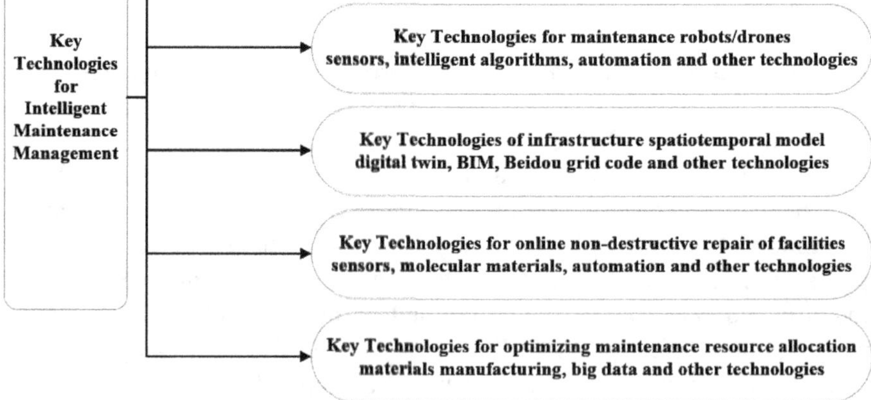

Figure 5.14. Key technologies of intelligent maintenance of infrastructure.

(1) **Business process management system:** It comprehensively manages information such as business requirements, business preconditions (material supply state, link-related physical state, etc.), and the physical state of each stage of business, associates and integrates each business, and forms a physical information model of business process management that comprehensively refers to the interaction between the physical state of each business. The facility state matches the business needs, associates with, and integrates the facility state physical information model with the business process information management system. It considers the equipment state to have the leading role in the business process and initially forms the infrastructure maintenance management information physical system. By matching all kinds of automatic intelligent maintenance equipment, it initially achieves high-quality, efficient, and low-cost condition repair. The system will gradually carry out business expansion and module optimization, and the maintenance management system will refine the division of labor for each business. Then, in-depth optimization on this basis is carried out in order to perform fine management of the whole business process and achieve the optimal decision for the business process, further improving efficiency and reducing costs.

(2) **Online adjustment of facility parameters:** It mainly realizes remote control of adjustable parameters to achieve global optimal decision-making for the linkage of various business areas. Based on the principles of facility structure in various business areas and the digital twin model, the key adjustable parameters of each point are sorted out, and then the adjustability is upgraded.

(3) **Maintenance robots/drones:** Maintenance robots/drones that perform maintenance operations during normal train operations use the surrounding environment to effectively avoid passing trains and reduce the window time for larger assignments. Single or modular operations can be completed between train passes to overcome the shortcomings of human-led maintenance methods. For example, arriving at the maintenance site takes a long time, work efficiency is greatly affected by external factors such as temperature, altitude, and light, and high labor costs are incurred.

(4) **Space-time model of infrastructure:** The space-time model, which integrates Beidou Grid, BIM, and GIS, can digitize, refine, and visualize

the location information of various infrastructure so as to provide a reliable data foundation for implementing more application services based on device location information.

Some of the functions realized by the application of intelligent maintenance technology based on BIM are the interactive closed-loop between static and dynamic data of the BIM model, multi-party collaborative operation, intelligent visual maintenance of facilities and equipment, intelligent data analysis, fault prediction, and full lifecycle management. It includes comprehensive maintenance based on BIM technology, the professional maintenance of railway lines, railway communication and signaling, power supply, and building construction, which can effectively improve the service level and efficiency of railway infrastructure operations and maintenance.

(5) **Online non-destructive repair of facilities:** There are a large number of vulnerable components in railway infrastructure, such as the contact wires of the contact network in power supply systems and the rails in railway lines. The existing solution for maintaining the contact wires is to repress or replace them, while the solution of maintaining rails is to polish or replace them. Studying online non-destructive maintenance techniques for such large and vulnerable components is beneficial to avoiding the consumption of a large amount of manpower and time and reducing the loss of resources.

(6) **Optimization of maintenance resources:** It fully considers the differences in the frequency of faults occurring in different sections of each line and studies the method of reasonably configuring robot/drone reception stations and maintenance material storage stations along the route so as to achieve optimization of maintenance resource allocation. By utilizing 3D printing technology to achieve the raw material configuration of spare parts, redundant requirements for single accessory configuration are reduced, and resource utilization is optimized.

5.7 Prognostic and Health Management and Intelligent Maintenance of EMU

China's high-speed railway covers more than 30 provinces, autonomous regions, and municipalities directly under the central government.

The operating conditions of high-speed trains are complex, with many lines having a range of over 1000 km. A wagon group can travel up to 3000 km per day and may encounter complex external environments, such as severe cold, high temperatures, high humidity, intense sandstorms, and high altitudes along the way. Based on the massive data accumulated during the operation of high-speed trains, it is extremely necessary to study the data-driven prognostic and health management (PHM) and intelligent maintenance of EMUs and promote the realization of digital and accurate state maintenance.

The PHM technology is developed based on fault monitoring, diagnosis, and early warning technologies. It mainly includes three aspects: enhancing diagnosis, which is the process of determining the functional status of a component with high fault diagnosis ability and low false alarm rate; prediction, which is achieved by modeling based on fault characteristics and failure mechanisms to achieve early warning of faults and residual life prediction; and health management, which makes decisions on the health condition, maintenance, and support activities of equipment or components based on monitoring, diagnosis, and prediction information, so that the equipment or components can meet the required health standards again. PHM is an important component of the intelligent equipment inspection and maintenance system. The maintenance of EMUs based on PHM is key to their intelligent maintenance and an important means of achieving planned preventive maintenance, leading to digital precision condition maintenance.

Model–data integration is an important driving force for the intelligent maintenance of EMUs. The data involved mainly include information concerning maintenance planning and operation, maintenance materials, crew operation, emergency response, technical data, decision support, alarms and early warnings, predicted maintenance, and health. The business models may involve EMUs' fault warning and prediction, health condition assessment, predictive maintenance decision-making, etc. The integrated fusion application based on model and data provides support for fault prediction, condition-based maintenance, predictive maintenance, etc., of EMUs.

5.7.1 *Overall framework*

The intelligent multiple-unit PHM and intelligent maintenance system is composed of vehicle-mounted systems, vehicle–ground data

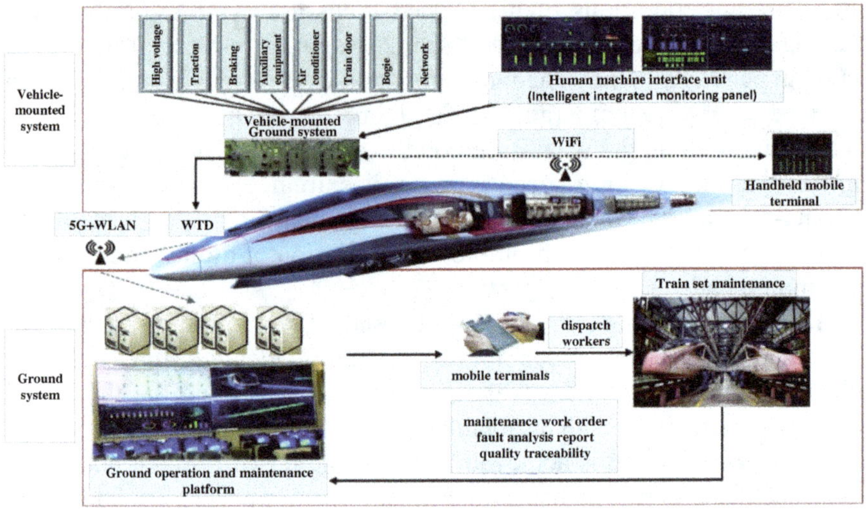

Figure 5.15. The overall framework of EMU PHM.

transmission systems, and ground systems. According to system functions, it is divided into subsystem PHM functions, vehicle-level PHM functions, and train-level PHM functions, comprehensively covering the PHM technology ecosystem of "vehicle–air–ground." The overall framework of EMU PHM is shown in Figure 5.15.

5.7.1.1 *Vehicle-mounted system*

The intelligent EMU vehicle-mounted system consists of a vehicle-mounted PHM unit, a human–machine interface unit, and a handheld mobile terminal. The vehicle-mounted PHM unit adopts a standard industrial control computer with third-party testing and certification, is equipped with high-performance processing units, and is capable of real-time data collection, data preprocessing, data storage, model calculation, and other functions. Through the train ethernet, a standard TRDP communication protocol is used to collect status parameters, fault parameters, and other information about key systems and components from the train bus. The human–machine interface unit achieves real-time status display, early warning and alarm information viewing, electronic interaction, manual querying, and other functions by reading the data from the vehicle-mounted PHM unit. The handheld mobile terminal utilizes a dedicated local area network with security measures within the train to interact with the PHM unit for data, providing mobile data query terminals for drivers and passengers within a certain range inside and outside the train. It greatly improves the efficiency of emergency handling during train operations, enhancing safety and reliability.

5.7.1.2 *Vehicle–ground data transmission system*

The data generated by the vehicle-mounted PHM system is compressed and processed through the vehicle–ground data transmission system. 5G equipment and antennas are used to send the data to the ground server through the specified protocol for centralized storage and processing, which is then transmitted back to each application unit through a dedicated channel for use by the ground PHM system platform. This provides a data basis for model development and exploration.

5.7.1.3 *Ground system*

The goal of the ground system is to optimize the design, manufacturing, operation and maintenance of multiple units, ensure safe vehicle operation, reduce vehicle operation maintenance costs, and improve passenger service satisfaction by improving the intelligence of EMUs. With a focus on maintenance and support, the ground PHM platform uses advanced sensor technology to obtain train operating status data. It uses intelligent algorithms and models to mine and analyze on-board big data to obtain quantitative knowledge of the health status of train equipment and conduct performance evaluation of key train systems and components. The platform can identify early faults, predict medium- and long-term faults, and provide support for condition-based balanced repair of trains. At the same time, while ensuring the quality of train operation service, it strives to achieve the goal of reducing maintenance costs. The ground PHM framework of intelligent EMUs is shown in Figure 5.16.

The intelligent EMU PHM includes application functions such as data access and storage, status monitoring, fault diagnosis, fault warning, health assessment, statistical analysis, maintenance decision-making, and system management. It is beneficial to accurately grasp the performance status of vehicles, handle faults early, and ensure the safety of vehicle

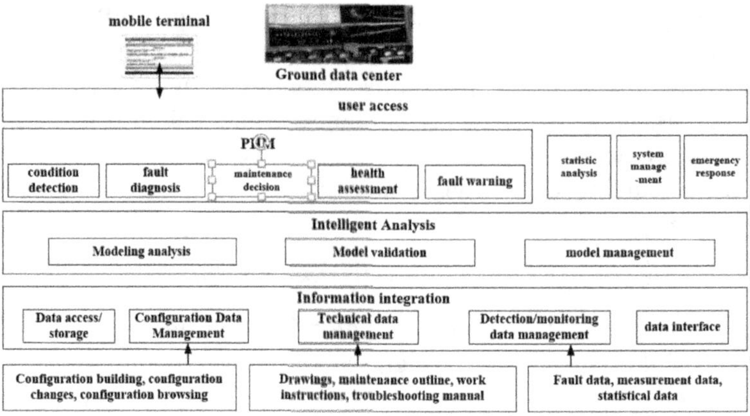

Figure 5.16. The ground PHM framework of intelligent EMUs.

operation. It optimizes spare parts inventory and maintenance resource planning based on performance, optimizes the repair process and system, and reduces maintenance costs throughout the entire lifecycle.

At the data processing level, the EMU PHM system can also be divided into several layers: data source, data collection, data processing, data storage, data analysis, and maintenance decision-making.

(1) **Data source layer:** This layer mainly includes the source data system and operation equipment required to support EMU PHM functional implementation, such as the EMU management information system, EMU vehicle-mounted information wireless transmission system, EMU operation fault image monitoring system, wheel set fault detection system, operation handheld terminal, mobile wheel rim and spoke detection equipment, as well as related equipment and systems, such as maintenance and railway communication and signaling.

(2) **Data collection layer:** This layer specifies the specific business data required for the intelligent EMU PHM system and classifies it based on the data structure obtained through data analysis. The classified data are then used to develop an extract–transform–load (ETL) extraction strategy based on the data update cycle. The data extraction process should ensure consistency between the extracted data and the source data, and a buffer layer should be installed in the collection layer to ensure the effectiveness and reliability of the data extraction process.

(3) **Data processing layer:** This layer uses distributed multi-type database storage to meet the requirements of large-scale storage and application scenarios for different types of data.

(4) **Data storage layer:** This layer provides batch processing methods for offline data processing and real-time processing methods for streaming data processing. Among their main functions are data cleaning, format conversion, and data aggregation from different dimensions. The real-time data processing method utilizes streaming data processing tools related to big data to achieve real-time computing and query processing functions.

(5) **Data analysis layer:** This layer utilizes data analysis methods, such as analysis algorithms, deep learning, and neural networks, to mine and analyze the stored data. At the same time, combined with the failure mechanism of relevant components of the high-speed train unit, it constructs a fault model for the components, summarizes the fault patterns, and obtains analysis and prediction results.

(6) **Maintenance decision-making layer (or data display layer):** This layer is responsible for visualizing the obtained data and providing scientific decision-making to management personnel.

5.7.2 *Key technologies*

The intelligent EMU PHM involves many aspects, such as data processing, condition monitoring, fault prediction, fault diagnosis, health management, and maintenance decision-making. It provides users with open fault prediction and health management services, mainly including the following key technologies.

(1) **Data collection and processing technology:** Obtaining parameter information that represents the health status of systems and components is a prerequisite and foundation for conducting PHM-related work, and the quality and quantity of state data directly affect the application effectiveness of PHM systems. The data acquisition and processing technology of the PHM system of multiple units mainly aims to solve two problems. First, it is necessary to use high-speed vehicle–ground communication technology to achieve stable and real-time transmission of

vehicle-mounted condition monitoring parameters of the EMU system and key components to the ground data center. This ensures the communication of comprehensive and complete train status information. Second, it is necessary to define a unified data processing format and specification, which facilitates the structuring, conversion, and integration of various types of data, such as vibration, electrical, and process parameters. These information and data mainly include vehicle-mounted condition, environmental, manufacturing, inspection, maintenance data, and historical data.

(2) **Early warning and prediction technology:** Fault early warning and prediction is based on the failure modes, characteristics, and mechanisms of components. It uses methods such as statistical reliability, data-driven, and mechanism research to predict the occurrence of faults, the deterioration trend of health indicators, and the remaining lifespan of components and to identify symptoms, locate faults, isolate faults, etc.

The research on fault warning and prediction technology for multiple units mainly comprises two aspects. First, in view of the fact that the existing scope of EMU condition monitoring is not comprehensive, it is necessary to carry out long-term simulation tests, fault simulations, real vehicle tracking, and other work. Then, the limited state parameters are used to research failure modes and failure mechanisms to establish corresponding early warning models. Second, we need to accurately identify fault characteristics, reasonably formulate early warning threshold ranges, and strictly control false alarm rate indicators, while ensuring the accuracy of early warning results.

(3) **Health condition assessment:** Equipment health condition refers to the ability to maintain a certain level of reliability and maintainability under specified conditions and for a specified period of time and to stably and continuously complete predetermined functions. The health condition evaluation of the EMU system and components has two primary aspects. The first is establishing a well-equipped health condition evaluation index system using qualitative or quantitative methods to define the subdivision level of health status and to achieve horizontal comparison and evaluation of components, systems, and vehicle models. Second, through real-time and long-term health condition evaluation, basic useful information can be provided for the study of the long-term service performance and evolution laws of the entire train and components of the multiple units. This can

achieve comprehensive and accurate evaluation and analysis of components, systems, and vehicle models.

(4) **Condition-based maintenance decision-making:** Condition-based maintenance refers to predictive maintenance that predicts the deterioration trend of a component based on its current state and determines the timing and method of maintenance. Condition-based maintenance is an important direction for solving the current problems of insufficient maintenance, excessive maintenance, and cost waste in China's EMU technology. It is also significant for improving the efficiency of the use of high-speed trains, saving maintenance costs, and improving the safety and reliability of EMUs.

Currently, there are three major issues that need to be solved for the condition-based maintenance of EMUs. The first is to continuously improve the monitoring range of key component status parameters and select effective feature parameters to characterize the performance status of the components. The second is to analyze the evolution law of component characteristic parameters and reveal any inferior trends in component performance. The third is to determine reasonable maintenance timings for components and systems based on the current maintenance system and eventually achieve the optimal ratio of maintenance production organization.

Review Questions

1. What are the key technologies included in a high-speed railway intelligent passenger station?
2. Briefly describe the innovative functions of high-speed rail intelligent ticketing.
3. What are the differences between the intelligent and traditional compilation technologies of train diagrams?
4. What are the functions of intelligent centralized traffic control and intelligent comprehensive dispatching?
5. What business models does intelligent maintenance of infrastructure include?
6. What is PHM? What are some innovations in PHM and intelligent maintenance of high-speed multiple units?

Chapter 6

Innovative Application Demonstration: Intelligent Beijing–Zhangjiakou High-Speed Rail

The Beijing–Zhangjiakou high-speed railway (HSR) played a crucial role as an important transport infrastructure during the 2022 Beijing Winter Olympic Games. Its main line is 174 km long, and its maximum design speed is 350 $km \cdot h^{-1}$. It is the world's first HSR to be designed based on the concept of an intelligent railway. The construction of the Beijing–Zhangjiakou HSR began on April 29, 2016, and the line was put into operation on December 30, 2019, decreasing the travel time between Zhangjiakou and Beijing from at least 3 h and 7 min to just 47 min. This HSR provided important support for the competitions at three venues in two cities during the 2022 Beijing Winter Olympic Games and promoted the connectivity and economic integration of Beijing, Tianjin, and Hebei. This chapter introduces the construction and operation practices of the intelligent Beijing–Zhangjiakou HSR in terms of its overall architecture design and intelligent construction, equipment, and operation.

6.1 Engineering Characteristics of Beijing–Zhangjiakou High-Speed Rail

6.1.1 *Project overview*

The Beijing–Zhangjiakou HSR is an important transport facility that served during the 2022 Beijing Winter Olympic Games. Its main line is

174 km long, and its maximum design speed is 350 km·h^{-1}. It is the world's first HSR to be designed based on the concept of an intelligent railways. The construction of the Beijing–Zhangjiakou HSR began on April 29, 2016, and the line was put into operation on December 30, 2019, decreasing the travel time between Zhangjiakou and Beijing from at least 3 h and 7 min to just 47 min. This HSR provided crucial support for the competitions at three venues in two cities during the 2022 Beijing Winter Olympic Games and promoted the connectivity and economic integration of Beijing, Tianjin, and Hebei. Ten stations are set along the main line of the Beijing–Zhangjiakou HSR: Beijing North, Qinghe, Shahe, Changping, Badaling Great Wall, Donghuayuan North, Huailai, Xiahuayuan North, Xuanhua North, and Zhangjiakou. In addition, Yanqing Station is set along the Yanqing Branch Line, while along the Chongli Railway, Taizicheng Station is set and Zhaochuan South Station is reserved. As shown in Figure 6.1, Beijing North is the originating and terminating station, and Qinghe and Zhangjiakou are both originating stations and through stations. The line passes through the core area of Beijing, the Badaling Great Wall Scenic Area, and the Guanting Reservoir. The construction of the Beijing–Zhangjiakou HSR overcame many difficulties, such as complex geology, challenging construction, and stringent environmental protection requirements. Other challenges included the variety of passengers, concentrated passenger flow periods, high travel service requirements, and significant international attention during the preparations for the Winter Olympics Games. To meet the requirements of the 2022 Winter Olympic Games in the construction and operation of the

Figure 6.1. Route map of the Beijing–Zhangjiakou HSR.

Beijing–Zhangjiakou HSR, the following tasks were both necessary and urgent: (i) to integrate and apply building information modeling (BIM), big data, artificial intelligence, BeiDou satellite navigation, 5G, and other new-generation information technologies; (ii) to achieve intelligent innovation in the construction, equipment, and operation technologies of the Beijing–Zhangjiakou HSR; and (iii) to realize comprehensive perception, ubiquitous interconnection, integrated processing, active learning, and optimized decision-making based on data driving.

6.1.2 *Technology system*

The technical system of the Beijing–Zhangjiakou HSR is an overall design of the core elements of the intelligent HSR and their interrelation at the systemic level. Based on the design concept of "carrying out top-level design and prioritizing the platform," the technical system is divided into five levels from top to bottom: sector, field, subject, innovation, and basic platform, as shown in Table 6.1.

There are three major sectors: intelligent construction, intelligent equipment, and intelligent operation. Under the framework of the three major sectors, there are 10 major fields: survey and design, engineering construction, construction management, mobile equipment, communication and signaling, traction power supply, inspection and monitoring, passenger service, transportation organization, and maintenance and repair. These 10 major fields cover 18 subjects: space–air–ground integrated engineering survey; BIM-based engineering design; intelligent construction of bridge, tunnel, subgrade and track works; intelligent construction of passenger station works; intelligent construction of communication, signaling, power and electrification works; BIM+GIS-based construction management; intelligent EMUs; intelligent comprehensive inspection trains; intelligent signals; intelligent communications; intelligent traction power supply; intelligent inspection and monitoring; intelligent passenger transport; intelligent ticketing; intelligent integrated dispatching; intelligent traffic dispatching; integrated operation and maintenance of the track, communication, signal, and power supply; and intelligent operation and maintenance of EMUs. Under the framework of these 18 subjects, there are numerous innovations, including space–air–ground integrated intelligent surveying and mapping; BIM building; intelligent beam plants; the intelligent construction of passenger stations; BIM-based virtual construction; EMU intelligent monitoring; intelligent inspection equipment;

Table 6.1. The technical system framework of the intelligent Beijing–Zhangjiakou HSR.

Technical system framework of intelligent HSR			
3 sections	**10 fields**	**18 directions**	**N innovations**
Intelligent construction	Survey and design	Space-air-ground integrated engineering survey	GIS-based intelligent exploration; space-air-ground integrated intelligent survey and mapping; digital survey delivery
		BIM-based engineering design	BIM modeling; collaborative design; digital design delivery
	Engineering construction	Intelligent construction of bridge, tunnel, subgrade, and track works	Intelligent beam yard; intelligent shield tunneling; intelligent subgrade filling; intelligent slab yard
		Intelligent construction of passenger station works	Intelligent construction of passenger stations; coordination for construction of integrated transport systems
		Intelligent construction of communication, signaling, power and electrification works	Intelligent construction of electrification works; intelligent construction of communication works; intelligent construction of signaling works; intelligent construction of information works
	Construction management	BIM+GIS-based engineering construction management	BIM-based virtual construction; full-process digital management; digital completion delivery
Intelligent equipment	Mobile equipment	Intelligent EMUs	Intelligent monitoring of EMUs; intelligent diagnosis of EMUs; intelligent services of EMUs
		Intelligent comprehensive inspection trains	Intelligent inspection equipment; analysis of intelligent inspection data

	Communication & signaling	Signaling	Automatic control of station route; train operation control; automatic train operation (ATO)
		Communication	Intelligent communication bearer network; intelligent mobile communication; intelligent dispatching communication
	Traction power supply	Intelligent traction power supply	Intelligent traction substation; simple, unified and standardized OCS; intelligent power supply dispatching system
	Inspection and monitoring	Intelligent inspection and monitoring	Inspection and monitoring of infrastructure; monitoring and warning of natural disaster; intelligent monitoring of perimeter intrusion; intelligent monitoring of environment
Intelligent operation	Passenger service	Intelligent passenger transport	Integrated production command and management of passenger transport; intelligent management and monitoring of passenger transport equipment; key station and train service; passenger safety of stations and trains; intelligent station and train service
		Intelligent ticketing	Electronic ticketing; intelligent product design and ticketing organization; travel planning; targeted marketing; integrated traffic information sharing

(Continued)

Table 6.1. (*Continued*)

Technical system framework of intelligent HSR			
3 sections	**10 fields**	**18 directions**	**N innovations**
	Transport organization	Intelligent comprehensive dispatching;	Transportation situation awareness; planning integration and coordination; intelligent plan adjustment; intelligent emergency dispatching
		Intelligent traffic control;	Automatic adjustment of train operation plans; route and command control; integrated simulation of train operation control
	Maintenance and repair	Integrated operation and maintenance of track, communication, signaling, and power supply;	Digital log management; intelligent fault diagnosis; condition-based intelligent integrated warning; intelligent decision-making on operation and maintenance
		Intelligent operation and maintenance of EMU	Fault prediction; health assessment; operation and maintenance analysis; decision-making support
1 platform	Basic platform of intelligent HSR	Master data service; Metadata service; Geographic information service; Big data analysis services; Artificial intelligence service	
	Supporting technology	Network security assurance; Beidou satellite navigation; Internet of Things	

automatic train operation (ATO); intelligent traction substations; natural disaster monitoring and warning; intelligent stations and train service; electronic ticketing; automatic adjustment of train operation plans; and intelligent operation and maintenance decision-making. Finally, there is one basic platform that provides common computing, storage, data, algorithms, and other services for all the systems and subsystems of the intelligent Beijing–Zhangjiakou HSR to integrate cutting-edge new information technologies.

This technical system has become the general framework guiding the construction of China's intelligent HSRs. It is at the forefront of railway technologies around the world.

6.2 Innovative Application of Intelligent Construction Technology

During the construction of the Beijing–Zhangjiakou HSR, technologies such as BIM-based collaborative design, digital manufacturing, and intelligent construction were developed, and a BIM + GIS-based engineering management platform was built. These advancements enabled the automatic collection of construction data and information interconnection; supporting the collaborative management and decision optimization among all parties involved in the construction.

6.2.1 *BIM-based multi-area collaborative design platform*

The design of the intelligent Beijing–Zhangjiakou HSR is a huge, complex, and systematic work involving many areas and requires difficult coordination. It requires close cooperation among different areas of expertise within the design team. At present, two-dimensional (2D) drawings are used by all railway areas in the design process to represent the final design results. Such drawings have weaknesses, including unclear design expression and understanding and poor communication among areas.

BIM technology enables digitalization, visualization, multiple dimensions, collaboration, and simulation for the entire life cycle of the railway, including design, construction, operation, and maintenance. Therefore, a multidisciplinary collaborative design platform based on BIM has been built for the design of the entire Beijing–Zhangjiakou HSR, as well as the detailed design of three stations and three tunnels.

Meanwhile, BIM collaborative design software for multiple areas — including surveying and mapping, tracks, bridges, tunnels, subgrades, overhead contact line systems (OCSs), and signaling — has been independently developed to improve design efficiency and methodology. The BIM model has been used to effectively optimize the design scheme and facilitate multidisciplinary visualized collaboration in the design of complex nodes, thereby breaking the restrictions imposed by the traditional way of design, in which 2D drawings are used to illustrate structures. Several technical problems have been solved, including the evolvement of map projection, data format fusion and conversion, data source organization and management, and the development of display strategies. Deliverables have been realized, such as rapid parametric design, the calculation of project workload, interdisciplinary fitting, the integration management of design results, and seamless information transmission among various railway sectors (Figure 6.2(a)). The BIM-based

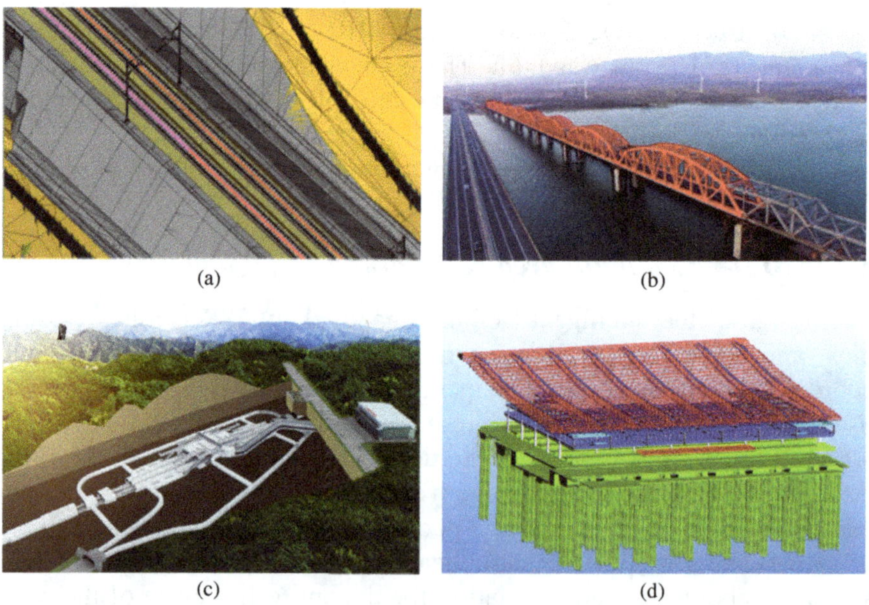

Figure 6.2. Intelligent construction of the Beijing–Zhangjiakou HSR: (a) BIM-based collaborative design platform; (b) simultaneous incremental launching of the Guanting Reservoir Super Major Bridge; (c) BIM-based intelligent construction of the New Badaling Tunnel; (d) intelligent construction of Qinghe Station.

collaborative design platform decreases coordination and liaison time by 8% and material costs by 3%.

6.2.2 *Intelligent plant for bi-block sleepers-enabled digital manufacturing*

The application of the Internet of Things (IoT), automation, intelligent image recognition, and other technologies enables the intelligent bi-block sleeper plant to realize automation and intelligence throughout the manufacturing processes. With all manufacturing data, the plant boasts an intelligent bi-block sleeper production line with the assistance of industrial robots and big data technologies, which decreases the number of workers by 50%. With the introduction of a production management information system based on the Industrial Internet, the plant provides centralized online management of all production elements, increasing the data collection efficiency by 20%. Thanks to laser measurement and deep learning for machine vision, the bi-block sleeper plant is capable of intelligent inspection and testing, with a five-fold increase in inspection precision and a 15-fold increase in inspection efficiency, thereby achieving the digital manufacturing of bi-block sleepers.

6.2.3 *Intelligent construction of subgrades, bridges, tunnels, passenger stations, and other infrastructure*

6.2.3.1 *New technologies for intelligent subgrade filling*

For better relevance and integrity of sampling conducted after the routine quality inspection of subgrade compaction, technologies based on BeiDou satellite positioning have been developed, such as unmanned compaction, automatic route planning, and automatic re-compaction in inadequate areas. In this way, automatic control of filler transport, automatic identification of filler quality, and real-time inspection of subgrade compaction quality have been realized. Automatic guidance and collaboration among bulldozers, pavers, and excavators have been realized, making the full-coverage and continuous inspection of vibration and compaction possible. Compared with conventional construction process, this new approach improves subgrade filling quality, compactness, and uniformity, as well as inspection efficiency and subgrade quality.

6.2.3.2 *Simultaneous incremental launching technology for the Guanting Reservoir Super Major Bridge*

With a total length of 9.08 km, the Guanting Reservoir Super Major Bridge is composed of simply supported curve-chord steel truss girders and registers with an opening number of eight and a span length of 110 m. This bridge, which is located in the Guanting Reservoir Water Source Protection Area, is one of the critical projects of the Beijing–Zhangjiakou HSR. In order to protect the water body to the greatest extent possible, a new construction method has been proposed in which the steel girders are assembled on the bank and then launched incrementally. A gantry-type automatic welding system based on welding robots has been developed, realizing the simultaneous welding of multiple stiffening ribs. Intelligent and simultaneous incremental launching at multiple points and automatic monitoring have been used, breaking new ground in the incremental launching of bridges with large-span steel girders and multiple openings. This method ensures accuracy, safety, improved construction efficiency, better welding, and better construction quality (Figure 6.2(b)).

6.2.3.3 *BIM-based intelligent construction of the New Badaling Tunnel at Badaling Great Wall Station*

With a total length of 12 km, the Badaling Tunnel on the Beijing–Zhangjiakou HSR is home to the Badaling Great Wall Station, the world's largest and deepest HSR station, whose construction has involved the use of subsurface excavation and an extremely complex structure of overlapped-layer tunnel clusters. With the help of a BIM model, design optimization and visualization for technology clarification can be realized for the stacked station entrance and exit passages, circular rescue galleries, parallel twin tubes, and three-dimensional (3D) multi-branch shafts. In this way, the time consumed at different interfaces has been shortened, and the onsite construction efficiency has been improved by more than 20%. Based on the BIM model, the IoT and big data technologies have been integrated to carry out real-time prediction and forecasting of the risks to adjacent buildings and structures. This provides support for the intelligent and visualized dynamic monitoring and management of the whole process of shield tunneling and effectively mitigates the impact of construction on the surrounding area (Figure 6.2(c)). A quality management system for tunnel sections based on high-speed laser scanning has been developed for the automatic, high-precision inspection of overbreaks, underbreaks, clearance, and cavity volume.

6.2.3.4 *Intelligent construction and detail-oriented management of passenger stations*

For the construction of large passenger stations, BIM technologies have been used to develop 3D fine and green designs of buildings and structures, including passenger station decoration, steel structures, mechanical and electrical pipelines, and customer service equipment rooms. A BIM + IoT-based construction management system for passenger stations has been established to realize the 3D visualized control of raw material traceability, deep foundation pit monitoring, high formwork monitoring, inspection lots, tower crane anti-collision, and steel structure welds. A BIM-based simulation system for construction organization optimization has been developed to support detail-oriented management and onsite safety and quality control (Figure 6.2(d)). With the use of intelligent construction technology, the utilization rate of nonconventional water sources has reached 22.14% and the proportion of recyclable building materials has reached 15.30%.

6.2.4 *BIM + GIS-based HSR engineering management platform*

The construction of the Beijing–Zhangjiakou HSR is characterized by large investments, a long project period, complex technology, multiple interfaces, decentralized management, and complex coordination. Thus, it is difficult to meet the requirements of high quality, high efficiency, and high speed for the construction of the Beijing–Zhangjiakou HSR by conventional management means. To successfully complete the construction of this project, information-based intelligent means have been adopted to move conventional offline management mode online; at the same time, standardized management has been adopted as an effective way to develop the HSR engineering management platform, which has BIM + GIS technology at its core and the construction project as its carrier. The platform is divided into three parts: namely, comprehensive management, process control, and site management: in four dimensions: namely, space, data, management, and application. It involves six applications: comprehensive management, progress management, material management, quality management, safety management, and investment control. This platform has shattered the restrictions in the development of BIM + GIS multisource data fusion technology and lightweight modeling technology; realized visualized and integrated component-level control over progress, quality, and safety; and improved the capability of construction data for multi-angle, multidimensional, and multiscale comprehensive analysis and management.

Since the platform has been put into use, railway quality and safety incidents have been significantly reduced. The platform plays a prominent role in demonstrating the effective delivery of social and economic benefits.

6.3 Innovative Application of Intelligent Equipment Technology

6.3.1 *Intelligent EMU system*

The intelligent EMU system integrates intelligent composite sensors, real-time train-wayside information transmission, big data mining and analysis, automatic control, intelligent information processing, fault prediction, health management, and other technologies, thereby, it realizes intelligent perception, diagnosis, and decision-making. As members of the Fuxing China standard EMU series, the intelligent EMU trains serving the Beijing–Zhangjiakou HSR are classified as standard trains and Olympic trains. The standard trains, which are built on the Fuxing model, are equipped with additional intelligent modules and dedicated functions specific to the Beijing–Zhangjiakou HSR, while the Olympic trains are equipped with additional Olympic modules as upgrades to standard trains. The intelligent EMU system in the Beijing–Zhangjiakou HSR is shown in Figure 6.3:

- The intelligent modules enable intelligent train operation, intelligent operation and maintenance (up to 2718 monitoring points), intelligent service, and safety monitoring (168 more running gear vibration monitoring points).
- The Olympic modules involve Olympic-themed painting, an Olympic-specific layout, storage areas for sports equipment, and live broadcasts of Olympic events.

(a)

(b)

Figure 6.3. Intelligent equipment in the Beijing–Zhangjiakou HSR: (a) an intelligent EMU train; (b) automatic operation at 350 km·h^{-1}.

- Dedicated functions specific to the Beijing–Zhangjiakou HSR include adaptability to severe cold weather (–40°C) in Beijing, operation ability (including traction and braking) for a grade of 30‰, and emergency self-running capacity.

The intelligent EMU trains serving the Beijing–Zhangjiakou HSR are able to run on a long, steep grade of 30‰, which is a significant improvement over all other trains. The EMUs are more environmentally friendly than other HSR trains: their water consumption has been decreased by 10%; they feature a 1–2 dB reduction in noise; up to 75% of their interior materials are recyclable; and 50% of their interior materials are degradable. Their energy consumption has been brought down by 7% through the application of aerodynamic optimization, resistance reduction, lightweight material (the weight of each train has been decreased by about 7 tons), and the multidimensional alignment of train weight, resistance, and power, along with other modifications. The intelligent technologies of environmental perception and adjustment translate into improved riding comfort.

6.3.2 *Automatic operation of a 350 km·h^{-1} high-speed EMU*

ATO technology is one of the core technologies applied to the intelligent Beijing–Zhangjiakou HSR. ATO-related equipment has been incorporated based on the Chinese Train Control System (CTCS)-3, thus enabling ATO. Accurate-positioning balises mounted on tracks at stations enable the trains to stop accurately at the signs. As a result, trains can perform automatic departures, automatic operation in sections, automatic stops, and automatic door opening; they can also realize linkages between train doors and platform screen doors.

The Beijing–Zhangjiakou HSR has set a world record for the first automatic operation of a 350 km·h^{-1} EMU train; it has also optimized the movement modeling and speed control of trains at high speeds and improved the strategy calculation and control accuracy for train operation in long sections. By combining the intelligent control algorithm with the train control strategy, the ATO train control algorithm has been optimized, and the HSR's performance, including train control comfort, stop accuracy, and energy efficiency, has been improved. Practice has shown that the automatic operation of high-speed EMU trains is safe and can reduce traction energy consumption, decrease the driver's workload, and improve the travel experience of passengers.

6.3.3 *Intelligent traction power-supply system*

At present, HSR traction power-supply systems are independent. There is limited real-time communication and information exchange between traction substations and posts, which makes it difficult to provide a reliable power supply around the clock in all scenarios for the Beijing–Zhangjiakou HSR. The development of the intelligent traction power-supply system of the Beijing–Zhangjiakou HSR is based on intelligent traction substations, section posts, auto-transformer (AT) posts, and other intelligent power-supply facilities. Digital and internet technologies are used to eliminate information blockages between traction substations and/or posts, thus enabling system integration and multisource data sharing. Wide-area protection of measurement and control, reconstruction of error self-recovery, error alarm and warning mechanisms, and a health assessment system are being established to improve the reliability, operation, and maintenance capacity of the traction power-supply system. In addition, intelligent traction substations featuring holographic perception, multidimensional integration, reconstruction and self-recovery, and intelligent operation and maintenance have been established, thereby enabling the realization of whole-station unmanned operation. The delay in wide-area protection has been shortened from 100 ms to 20 ms, and the primary and secondary equipment of the substations has reached an optical fiber isolation level of 100%.

6.3.4 *Intelligent safety monitoring and emergency response system*

In terms of safety and emergency response, the Beijing–Zhangjiakou HSR is characterized by numerous disaster-monitoring devices (for wind, rain, snow, earthquakes, and so on) along the line, the ability to analyze warnings in an extremely timely manner, and wide-range linkage coverage for emergency response. Based on these features, a whole-process intelligent safety monitoring and emergency response system has been established, with the functions of continuous safety data aggregation, risk prediction, real-time risk warning, and the temporal linkage of emergency response. This system can realize the analysis of the time and spatial distribution of disasters based on big data, intelligent warning and comprehensive judgment of risks, and the real-time transmission of safety monitoring information between trains and the ground. It connects other systems, such as disaster-monitoring, traffic control, and integrated video systems, and coordinates the emergency response of the railway with that of relevant local departments, thus improving the timeliness and coordination of safety warnings and emergency response.

6.4 Innovative Application of Intelligent Operation Technology

6.4.1 *Intelligent travel service*

A whole-process travel ticketing service system has been established which provides important services, such as e-tickets based on passenger service records (PSRs), e-ticket checking under high concurrent conditions, online–offline integrated facial recognition, and onboard inspection based on station–train wireless interaction systems. Breakthroughs have been made in the technology of having a single ticket for all travel modes, thus accelerating the process of ticketing internationalization for the Beijing Winter Olympic Games and providing paperless, self-service, and integrated travel services for passengers.

An intelligent passenger station system has been built which manages all business and production elements and provides integrated production organization, efficient emergency response, full life cycle equipment management, and accurate safety control. It can also provide self-service and intelligent services for passengers, such as travel reminders, station entry by face recognition, integrated ticket checking and ID authentication, broadcast guiding, e-navigation, intelligent information desk services, self-service ticket refunds, self-service document preparation, and remote ticket sales. Technical support is also available, such as passenger flow information sharing, transfer recommendations, and tourism information for integrated transport hub stations. Examples of intelligent travel service are shown in Figure 6.4(a).

6.4.2 *Intelligent infrastructure operation and maintenance system*

Based on integrated BIM and GIS technology, a seamless transition from the construction stage to the operation stage and full life cycle

(a)

(b)

Figure 6.4. Intelligent operation of the Beijing–Zhangjiakou HSR: (a) intelligent travel service; (b) intelligent integrated operation and maintenance system.

management of infrastructure data, such as track maintenance, communication and signaling, and power supply, have been achieved. Data acquisition and processing have been standardized for track maintenance, communication and signaling, and power supply. The data access and processing capacity of mobile inspection, fixed monitoring, and static inspection systems has also improved. A planning system for producing analysis of testing and inspection status that integrates multiple areas, such as track maintenance, communication and signaling, and power supply, has been established to support integrated operation and maintenance. The application of this system improves the utilization efficiency of the HSR maintenance window and reduces costs in the operation and maintenance of infrastructures. The intelligent integrated operation and maintenance system is shown in Figure 6.4(b).

6.4.3 *Intelligent centralized traffic control system*

As the nerve center of HSR transport command, the centralized traffic control system performs the important task of organizing and commanding the safe, punctual, and efficient operation of high-speed trains. The development of such a system has become a trend, both in China and abroad.

Based on emerging technologies, such as artificial intelligence and big data, the HSR intelligent centralized traffic control system was developed based on the Centralized Traffic Control (CTC) 3.0 software and hardware platform, with critical functions such as the automatic adjustment of train operation plans, safety control of train routes and commands, and ATO. This system has generally improved the train dispatching and command capacity, the capability of ensuring train operation safety, and enhanced the intelligence level of the Beijing–Zhangjiakou HSR.

Index

W

X

www.ingramcontent.com/pod-product-compliance
Lightning Source LLC
LaVergne TN
LVHW010326220926
848484LV00004BB/849

* 9 7 8 9 8 1 1 2 8 7 4 4 2 *